4

RECONSTRUCTING QUATERNARY ENVIRONMENTS

Reconstructing Quaternary Environments

J. J. Lowe
Senior Lecturer in Geography
City of London Polytechnic

M. J. C. Walker
Senior Lecturer in Geography
St David's University College
Lampeter

Longman
London and New York

For Jeanette and Margaret

Longman Group Limited
Longman House, Burnt Mill, Harlow
Essex CM20 2JE, England
Associated companies throughout the world

Published in the United States of America
by Longman Inc., New York

© Longman Group Limited 1984

First published 1984

British Library Cataloguing in Publication Data

Lowe, J. J.
    Reconstructing quaternary environments.
    1. Paleoecology   2. Paleontology –
    Quaternary
    I. Title   II. Walker, M.J.C.
    560'.178   QE741
    ISBN 0-582-30070-3

Library of Congress Cataloging in Publication Data
Lowe, J. J.
    Reconstructing Quaternary environments.
    Bibliography: p.
    Includes index.
    1. Geology, Stratigraphic – Quaternary. 2. Paleo-
geography. 3. Geomorphology. I. Walker, M. J. C.
II. Title.
    QE696.L776   1984   551.79   83-13535
    ISBN 0-582-30070-3

Set in 10/12 pt Linotron 202 Palatino
Printed in Hong Kong by
Wing Lee Printing Co. Ltd

Contents

List of Figures

List of Tables

Preface

The Quaternary is the most recent geological period, spanning the last two million years or so of the earth's history, and extending up to the present day. Despite the fact that it is one of the shortest formal episodes of geological time, the Quaternary has attracted the interest of a great many scientists from a wide range of disciplines, and there is now a large and rapidly-expanding literature on topics as diverse as glacial geology, climatic history, oceanic circulation and sedimentation, floral and faunal changes, and human evolution. Surprisingly, however, in spite of the wealth of material that is available, relatively few texts have so far been concerned specifically with landscapes of the Quaternary, and with the way in which different forms of evidence can be integrated to provide an insight into both spatial and temporal changes in Quaternary environments. That is the aim of this book, for it contains a description and assessment of the principal methods and approaches that can be employed in the reconstruction of Quaternary environments. The book is designed for undergraduate and first-year postgraduate students who may have been introduced to certain aspects of Quaternary studies, but whose training has not focused specifically on palaeoenvironmental reconstruction. The text is very wide-ranging and reflects the multi-faceted nature of Quaternary research. Although written by geographers, therefore, it is anticipated that students of archaeology, anthropology, botany, geology and zoology will find within the following pages material that is of use to them.

A large number of people have contributed either directly or indirectly towards the writing of this book. We owe a considerable debt to Dr J. B. Sissons who was our research supervisor in the Department of Geography, University of Edinburgh, for not only did he stimulate our interest in the Quaternary, he also taught us to seek out the flaws in arguments and to think clearly and argue logically. Moreover, he established a research school with whose members we have continued to enjoy rewarding academic contact. We are particularly grateful to Dr R. Cornish, Dr R. A. Cullingford, Dr A. G. Dawson, Dr J. M. Gray, Dr D. E. Smith and Dr D. G. Sutherland for discussing material with us, and for their assistance and companionship in the field over a number of years. We should also like to thank those who have read and commented upon drafts of the chapters of this book. Dr R. A. Cullingford (University of Exeter), Dr B. D'Olier (City of London Polytechnic), Dr P. Gibbard (University of Cambridge), Professor F. Oldfield (University of Liverpool), Dr J. D. Peacock (Institute of Geological Sciences), Dr R. C. Preece (University of Cambridge), Dr. J. E. Robinson (University of London), Mr J. Rose (University of London), Dr D. E. Sugden (University of Aberdeen) and Dr D. G. Sutherland (University of Edinburgh) all provided useful suggestions, and large sections of the text have been much improved as a result of their constructive critical appraisal. Professor G. I. Meirion-Jones made the cartographic facilities of the Geography Section, City of London Polytechnic, available to us, and we are grateful to Don Shewan, Susannah Hall, Connie Flewitt, Ed Oliver, Mavis Teed and Clare Terry for their efficient production of the line drawings.

We also wish to express our gratitude to colleagues who have kindly supplied photographs: Dr N. F. Alley (Figs 2.15, 2.16, 2.21, 2.22, 2.23, 3.8, 5.14, 5.15 and 5.24); Dr J. Crowther (Fig. 3.22); Dr J. M. Gray (Fig. 2.7a); Mr S. Lowe (Fig. 3.1); Dr J. A. Matthews (Figs 2.1 and 5.19); and Mr R. Tipping (Fig. 3.15).

Overall, however, our greatest debt is to our wives. Not only have they helped with the usual routine chores of typing, correcting, editing and proof-reading, but they have provided encouragement when we began to despair of the book ever seeing the light of day, and they have usually managed to keep Kathy and Stephen away from their harassed fathers when the mountains of typescript threatened to engulf us. They, probably more than we, will be glad to see the manuscript safely in the hands of the publishers. This book, therefore, is for them.

John Lowe
Mike Walker
October 1983

Acknowledgements

We are grateful to the following for permission to reproduce copyright material:

Academic Press Inc (London) Ltd & the authors for figs 4.14 from fig 1 (Sparkes 1961) Copyright 1961 Linnean Society of London, 4.15 from fig 116 (J. G. Evans 1972), 5.7 from fig 1.9 (Fritts 1976) & Table 4.6 from Table p 338 (Kerney 1977b); George Allen & Unwin Ltd for figs 4.17, 4.20 from parts of figs 13.27, 8.3, 12.6 (Brasier 1980); American Association for the Advancement of Science & the authors for figs 4.21 from fig 2 (Ericson & Wollin 1968), 4.24 from fig 2 (McIntyre, Kipp et al 1976); Edward Arnold (Publishers) Ltd for Table 3.2 from Table p 371 (Embleton & King 1975); Associated Book Publishers for fig 2.33c from fig 9.6 p 130 (Sissons 1976); Blackwell Scientific Publications Ltd for Tables 4.2 from Table 2 (Walker & Lowe 1979), 4.5 from Table 1 (Kenward 1976); the Editor 'Boreas' for figs 4.10 from figs 7, 9, 12, 13 (Coope & Brophy 1972), 4.19 from fig 1 (Stuart 1979), 5.18 from fig 3 (Mörner 1980b); Cambridge University Press for fig 3.29 from fig 6 (Pennington 1970) & Table 3.6 from Table 2 (Godwin 1975); the author, Dr. G. R. Coope for fig 7.9 from fig 1 (Coope 1975a) & *Phil. Trans. Roy. Soc. B280* 313–340 1977; Edinburgh University Press for Tables 5.1, 5.2 from Tables 1, 2 (Harkness 1975); Geo Books, Norwich for figs 6.11 from fig 5 (Morrison 1978), 7.3 from fig 20 (Worsley 1977); The Geological Association for fig 2.38 from plate 13 (Hare 1947); The Institute of British Geographers for figs 2.12 (Sissons 1974), 2.32 (Cullingford & Smith 1966), 2.33a (Andrews 1970), 7.16b (Sissons et al 1966) & Table 2.1 from Table p 109 (Sissons 1974c); Institute of Geological Sciences for fig 4.18 based on (Peacock et al 1978); International Glaciological Society & the authors for fig 2.18 (Sissons & Sutherland 1976); the author, Dr. T. Kellogg for fig 4.23 from fig 23 (Kellogg 1976); Longman Group Ltd for figs 1.2b, 2.3, 3.10 from figs 11.3, 12.3, 3.2 pp 257, 280, 23 (R. G. West 1977a); the author, Andrew McIntyre for fig 7.2 from fig 17 (Ruddiman & McIntyre 1976); Macmillan Journals Ltd & the authors for figs 2.27 from fig 3 (Devoy 1977), 3.27 from fig 3 (Aaby 1876a), 3.33 from fig 6 (Paterson et al 1977), 3.34 from fig 2 (Dansgaard et al 1975),

5.4 (Wintle 1981), 5.13 from fig 1 (Pearson et al 1977), 7.14 from fig 1 (S. Lowe 1981) & Table 5.6 from Table 1 (Miller et al 1979); Masson Editeur s.a. for fig 4.11 from fig 2 (Coope 1970b); The MIT Press for Table 2.3 from Table X (Mabbutt 1977); Mouton Publishers for figs 6.10 from fig 6 (Kukla 1975), 6.12 from fig 3 (Butzer 1975) & Tables 6.3, 6.4 (Butzer 1975); Munksgaard International Booksellers & Publishers for Table 4.3 from Table p 107 (Faegri & Iversen 1975); The Editor 'New Phytologist' for figs 4.2, 4.3, 4.4 from figs 2, 4, 6 (M. J. L. Walker 1982), 4.8 from fig 4 (H. H. Birkes 1972), 5.10 from fig 9 (Pitcher et al 1977); the Editor, 'Norsk Geografisk Tidsskrift' for fig 5.20 (Matthews 1974); Oxford University Press for figs 2.10, 2.11, 3.18 from figs 17.1, 17.2, 16.1 (Boulton et al 1977), 4.6 (du Saar 1978); Pergamon Press Ltd for fig 7.20 from fig 4 (Lowe et al 1980); the Editor, 'Pollen et Spores' for Table 4.4 from Tables 2, 3 (Havinga 1964); the Editor, 'Polskie Archium hydrobiologii' for Table 5.4 from Table 1 (R. Thompson 1978b); Princeton University Press for fig 3.19 from fig 3 (Ruhe 1965); the Editor 'Quaternary Research' for figs 2.13 from fig 4 (S. C. Porter 1975), 2.42 from fig 1 (G. I. Smith 1974), 3.31 from fig 12 (Street & Grove 1979), 5.23 from fig 2 (Berggren et al 1980), 6.8 from fig 9 (Shackleton & Opdyke 1974) & Tables 7.2, 7.4 from Tables 4, 5 (Lockwood 1979); the Editor, 'Quaternary Research' & The Association of American Geographers for fig 3.20 from fig 3 (Sorensen 1977); Royal Geographical Society & the author for fig 2.41 from p 200 (Grove & Warren 1968); Royal Meteorological Society for fig 2.26 from fig 4 (Jelgersma 1966); The Royal Society & the authors for figs 4.13 from fig 12 (Kerney 1963), 7.4 from fig 3 (A. V. Morgan 1973) & Table 7.1 from Table 4 (West et al 1974); the author, Dr. N. J. Shackleton for fig 6.7 from fig 1 (Shackleton & Opdyke 1973) & Table 6.2 from Table 2 (Shackleton & Opdyke 1976); the author, Dr. J. B. Sissons for fig 3.9 from fig 28 (Sissons 1967); Societe Nationale des petroles d'Aquitaine for Table 4.7 from Table 2 (Delorme 1971); the author, Dr. J. A. Sutcliffe for fig 3.23 from fig 1 (Sutcliffe 1981); the author, Dr. R. Thompson for figs 5.21 (R. Thompson 1978b), 5.22 based on data (Thompson & Turner 1979); University of Chicago Press for fig 6.6 from fig 2 (Emiliani 1955); the Managing Editor for Regents of University of Colorado for fig 2.17 from fig 2 (Loewe 1971); University of Utah Press for fig 3.32 from fig 8 (G. I. Smith 1968); the author, Dr A van der Werff for fig 4.5 from (van der Werff & Huls 1958–74); John Wiley & Sons Ltd for figs 2.33b from fig 6 (Morner 1980a), 2.39 adapted from fig 17.3 (Flint 1971), 3.30 from fig 5.16 (Edwards & Rowntree 1980), 5.11 from (Suess 1970a), 5.17 & Table 5.3 from fig 1, Table 3 (Tauber 1970), Tables 3.3 from Table 13.1 (Bull 1980), 6.1 from (Hedberg 1976); the author, Dr. R. B. G. Williams for fig 3.16 from fig 2 (Williams 1975); the author, W. H. Zagwijn for fig 1.4 from fig 8 (Zagwijn 1975).

We are also indebted to the following for permission to reproduce photographs:

Dr. N. F. Alley for figs 2.8, 2.15, 2.16, 2.21, 2.22, 2.23, 3.8, 5.14, 5.15, 5.24; Dr. J. Crowther for fig 3.22; Dr. J. A. Matthews for figs 2.1, 5.19; Dr. J. M. Gray for fig 2.7a & Richard Tipping for fig 3.15.

CHAPTER 1

The Quaternary period

INTRODUCTION

The **Quaternary** is the most recent major subdivision (**period**) of the geological record, and it extends up to, and includes, the present day (Fig. 1.1). Together with the **Tertiary** it forms the **Cenozoic**, the fourth of the great geological **eras**. In the geological time-scale, periods are conventionally divided into **epochs**, and the Quaternary is often held to include two intervals of epoch status: the **Pleistocene**, which ended around 10 000 years ago, and the **Holocene**, which is the present warm interval within which we live.[1] However, since there are good grounds for believing that the current warm period is part of a long-term climatic cycle (see below) and is comparable to previous warm episodes of the Quaternary, the last 10 000 years can be seen as part of the Pleistocene epoch (West 1977a), and the Pleistocene can therefore be considered to extend up to the present day. This interpretation is adopted here and throughout this book the terms 'Quaternary' and 'Pleistocene' are used interchangeably, since they refer to the same interval of geological time.

THE CHARACTER OF THE QUATERNARY

The Quaternary has long been considered to be synonymous with the 'Ice Age', a view that can be traced back to the writings of Sir Edward Forbes who, in 1846, equated the Pleistocene with the 'Glacial Epoch'. It is now known, however, that the Quaternary was characterised not only by periodic widespread glacier activity (**glacials**), but also by intervening warm episodes (**interglacials**) during which temperatures were experienced that may even have been higher than those of the present day. During the last interglacial in Britain, for example, such tropical creatures as the hippopotamus swam in the River Thames, while lions and elephants roamed the present site of Trafalgar Square in central London! However, the hallmark of the Quaternary is not simply the occurrence of either warm or cold phases, but rather the high amplitude

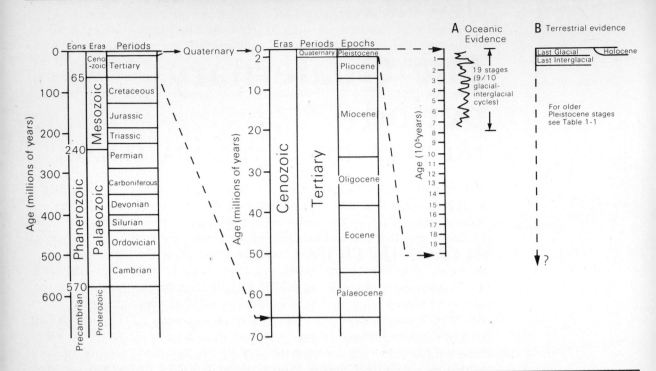

Fig. 1.1: The Quaternary in relation to the geological time-scale. The Quaternary is divided into stages or climatic episodes (glacials and interglacials). Nine or ten glacial/interglacial cycles have been identified in ocean sediments spanning the last 700 000 years, and as many as twenty such cycles may have occurred throughout the entire Quaternary period. Fewer glacial/interglacial oscillations can be detected in the terrestrial record (see text).

and frequency of climatic oscillations within what is, in geological terms at least, an extremely short time-span. In some parts of the world, temperatures may have fluctuated through more than 15 °C between warm and cold episodes, and it now appears that there were at least twenty glacial/interglacial cycles within the Quaternary period.

The effects of these climatic changes were dramatic. In the mid- and high-latitudes, ice sheets and valley glaciers waxed and waned, and the areas affected by periglacial (cold climate) processes expanded and contracted. In low-latitude regions, the desert and savannah margins shifted through several degrees of latitude as phases of aridity alternated with periods of increased precipitation. Throughout the world, weathering rates and pedogenic processes varied with changes in temperature and precipitation, river régimes fluctuated markedly, sea-levels rose and fell through perhaps 150 m, and plant and animal populations were forced to migrate and adapt as these environmental changes were thrust upon them.

The restless nature of the Quaternary environment is reflected in the complex mosaic of landforms, sedimentary sequences, faunal and floral remains and assemblages of human artefacts that date from this latest chapter of earth history. The record is often highly fragmented; evidence is absent from many areas, while detailed sequences are only locally preserved. The seemingly cyclic nature of climatic change has produced similar environments at different times, and because few sequences can be dated accurately, the linking of individual successions (**correlation**) becomes a hazardous exercise. In a single exposure of Quaternary sediments, there may be much to perplex the geologist, the geomorphologist, the botanist, the zoologist or the archaeologist, and an explanation of observed geological changes will often require the combined expertise of all of these disciplines. The purpose of this book is to illustrate the very wide range of methods that are currently employed in Quaternary research, and to demonstrate that both a **multidisciplinary** and an **interdisciplinary** approach are required if a proper understanding of the complexities of the Quaternary environment is to be achieved.

THE DURATION OF THE QUATERNARY

The beginning of the Quaternary is very difficult to define. A view that long held sway was that the Quaternary lasted for approximately one million years (a figure derived from extrapolations from weathering profiles), and that it could be differentiated from the preceding Tertiary on the basis of evidence for widespread glaciation. It is now apparent, however, that many areas of the world, particularly in the high-latitude and high-altitude regions, supported glaciers long before the onset of the Quaternary (Andrews 1975), reflecting the fact that although global temperatures had oscillated, there had been a gradual world-wide cooling throughout the Tertiary period. In the geological column, therefore, the boundary between the Pliocene and the Pleistocene cannot easily be drawn on the basis of glacial evidence but, in accord with conventional geological procedure, its position is best determined by changes in the fossil record (Flint 1965a). Accordingly, the Pleistocene is considered to begin at that point in the stratigraphic column where faunal and floral elements indicate an abrupt change from warm to cold conditions, marking the first major cold pulse of the Quaternary. However, because different indicator fossils have been used in different areas, correlations between individual sequences are difficult to effect, and because precise dating is seldom possible, general agreement on the position and age of the Pliocene-Pleistocene boundary has yet to be achieved. Age estimates range from *ca*. 1.6 to more than 2.4 million years, depending on locality and on the biostratigraphic evidence upon which the boundary is based. Clearly, however, the Quaternary spans considerably more than the one million years that is often quoted, although even then it covers no more than about 0.04 per cent of the total age of the earth, or approximately 0.3 per cent of the **Phanerozoic**, the period of geological time during which fossils have been found in rocks.

THE DEVELOPMENT OF QUATERNARY STUDIES

The term 'Quaternary' can be traced back to the work of the French geologist Desnoyers who, writing in 1829, differentiated between the strata of 'Tertiary' and 'Quaternary' age in the rocks of the Paris Basin. The Quaternary was redefined by Reboul in 1833 to include all strata characterised by the remains of flora and fauna whose counterparts could still be observed in the living world. The term 'Pleistocene' (most recent) was first used by Lyell some six years later to refer to all rocks and sediments in which over 70 per cent of the fossil molluscs could be recognised as living species. Only after the writings of Forbes in the 1840s did the term Pleistocene become synonymous with the glacial period.

Quaternary studies represent one of the youngest branches of the geological sciences, with a history that goes back less than 200 years (Chorley *et al.* 1964; Davies 1968). Prior to that it was widely believed that the earth had been created in 4004 BC, a figure derived from the biblical calculations of Archbishop Ussher of Armagh published in 1658, and hence early views on geological and environmental changes were confined within the limited time-scale of 6000 years. As a consequence, a **Catastrophist** philosophy held sway in which the form and character of the earth's surface were explained largely through the operation of great floods and other cataclysmic events. Around the turn of the eighteenth century, however, the work of the famous Edinburgh geologists Hutton and Playfair indicated that the features of the earth's surface could more reasonably be explained by the operation, over a protracted time-span, of processes similar to those of the present day. This significant departure in geological thinking gave rise to the principle of **uniformitarianism**, first expounded by Hutton but subsequently popularised by Lyell in his famous dictum 'the present is the key to the past'. Uniformitarian reasoning, in which present-day analogues are used as a basis for the interpretation of observed features within the stratigraphic record, is still fundamental to many aspects of palaeoenvironmental reconstruction, and is perhaps most widely employed in the analysis of biological evidence (Rymer 1978).

The nineteenth century saw a number of significant advances in Quaternary studies, many of which stemmed directly from the introduction and gradual acceptance of the glacial theory. Although there had been speculation for some time that certain Swiss and Norwegian glaciers had formerly been more extensive, it was not until the 1820s that credence was given to the notion of a glacial epoch. The work of Esmark in Norway, Bernhardi in Germany, and particularly the investigations of the two engineers de Venetz and Charpentier in Switzerland, produced evidence for former glacier activity far beyond the limits of present-day glaciers, but it fell to the Swiss zoologist Agassiz to expound, in 1837, the first coherent theory of 'the great ice period' involving world-wide climatic changes. Subsequently, Agassiz visited both Britain and North America and in both areas demonstrated that surficial deposits that had previously been interpreted as the products of marine inundation during the flood (**diluvium**) could more reasonably be regarded as the results of extensive glaciation in the relatively recent past.

Although the glacial theory did not immediately gain widespread acceptance, its adherents rapidly refined and developed the concept. By the 1850s evidence was beginning to emerge for two glaciations in parts of Britain and Europe, and as early as 1877, Geikie was describing evidence for four separate glaciations in East Anglia. The strata between the glacial deposits (**drift**) were referred to as interglacial, and hence the idea of oscillating warm (interglacial) and cold (glacial) episodes emerged. By the end of the nineteenth century, drift sheets of four separate glaciations (named Nebraskan, Kansan, Illinoian and Wisconsinan) had been identified in North America, while evidence began to emerge for multiple glaciations in different parts of Europe. Probably the most influential work in this respect, however, was that of Penck and Bruckner (1909) who resolved the river terrace sequences in the valleys of the northern Alps into four separate series, each related to a glacial episode. The phases of glaciation were named, in descending order of age, Gunz, Mindel, Riss and Wurm after major rivers of southern Germany. In both Europe and North America, the maximum limits of Quaternary glaciation were first mapped around the turn of the twentieth century and have subsequently been modified only in detail (Fig. 1.2), although views on the terminology adopted, and on the number of glacial/interglacial stages experienced during the Quaternary have changed dramatically (see below).

Other effects of glacier expansion and contraction were also recognised at a relatively early stage. The relation between glaciers and sea-level was discussed by MacLaren who, in 1841, reasoned that at times of glacier build-up sea-levels would fall as water was extracted from the ocean basins and locked up in the ice sheets, whereas following ice melting, sea-levels would rise as water was returned to the oceans. This was the first statement of the glacio-eustatic theory of sea-level change (Ch. 2). MacLaren suggested that sea-levels would fall by by 350–400 ft (ca.

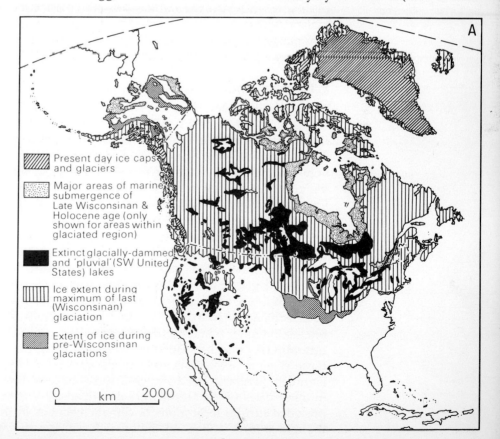

Fig. 1.2A: Glacier limits, major ice-dammed lakes and the principal 'pluvial' lakes in North America (after *USGS Nat. Atlas of USA* 1970).

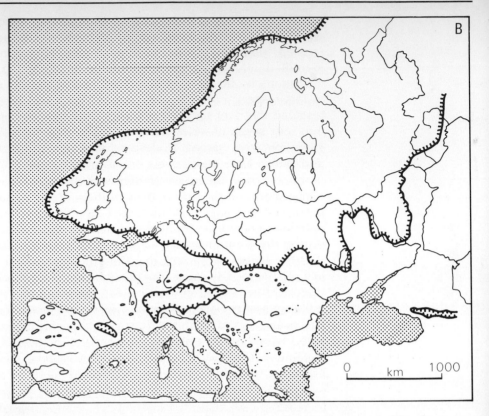

Fig. 1.2B: Maximum extent of glaciation in Europe (after West 1977a).

110–130 m) during a glacial phase, a figure that is in remarkably close agreement with many more recent estimates. In addition to its effects on global sea-levels, the results of the build-up of ice on the earth's surface were also noted. A number of workers, including Playfair and Lyell, had described the raised shoreline sequences in Scandinavia and around the coasts of Scotland, and had inferred that in both regions crustal uplift had occurred. The mechanism involved in crustal warping, however, remained unclear. In 1865, the Scottish geologist Jamieson finally made the link between the raised shoreline evidence and the glacial theory when he deduced that crustal depression would result from the build-up of glacier ice and that uplift would follow deglaciation as the crust returned to its pre-glacial state. This was the first clear statement of what are now referred to as glacio-isostatic effects (Ch. 2).

During the later years of the nineteenth century, evidence began to emerge for major environmental changes in areas beyond those directly affected by glacier ice. In the semi-arid south-west of the United States, for example, work by Russell and Gilbert in particular showed that extensive lakes had existed at some time in the past (Fig. 1.2) and that phases of higher rainfall (**pluvials**) had alternated with more arid **interpluvial** episodes. Moreover, a relationship was postulated between these climatic oscillations and the glacials and interglacials at higher latitudes. Similar relict drainage features in desert and savannah regions in other parts of the world were described by Victorian explorers and provided further indications of climatic changes in the low-latitudes. In the mid-latitude zones, on the other hand, it was gradually recognised that phases of glacier expansion would, in turn, be accompanied by an extension of the tundra belt in which cold-climate (although non-glacial) processes would predominate. The term **periglacial** was first used to describe such regions by the Polish geomorphologist von Lozinski in 1909.

Biological evidence for Quaternary environmental change also began to emerge soon after the introduction of the glacial theory. The writings of Forbes (1846) in which various geographical components of the British flora and fauna were related to successive migrations into the British Isles under different climatic conditions, and Heer (1865) wherein ecological changes in Switzerland were discussed in terms of Quaternary climatic changes, were particularly important milestones. In the later years of the nineteenth century, the work of the Scandinavian botanists Blytt and Sernander demonstrated the wealth of information on climatic and vegetational change that could be derived from the stratigraphy and macrofossil content of peat bogs. The scheme of postglacial climatic changes constructed by Blytt and Sernander from Scandinavian peat bog records (Table 3.4) was subsequently refined by the results of pollen analysis, a technique developed in Sweden by von Post (1916), and which is still one of the most widely used and successful methods in palaeoecology.

The twentieth century has seen many important developments in Quaternary studies, but three aspects in particular merit attention. The first is the methodological advances that have been made in, and the widespread application of, a range of field and laboratory techniques. Increasingly sophisticated methods of sedimentological analysis have offered new insights into the nature of Quaternary depositional environments, while the interpretation of Quaternary stratigraphy has been greatly assisted by the development of coring equipment and by the availability of borehole records. Landform analysis has been markedly improved by the use of remote sensing techniques, and especially by the improvements that have been made in recent years to the range and quality of photographic images obtained from earth satellites. Palaeoecological investigations have likewise benefited from general technological advances, particularly in the extraction of fossil remains, and in the field of microscopy. These techniques are considered in more detail in Chapters 2–4.

The second major development has been in the dating of Quaternary events. In the nineteenth century, notions of time were founded largely on estimates of the rates of operation of geomorphological processes. Hence, delta-construction, cliff retreat, stream dissection, weathering profiles and soil development were all used as the basis for a Quaternary time-scale (Flint 1971). The first, and for many years the only, method for estimating the passage of time was the varve chronology developed around the turn of the century by the Swedish geologist de Geer (1912). The major breakthrough came in the years following the Second World War with the discovery by Libby of the technique of radiocarbon dating. Other radiometric methods, notably potassium/argon and uranium-series dating, were developed in the 1950s and 1960s, along with the techniques of dendrochronology and palaeomagnetism. The 1970s have seen the refinement of a number of these methods, and also the development of new dating techniques including the highly promising method of amino-stratigraphy. The principles of these and other dating methods are discussed in Chapter 5.

The third important development in Quaternary studies during the twentieth century has been the investigation of sedimentary sequences on the deep-ocean floors. Indeed, it would not be overstating the case to suggest that the results of research into ocean sediments have revolutionised our view of the Quaternary. In one sense, trying to reconstruct environmental changes from terrestrial evidence alone is like trying to assemble a jigsaw puzzle with many of the pieces missing, for much of the evidence has been removed by sub-aerial weathering processes and, in mid- and high-latitudes, by glacial erosion. In the deep oceans of the world, however, sediments have been accumulating in a relatively-undisturbed manner for thousands, or even millions, of years, and deposits on the ocean floors therefore often form a continuous record spanning the whole of the Quaternary period.

Although the investigation of deep-sea sediments actually began in the nineteenth century with the voyage of the British government research vessel HMS *Challenger* in 1872 (Deacon 1973), detailed work on the fossil content of core samples from the ocean floors was first undertaken by the German palaeontologist Schott in the 1930s. Prior to the Second World War, only short sediment cores (less than a metre in length) could be raised from the sea bed. The development of a piston corer by the Swedish oceanographer Kullenberg opened up the modern phase of deep-sea research, for with the Kullenberg corer and specially-equipped research ships, it became possible to take undisturbed sediment cores 10–15 m in length (Kullenberg 1947, 1955). The changing fossil content of these cores has provided a remarkable record of changes in ocean water temperatures, and, by implication, in atmospheric temperatures during the Quaternary (Ch. 4). Many fossils, however contain another index of environmental change, namely variations in oxygen isotope content. Pioneered by Emiliani (1955), oxygen isotope analysis is now regarded as one of the principal techniques in Quaternary stratigraphy and palaeo-environmental reconstruction (Ch. 3). A highly-readable account of the major developments in the study of ocean sediments can be found in Imbrie and Imbrie (1979).

THE FRAMEWORK OF THE QUATERNARY

The conventional division of the Quaternary is into **glacials** and **interglacials**, with further subdivision into **stadial** and **interstadial** episodes. These terms are not always easy to define precisely however. Glacials are generally considered to be lengthy cold phases during which the major expansions of ice sheets and glaciers took place, whereas stadials are regarded as shorter cold episodes in which local ice advances occurred. Interglacials are usually considered to be warm intervals during which temperatures at the thermal maximum were as high or even higher than those experienced during the Holocene, and which were characterised in the mid-latitudes by the development of mixed woodland.

Table 1.1: Quaternary stratigraphic schemes for the Northern Hemisphere based on terrestrial evidence. Warm phases (interglacials) are shown in italics. Note (a) no direct evidence of glaciation has been found in the pre-Anglian of Britain, and (b) no correlation is implied between regions.

European Alps*	Central North America†	Northern Europe‡	Britain§
Postglacial	*Holocene*	*Holocene*	*Flandrian*
Wurm	Wisconsinan	Weichselian	Devensian
Riss-Wurm	*Sangamon*	*Eemian*	*Ipswichian*
Riss	Illinoian	Saalian	Wolstonian
Mindel-Riss	*Yarmouthian*	*Holsteinian*	*Hoxnian*
Mindel	Kansan	Elsterian	Anglian
Gunz-Mindel	*Aftonian*	*Cromerian*	*Cromerian*
Gunz	Nebraskan	Menapian	Beestonian
**Donau		*Waalian*	*Pastonian*
**Biber		Eburonian	Pre-Pastonian
		Tiglian	*Bramertonian*
		Pretiglian	Baventian
			Antian
			Thurnian
			Ludhamian
		Reuverian	*Reuverian*
		(= Pliocene)	

* From Penck and Bruckner (1909).
** Stages added by Wolstedt (1958).
† From Flint (1971).
‡ From West (1977a).
§ After Mitchell *et al.* (1973) and West (1980)

Interstadials, on the other hand, were relatively short-lived periods of thermal improvement during a glacial phase, in which temperatures did not reach those of the present day and in lowland mid-latitude regions an interglacial temperate woodland did not develop. This type of categorisation, based essentially on inferred climatic characteristics is referred to as a **climatostratigraphic** scheme, and is considered further in Chapter 6.

The traditional stratigraphic subdivisions of the Quaternary in Europe and North America are shown in Table 1.1. These are provided primarily as a guide to terminology used throughout this volume, although it must be emphasised that no correlations are implied by the table, and that no concepts of time are involved. However, it is generally agreed that the **Flandrian** of the British sequence can be equated with the Holocene of the European and North American schemes and that the last glacial stage identified in Britain (**Devensian**), northern Europe (**Weichselian**), North America (**Wisconsinan**), and perhaps also in the Alps (**Wurm**), can be considered as broad correlatives. General agreement also exists in north-west Europe over the short climatic oscillation that occurred towards the close of the last glacial stage (termed the **Devensian Lateglacial** in Britain and the **Weichselian Lateglacial** in continental north-west Europe), for this period can be more precisely dated than older parts of the succession. However, opinions differ over the extent to which the 'Lateglacial' period can be subdivided; in Britain, most follow a twofold subdivision of a **Lateglacial Interstadial** followed by a **Loch Lomond Stadial**, whereas a rather more complex sequence is envisaged on the European mainland (Fig. 1.3). For reasons that are not currently understood, there appears to be no clear record of a 'Wisconsinan Lateglacial' climatic oscillation in North America.

Beyond that, however, there is little general agreement, for large parts of the Quaternary terrestrial record are undated, and therefore correlations

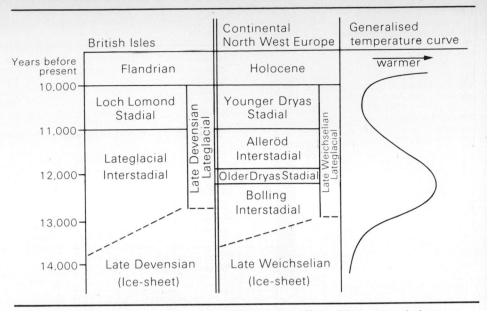

Fig. 1.3: The Lateglacial in Britain and North-West Europe (after Mangerud *et al.* 1974; Gray and Lowe 1977).

between individual sequences are difficult to effect. Moreover, it is now apparent that there are major gaps in the terrestrial stratigraphic record. For example, a comparison between the sequences in south-east England and the coastal areas of the Netherlands (Fig. 1.4) suggests that over one million years of early and middle Quaternary sedimentary history may be missing from the former area. Evidence from the deep-ocean floors where relatively-uninterrupted sedimentation has been in progress throughout the Quaternary indicates that there were nineteen stages or major climatic episodes in the past 700 000 years alone (Shackleton and Opdyke 1973), and that within the Quaternary as a whole, twenty-one full glacial and interglacial cycles may have occurred (van Donk 1976), many more than are represented in Table 1.1. The stages shown in Table 1.1 must therefore be regarded as no more than a partial record of the Quaternary glacial/interglacial sequence; continuity from one stage to another should never be assumed, and correlations that have been made between stages from different stratigraphic schemes should be treated circumspectly (Bowen 1978).

THE NATURE OF CLIMATIC CHANGE

Although a detailed consideration of climatic change lies beyond the scope of the present volume, some reference is necessary to the controlling influences on Quaternary climate and, in particular, to those factors that may have combined to produce the sequence of glacial and interglacial episodes. It has long been recognised that the climatic fluctuations of the Quaternary followed a cyclic pattern, and numerous suggestions have been proposed to account for both the regularity and frequency of Quaternary climatic fluctuations (see reviews in e.g. Flint 1971; Sparks and West 1972; Goudie 1977; West 1977a). One hypothesis that has attracted

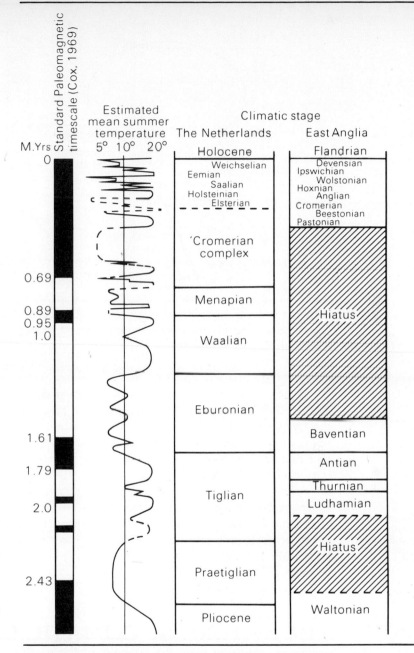

Fig. 1.4: Comparison of Quaternary stratigraphies in the Netherlands and East Anglia (after Zagwijn 1975).

much critical attention is the **'Astronomical Theory'**, developed by Croll a little over 100 years ago, and subsequently elaborated by the Yugoslavian geophysicist Milankovitch in the early years of the twentieth century. The theory was based on the assumption that the surface temperatures of the earth would vary in response to periodic changes in the earth's orbit and axis. Over approximately 96 000 years, the shape of the earth's orbit is known to change from circular to elliptical and back (Fig. 1.5A), while the axis tilts from about 21.5° to 24.5° and back over the space of around 42 000 years (Fig. 1.5B). A third variable is the 'precession of the

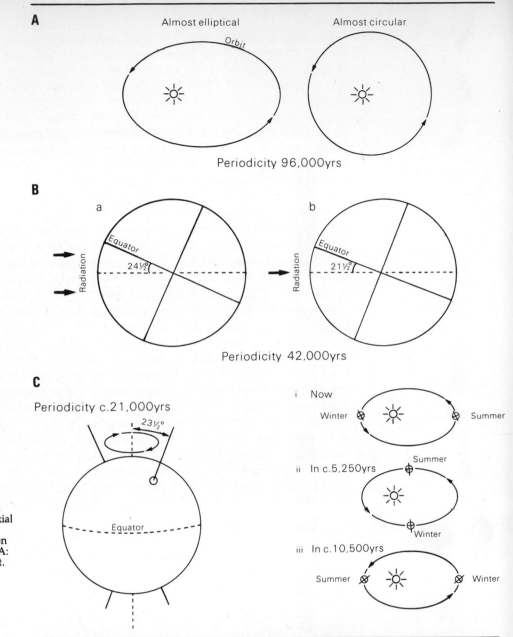

Fig. 1.5; Orbital and axial variations affecting receipt of solar radiation at the earth's surface. A: Eccentricity of the orbit. B: Axial tilt. C: Precession of the equinoxes (right) resulting from the 'wobble' of the earth's axis (left). For further explanation see text.

equinoxes' resulting from the 'wobble' of the earth's axis. This means that the time of year at which the earth is nearest to the sun (**perihelion**) varies. At present, the northern hemisphere winter occurs in perihelion (Fig. 1.5Ci), while the summer occurs at the furthest point on the orbit (**aphelion**). In about 10 500 years time the position will be reversed (Fig. 1.5Cii), while 21 000 years hence the cycle will be complete and the present situation will once again obtain (Fig. 1.5Ciii). These variables in combination will affect the amount of radiation received at the earth's surface, and patterns of change can be calculated from astronomical data. Using these data Milankovitch was able to obtain estimates for radiation inputs at different latitudes and hence infer temperature changes through time.

The theory was published in 1924, but because the predicted pattern of climatic changes did not appear to accord with the number and frequency of glacial/interglacial cycles observed at that time in the terrestrial record it was widely rejected. More recently, however, an appreciation of the regularity and periodicity of Quaternary climatic change that has arisen from work on sea-level variations (e.g. Mesolella *et al.* 1969), and particularly from the continuous oxygen isotope traces from deep-ocean sediments (Ch. 3 and 6), has led to a re-establishment of the Milankovitch hypothesis (Imbrie and Imbrie 1979). Detailed analysis of ocean core sequences by Hays *et al.* (1976) suggests that climatic oscillations operate over 100 000, 43 000, 24 000 and 19 000 year cycles. The longest cycle appears to drive the glacial/interglacial sequence, while combinations of the others introduce minor perturbations onto the main curve, and these may account for the short-term climatic episodes (e.g. the stadial and interstadial phases) observed in the stratigraphic record. The possible climatic effects of these variables are represented schematically in Fig. 1.6.

Variations in the earth's orbit and axis, therefore, may offer a coherent explanation of Quaternary climatic change that accords with the evidence for climatic fluctuations preserved in the ocean sediment records. It should be stressed, however, that debate continues about the precise climatic effects of the individual cycles, and about the other factors that need to be taken into consideration in the development of a comprehensive theory of climatic change. In this respect, the gradual northward movement of the continents, mountain-building along the margins of the continental plates, changing patterns of ocean-water movement and variations in the output of solar radiation may all have had an influence on global temperatures during the Quaternary, and these effects would therefore have been superimposed on the temperature variations resulting from astronomical changes (Wigley 1981). However, the degree to which these factors affected Quaternary climates has yet to be assessed or quantified.

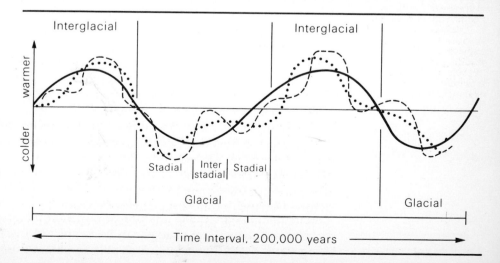

Fig. 1.6: Schematic representation of possible Quaternary mean temperature variations according to 100 000 (solid line), 43 000 (dotted line) and 24 000 (dashed line) year cycles of Hays *et al.* (1976).

THE SCOPE OF THIS BOOK

The aim of this book is to provide a critical assessment of the methods and approaches that are currently employed in the reconstruction of Quaternary environments. The work does not, however, claim to be exhaustive. Indeed, in view of the wide range of disciplines involved in Quaternary research, a comprehensive treatment would run far beyond the space of a single volume. Some aspects are therefore considered only briefly, while others (which some will no doubt believe to be important) are omitted altogether. To some extent the choice of material reflects the interest of the authors, but an attempt has nevertheless been made to present what is considered to be a balanced view of the methods and sources of evidence that form the basis for Quaternary environmental reconstructions. Some temporal bias is inevitable, as far more is known about the later part of the Quaternary than about the earlier parts of the period, and therefore the majority of examples are drawn from the last interglacial and last glacial stages. The methods, approaches and principles, however, are equally applicable to the analysis of early and middle Quaternary environments. Further, although evidence has been presented from many parts of the world, there is inevitably an emphasis on British material, as it is within the British Isles that the majority of the authors' own research experience lies. Nevertheless, it is hoped that readers in Europe and North America will find material here that is of interest to them also.

The book falls naturally into three parts. In Chapters 2, 3 and 4, the morphological, lithological and biological evidence that forms the basis for environmental reconstruction is outlined. Although these are useful general categories within which to describe particular techniques and approaches they are, to some extent, artificial and there are considerable overlaps between them. Hence, in Chapter 2 where the emphasis is on morphology, certain aspects of the stratigraphy of river terraces and raised shoreline sequences need to be considered also, while in Chapter 3, where sedimentological evidence is being discussed, reference is frequently made to landform evidence as, for example, in the analysis of sand dunes formed in loess and coversand deposits. In all three chapters, field and laboratory techniques are introduced in order to give an indication of the procedures that are involved in obtaining the basic data.

Chapters 5 and 6 comprise the second part of the book. The various dating methods that are currently employed in Quaternary studies are described and evaluated in Chapter 5, while the principles of stratigraphy and correlation which enable the researcher to build up meaningful spatial and temporal sequences from often fragmentary evidence are outlined in Chapter 6. The final part (Ch. 7) consists of a reconstruction of changing environmental conditions in the British Isles during the last cold stage (Devensian), and exemplifies the methods and approaches that have been described in earlier chapters. It illustrates how often diverse evidence can be synthesised into a coherent picture of environmental change and, in so far as it highlights the gaps in our present state of knowledge, it also serves as a basis for future investigations.

NOTES

1. Because the end of the Pleistocene coincided with the close of the last glacial phase, the period of time since then is often referred to as the 'Postglacial'.

CHAPTER 2

Morphological evidence

INTRODUCTION

The pronounced oscillations in global climate that occurred repeatedly during the Quaternary led to major changes in the types and rates of operation of geomorphological processes in many parts of the world. Undoubtedly the most spectacular manifestations of climatic change were the great ice sheets whose passage resulted in widespread modification of the land surface of the mid- and high-latitude regions. The growth and decay of the ice sheets were accompanied by the expansion and contraction of areas affected by periglacial activity, there were fundamental changes in the régimes of many of the major rivers, and the nature and effectiveness of geomorphological processes were strongly influenced by the changes that were continually taking place in the distribution and type of vegetation cover. In the low latitudes, phases of aridity were interspersed with periods of wetter climatic conditions so that desert regions were often more extensive during the Quaternary, water levels in pluvial lakes rose and fell, and alluvial processes fluctuated markedly. On the global scale, sea-level during the glacial phases was over 100 m lower than that of the present day, but rose to a position above that of the present during the interglacials.

In many parts of the world, the characteristic landforms that developed under a previous climatic régime have often survived, albeit in a much denuded form as 'relict' or 'fossil' features. Careful analysis of these landforms, and particularly of landform assemblages, can often provide information on the nature of the climatic régime under which they evolved, and also on other aspects of the environment in which they were formed. The use of morphological evidence in this way, however, requires a proper understanding of the relationships between geomorphological processes and land forms. Moreover, it must be emphasised that there is frequently a close relationship between morphological evidence and lithological evidence (Ch 3) and that, wherever possible, the two should be used in conjunction in the reconstruction of Quaternary environments.

METHODS

Field methods

Mapping

The production of a map illustrating the distribution of the principal landforms is often the first stage in the investigation of the Quaternary history of an area. In some types of analysis, for example the interpretation of drainage characteristics or variation in pedological development, it may be necessary to construct a map showing specific aspects of landforms and landscape type. This can be achieved by the technique of **morphological mapping** (Waters 1958; Savigear 1965), which is concerned specifically with the recognition of individual slope elements in the landscape and the nature of the junctions between them (Crofts 1981). A simple instrument such as an Abney Level or a clinometer can be used for slope measurement. Typical morphological maps are produced at scales of 1: 10 000 or larger for even the most subtle changes in the shape of the land surface are being recorded. This approach has been widely employed in geomorphological investigation, but because it is not specifically concerned with landscape evolution, it has found less favour with Quaternary workers. **Geomorphological mapping**, on the other hand, is one of the most important techniques in Quaternary research, for the maps produced contain not only information on morphology, but also on genesis and, in some cases, age of the landforms. This type of mapping can be carried out at a variety of scales ranging from very detailed maps of small areas (typically 1: 10 000) to maps at the national scale (e.g. IGS Quaternary Map of the British Isles 1977, scale 1: 625 000). Geomorphological mapping is essentially interpretative and therefore requires both an appreciation of the complexity of landform assemblages and a detailed knowledge of their genesis. It also needs an eye for detail, a grounding in mapping and survey techniques (Gray 1981), and a knowledge of the properties of aerial photographs, as the mapping of landforms on aerial photographs usually precedes work in the field. Geomorphological mapping, particularly at large scales (*i.e.* 1: 10 000 or greater), has been most effectively employed in the analysis of glacial landscapes, particularly those resulting from the passage of the last ice sheets and from more recent phases of glacier activity (see, for example, Sissons 1967a, 1976 and refs therein).

Further discussion on mapping techniques in geomorphology can be found in King (1966), Cooke and Doornkamp (1974) and Crofts (1974, 1981).

Instrumental levelling

In reconstructing the Quaternary history of an area, it is often necessary to determine the precise altitude of, and differences in altitude between, particular landforms. Altitudinal data can aid in the interpretation of landform assemblages, and may also enable landforms of different age to be identified. For example, only fragments of former river terraces may be preserved in a particular area, and it may be impossible to identify and correlate terrace fragments of similar age, and to develop a chronology of terrace development simply on the basis of field mapping. By obtaining precise altitudinal measurements on each terrace fragment, however, formerly continuous features can be reconstructed, gradients can be deduced, and the temporal relationships between individual terraces can be established. The same applies also to the investigation of abandoned shoreline features (see discussion on sea-level changes below). Where only a general impression of altitude is required and where the mapping is being carried out at a relatively small scale, it may be sufficient to obtain the altitudinal data from spot heights and contours on the relevant base maps. Where a more detailed investigation is being conducted, however, altitudes and landform surface gradients must be obtained by instrumental measurement in the field.

The comparison of altitudes of landforms, especially from widely-separated localities, requires a common **datum**, a plane of known altitude to which all subsequent measurements can be referred. A frequently employed datum has been sea-level, but as this can be highly variable (Lisitzin 1974), altitudinal data are more reliable when related to national survey bench marks (although clearly, this is difficult in some of the more remote high-latitude regions). In Britain, these are known as **Ordnance Survey** bench marks, points of known altitude above a common **ordnance datum** (OD, formerly OD Liverpool and now OD Newlyn, Cornwall).

Surface altitudes have been obtained in the field using: (a) an **aneroid barometer**; (b) hand-held (e.g. **Abney**) levels; and (c) a **surveyor's level**. Atmospheric pressure decreases with increase in altitude, and since it is possible to establish relationships between pressure and altitude, measurements of atmospheric pressures using an aneroid barometer can provide an indication of ground altitude. Comparison of altitudes for a series of stations using this method is termed **barometric levelling.** The method, however, is frequently inaccurate as a consequence of relatively rapid pressure and temperature changes during variable weather conditions. For similar reasons, unreliable results are often obtained in areas of hilly terrain. Hand-held levels are useful for rapid surveys, but also tend to produce variable results due principally to operator errors. It is, for example, very difficult to maintain a horizontal line between ground stations with an instrument which is only hand-held and not supported. In most geomorphological fieldwork in the mid-latitude regions therefore, particularly where precise altitudinal data are required, surveyor's levels are used and traverses are closed to national survey bench marks. However, in areas where no such bench marks are available, such as Arctic Canada, Greenland and Antarctica, sea-level must serve as the datum, and traverses are usually closed to measured sea-level in each local area. Levelling methods are outlined in detail in Higgins (1970) and Bannister and Raymond (1972).

Remote sensing

Remote sensing refers to the acquisition of images of earth surface features (and, to a limited extent, sub-surface features) by a variety of devices that receive radiation (electromagnetic) or sonar information reflecting variations in the topography, density or other aspects of the earth's surface. This includes conventional **photographic** images obtained from aircraft or satellites, using the visible and non-visible (e.g. infra-red) light spectra, **radar** sensors, **sonar** techniques (echo sounders), and recently (although not yet in common use) **lasers**. Some of the more widely-used techniques in Quaternary research are considered briefly in this section.

Aerial photography

Since the First World War, aerial photographic reconnaissance has increased both in frequency of use and in degree of sophistication. Good quality aerial photographs are now available even for the most inaccessible areas allowing at least a preliminary map to be made of any prominent or strongly-patterned landforms (e.g. Karlén 1973; Hjort 1979). A system of grid corrections can be used for transferring details from photographs to maps where scales differ, or where the photographs contain serious distortions (see e.g. Kilford 1963; Bannister and Raymond 1972; Thompson 1966; or Wolf 1974, for relevant techniques). Aerial photographs are particularly useful in the mapping of landforms in that:

(a) they direct attention to areas where landforms are most evident or abundant, thus avoiding much wasted ground reconnaissance;
(b) they illuminate ground detail that is not obvious at ground level;
(c) they reveal larger scale landform patterns that are often not visible on the ground, such as shorelines of lakes in semi-arid areas;
(d) they may record morphological features since obscured by afforestation programmes or urbanisation;
(e) repeated surveys allow the monitoring of, for example, changing landscapes, changing landform assemblages or changes in the position of glacier termini through time.

Disadvantages of aerial photographs include distortions due to camera tilt or variation in camera altitude, loss of detail due to cloud cover or shadow effects, poor tonal contrasts so that, for example, drift cannot easily be separated from bedrock surfaces, and difficulty in detection of small-scale, yet often geomorphologically significant landforms. Field mapping therefore remains essential, and even where mapping is based on large-scale, good quality aerial photographs, the results must be viewed as a *provisional* map of the Quaternary geomorphology of a region until the interpretations can be checked on the ground. For areas where good topographic maps are unavailable, aerial photographs provide the only reliable basis for analysing landform distributions, and enlargements can be made for this purpose (Gordon 1981).

Satellite imagery

A number of spacecraft have produced images of the earth that can form the basis for terrain evaluation, but the most useful so far have been the American satellites (LANDSAT-1 *formerly Earth Resources Technology Satellite*, or ERTS-1, launched 1972), LANDSAT-2 (launched 1975) and SKYLAB-1 (launched 1973). LANDSAT makes a circular orbit of the earth fourteen times each day, and transmits images continuously to receiving stations in Maryland, California and Alaska. The image sensors in each LANDSAT can cover the entire globe, with the exception of the poles, every eighteen days. The great advantages of satellite imagery over aerial photographs are that distortions are minimised[1], the process is much more rapid, and repetitive images of large parts of the earth's surface can be obtained. Conventional photographs are obtained by the simultaneous recording of all features viewed through the lens, whereas **scanning sensors** operate by sensing one spot at a time as they sweep across the area in view. The incoming radiation is focused onto a detector which transforms radiation intensity into an electronic signal and an image is composed by assembling a large number of spots (**pixils**), in a similar manner to the formation of a television picture (Verstappen 1977). In **multi-spectral scanning** (MSS) the energy received from different narrow wavelengths is recorded simultaneously, and images can be produced using only one spectral band, or, all of the information can be combined in a 'colour composite' image (NASA 1976; Townshend 1981). This technique has opened up new areas in remote sensing, for the use of infra-red or near infra-red wavelengths permits image sensing of features not observable in the visible part of the spectrum where, for example, objects are obscured by cloud, haze, smoke or even vegetation. The potential of satellite imagery in the analysis of Quaternary landforms is considerable, and present applications range from the mapping of former lake shorelines in Africa (Verstappen 1977; Ebert and Hitchcock 1978), to the investigation of landform assemblages in presently glacierised areas (various photographs in Sugden and John 1976).

Radar

This operates on the principle of the emission of pulsed signals, usually in the microwave and higher radio frequencies, from a transmitter, and the recording of the 'echos' of these signals as they are bounced back from the ground surface. The returning signals are affected by ground surface roughness, by the orientation of upstanding features, and by the density and electrical properties of ground materials. In dry sediments or cold ice, boundaries between stratigraphic units or ice layers can often be determined. Airborne radar equipment ('echo sounders') have been developed that automatically transform received signals into images (**imaging radar**), usually referred to as **sideways looking airborne radar** (SLAR). As in satellite scanners, the pulsed signals scan the terrain to one side (sometimes to both sides) of the aircraft, and the received signals are converted into electrical impulses that are recorded directly onto magnetic tape or transformed into a photographic image. Once again, the great advantage of this technique in geomorphological mapping is that data can be obtained even in cloudy or adverse weather conditions. Radar has proved particularly useful in the investigation of ice thickness, and of the nature of subglacial topography in presently glacierised regions (Morgan and Budd 1975; Whillans 1976; Robin et al. 1977).

Sonar and seismic sensing

A number of techniques have been developed that are based on the gravitational, magnetic or electrical properties of the earth. Movements of **acoustic** or **sonic** waves are affected by the density and other characteristics of different materials, and these have formed the basis for seismic surveys that have been particularly widely-used in geophysical exploration. Recently, Quaternary sediments and landforms have been investigated with seismic equipment. For example, the **Sparker** sound wave emitter has been employed in offshore areas around the coasts of the British Isles to locate submerged landforms and to establish the thickness and extent of Quaternary sediments (Garrard and Dobson 1974; Boulton *et al*. 1981). Similarly, **acoustic-reflection profiling** has been used to map offshore landforms near the Connecticut coast (Flint and Gebert 1976), submerged moraines in Lake Superior (Landmesser *et al*. 1982) and to investigate submarine terraces and morphology in the western Baltic (Healy 1981).

GLACIAL LANDFORMS

Ever since the general acceptance by scientists of the 'Glacial Theory' in the middle years of the nineteenth century, it has been recognised that landforms of glacial erosion and deposition are important palaeoenvironmental indicators. When mapped carefully, they reveal a great deal about the extent and thickness of former ice masses, the directions of ice movement at both local and regional scales, and the manner of glacier retreat. In certain circumstances, the evidence may be used to reconstruct the configuration of the former glacier surface. Where this can be compared with present-day glaciers for which certain controlling climatic parameters have been established, there is a basis for palaeoclimatic reconstruction. The first stage in this process, however, is the production of an accurate map of the extent of the former glaciers.

Extent of ice cover

Establishing the maximal extent of the Pleistocene ice sheets and glaciers has long been regarded as one of the most challenging aspects of Quaternary research. In North America, the systematic field mapping of the outer limit of Pleistocene ice began soon after 1860. The maximal extent of ice was based largely on the recognition of conspicuous 'end moraines' or on the termination of the cover of glacial drift ('Drift Border'), and by 1878 a map had been constructed showing the position of the drift border from Cape Cod to North Dakota (Flint 1965b). Similar investigations of glacial drift cover and end moraines were under way in Europe, Asia and parts of the southern hemisphere, so that by 1894, Geikie was able to compile maps of the worldwide distribution of glaciers during the 'Last Ice Age'.

The principal morphological evidence used in the reconstruction of former ice-marginal positions are lateral and terminal moraines, outwash spreads and sandar[2], marginal meltwater channels, and valley-side or downvalley limits of stagnation moraine or boulder spreads (boulder limits). Lateral and terminal moraines and meltwater channels mark the position of the ice margin at its maximum extent, whereas within those limits, linear morainic ridges will reflect subsequent

recessional stages (Fig. 2.1) as the ice becomes temporarily stabilised during deglaciation, while widespread moundy topography (**dump** or **hummocky moraine**) will result from glacier stagnation *in situ*. The types of deposits and landform assemblages produced during ice wastage will be determined by a range of often interconnected factors, including manner (e.g. rate) of glacier retreat, debris content of the ice, position of entrainment of debris in the ice, and topographical influences. In general, however, the overall distribution of a variety of glacial landforms will broadly define the extent of the formerly glaciated area, with both lateral (ice marginal) and vertical (valley-side) limits.

The land lying beyond or above the area directly affected by glacier ice will have experienced periglacial activity, with the shattering of exposed rocks by freeze-thaw processes, the development of solifluction features (lobes, terraces etc), and the formation of structures associated with the action of ground ice (wedges, patterned ground etc). The distribution of these periglacial features can, in certain cases, provide further evidence of the former extent of glacier ice. In upland areas, for example, the boundary (or, more commonly, the zone of transition) between glacially-scoured and frost-shattered bedrock is referred to as the **trim line** and indicates the approximate positions of the former ice margins. The lateral extent of the area affected by frost action can be used in a similar manner to delimit formerly glacierised areas.

A number of difficulties arise in mapping the former extent of Pleistocene glaciers on the basis of morphological evidence. First, at the height of the last glaciation in both Europe and North America, ice masses submerged many upland areas. Hence, although morphological evidence can be found in many places to mark the outer limits of the ice sheets and glaciers, the vertical extent of ice is more difficult to establish from field evidence alone. In order to obtain estimates of ice thickness, therefore,

Fig. 2.1: Recessional moraines formed after the 'Little Ice Age' glacier advance, Storbreen, Norway.

recourse must usually be made to models of former ice sheets (see below). Secondly, successive ice sheets covered broadly the same areas, except towards the outer margins (Figs 1.2a and 2.3), so that the morphological evidence from earlier glacial phases has usually been destroyed by later ice advances. As a consequence, in many areas of the mid-latitudes that were affected by Pleistocene glaciers, the majority of the landforms that have been preserved date only from the later stages of the last glaciation. Thirdly, many glacial landforms have been considerably modified by periglacial activity both during and after regional deglaciation, and this often poses problems in field mapping and interpretation. Fourthly, some glacial landforms may resemble ice-marginal features but may not, in fact, be so. Glaciofluvial landforms (kames, eskers etc), for example, often exhibit a linear trend and have, on occasions, been used as evidence for former ice limits (e.g. Charlesworth 1926). Such features may, however, form in a number of different glacial situations and are poor ice-marginal indicators. A proper understanding of the nature and origin of glacial landforms is therefore necessary if this type of evidence is to be used to determine the former extent of Pleistocene glacier ice. Finally, it should be noted that in many areas the outer margins of drift sheets have no distinctive morphological expression, and end moraines in particular are often absent. In some cases, the evidence has been destroyed, either by meltwater activity during deglaciation, or by postglacial sub-aerial weathering. In other situations, the glaciers either did not carry sufficient debris, or did not maintain a steady-state position for the length of time necessary for the construction of an end moraine.

Where end moraines are found within the areas formerly covered by glacier ice, further problems of interpretation are encountered. Much discussion has obtained over whether such moraines are, strictly speaking, recessional, in that they formed during a stillstand of the ice margin in a phase of overall glacier retreat, or whether they have been produced by a renewed phase of glacier activity and therefore reflect a **glacier readvance**. Where prominent end moraines occur, the latter interpretation has usually been adopted as it has been considered unlikely that large constructional forms would have developed simply during a stillstand of the ice margin. The morphological evidence, however, is frequently equivocal, and it must be emphasised that conclusive proof of glacier withdrawal and a subsequent readvance can only be obtained from stratigraphic evidence (e.g. where organic sediments are found between two glacigenic deposits) or, in some instances, from other morphological features, such as independent indicators of changing directions of ice flow between successive glacial episodes (e.g. Robinson and Ballantyne 1979). An additional complication in the interpretation of end moraine evidence is that it is frequently very difficult to distinguish between those landforms that have developed from a glacier readvance induced by a deterioration in climate, and constructional forms that have been produced by a **glacier surge** resulting from short-lived instability within the former glacier system which may or may not have been climatically determined. Although differences have been detected both in the morphology and internal composition of moraines produced by recent normal and surging glaciers (Rutter 1969), a distinction between older forms is more difficult, and this clearly poses problems both in glacier modelling and in palaeoclimatic interpretations based on reconstructed Pleistocene glaciers and ice sheets.

Morphological evidence and the extent of the last ice sheets and glaciers

Northern Europe

In many parts of northern Europe, conspicuous end moraines mark the outer limits of the last Scandinavian ice sheet, and also important recessional stages during glacier decay, while in the mountains of Norway and Sweden, there is abundant morphological evidence for more recent glacier activity. The southernmost extent of Weichselian ice is generally considered to be delimited by the Brandenburg moraine which can be traced intermittently for some 500 km across the North German Plain (Fig. 2.2), while its correlative (the Zyrian limit) trends north-eastwards for over 2000 km across European Russia. In northern Germany and Poland, recessional positions of the ice margin are marked by a number of morainic assemblages, including the Frankfurt, Pomeranian and Velgast moraines. These features, however, are complex and may have resulted, at least in part, from readvances of the ice margin rather than stillstands during overall deglaciation. For example, the younger Frankfurt moraine overlaps the older Brandenburg moraine in both the east and west, and similar indications of oscillations of the ice margin have been detected in Denmark (Berthelsen 1979) and in southern Sweden where large numbers of 'recessional moraines' are found (Berglund 1979; Mörner 1979). Sequences of submerged moraines have been discovered by echo sounding both off the Norwegian coast (Holtedahl and Sellevol 1972) and in the Gulf of Bothnia in the northern Baltic (Winterhalter 1972).

The best-known and most prominent moraines of the Fennoscandian ice sheet have been mapped as almost continuous belts across Norway (**Ra Moraines**), Sweden (**Central Swedish Moraines**) and Finland (**Salpausselkä Moraines**). All of these are believed to have formed as a result of a readvance during the Younger Dryas Stadial (Fig. 2.2). Morphologically and genetically, however, the moraines are very different. In many parts of Norway, for example, constructional forms are small (1–5 m) or are poorly developed (Aarseth and Mangerud 1974), whereas the Salpausselkä moraines are prominent features over 100 m in height and up to 2 km wide in places (Virkkala 1963). All were at some time considered to be true end moraines, but the Salpausselkä have been found to be composed almost entirely of stratified drift, and are now believed to be features that formed in fracture zones parallel to, but perhaps some distance away from the actual ice front (Virkkala 1963; Hyvärinen 1973). As such, they cannot be regarded as reliable indicators of former ice-marginal positions. Only one major moraine appears to have formed in Norway (Ra Moraine), at the same time as an oscillating ice front in Sweden and Finland produced several moraines. In addition, radiocarbon dates associated with the Ra Moraine indicate that the feature is not of the same age throughout its length but that the ice margin was advancing in some areas while retreat was underway elsewhere (Mangerud 1980).

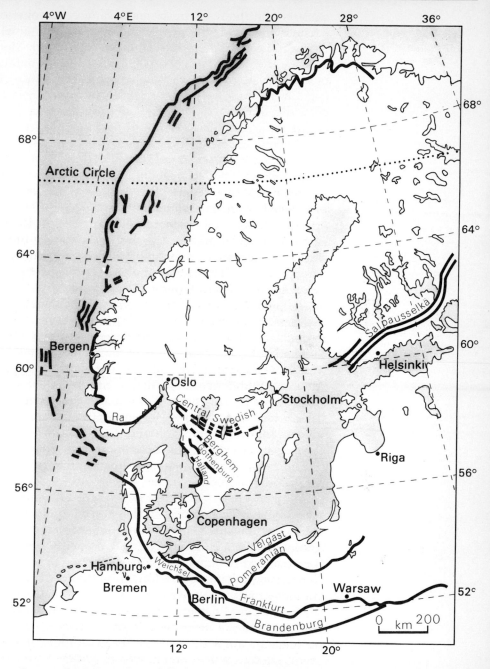

Fig. 2.2: The limit of the Weichselian ice sheet in Northern Europe, and inferred recessional stages, based principally on morphological evidence (after Bowen 1978; Andersen 1979; Berglund 1979; Hillefors 1979; and Mangerud 1979).

These contrasts reflect differing glaciological responses to topographic and climatic factors at different points around the ice sheet and suggest that, on the continental scale at least, former ice limits as shown by morphological evidence are more likely to have been metachronous than synchronous. This has obvious implications for ice sheet models derived from such data (see below).

By shortly after 9000 BP (yrs before present), the Scandinavian ice sheet had virtually disappeared (Andersen 1980; Ignatius *et al.* 1980), but in many high mountain areas of Norway and Sweden there is widespread morphological evidence for renewed glacier activity after that time. This consists principally of 'fresh' and relatively unvegetated terminal and lateral moraines, many of which occur in bands downvalley from present day active glaciers (e.g. Karlén and Denton 1976; Mottershead and Collin 1976; Griffey and Matthews 1978). Radiocarbon dating of soils beneath the moraines, associated with lichenometric and dendrochronological evidence (see Ch. 5), indicates that glacier advances occurred on a number of occasions during the Holocene but particularly during the **Neoglacial** period (*ca*. 3000–2000 BP) and during the **Little Ice Age**, which has been dated about AD 1500 to 1920 depending on locality.

British Isles

In Britain, the maximum extent of the last (Devensian) glaciation has traditionally been placed where a morphological distinction can be made between 'Older' and 'Newer Drift' (Fig. 2.3). In the borderland of south-east Wales, for example, contrasts in the degree of drift dissection and the relative preservation of the glacial landforms provide a basis for the mapping of the limit (Luckman 1970), while in the Irish lowlands the 'Southern Irish End Moraine' (Charlesworth 1928) which can be traced intermittently for over 200 km forms a prominent morphological divide between the relatively fresh and less-weathered drift to the north and the more deeply dissected surficial material to the south. Elsewhere, however, the morphological evidence is often equivocal, and in a number of places it is clear from the stratigraphic evidence that morphological features which have long been regarded as marking the outer limits of the 'Newer Drift' do not, in fact, do so. For example, the Escrick Moraines have usually been held to mark the maximum extent of the last ice sheet in the Vale of York (e.g. Embleton and King 1975a), but it now appears that the Devensian ice front extended some way to the south of these landforms (Gaunt 1974). Similarly, there is unequivocal stratigraphic evidence to indicate that, at its maximum, Devensian ice occupied a considerable area of south and west Wales beyond the 'South Wales End Moraine' (Bowen 1981). In both of these cases, therefore, the morphological evidence relates not to the maximal extent of the last ice sheet, but to recessional stages during ice wastage. Moreover, in a number of areas what had previously been mapped as 'moraines' are now known to be glaciofluvial features which may therefore have formed within, and not necessarily at, the margins of Devensian ice.

Morphological evidence has frequently been cited in support of readvances of the Devensian ice sheet. In many cases, however, postulated readvances have not been substantiated by stratigraphic investigation while in other cases (e.g. the 'Aberdeen-Lammermuir' and 'Perth' readvances – Fig. 2.3), the morphological evidence for a readvance has been shown to be capable of an alternative explanation (Sissons 1976). More convincing evidence for a renewed period of glacier activity following retreat from the Late Devensian maximum can be found in the Scottish Highlands, the Southern Uplands, and the hills of the Lake District, Wales and Southern Ireland. Terminal, lateral, and extensive spreads of hummocky moraine, clear trim lines on the mountain sides, intricate meltwater channel systems and the distribution of periglacial features define limits of the **Loch Lomond Readvance**[3] which is broadly

Fig. 2.3: Limits of glaciation in Britain during the Anglian, Wolstonian and Devensian (last) glacial stages, and readvances of the last ice sheet. Bold lines represent limits that are still generally accepted. Dashed or dotted lines mark readvance limits that are no longer considered to be valid (modified after West 1977a).

equivalent with the Younger Dryas of Scandinavia (Sissons 1979a). The 'freshness' and relatively unweathered nature of the landforms have meant that the extent of the 'readvance' can be mapped in great detail, and this has not only enabled the ice limits to be established accurately, but has also allowed the individual cirque and valley glaciers to be reconstructed. The application of this work in palaeoclimatic reconstruction is considered below.

The limits of the last ice sheet in Britain and the readvances that occurred during ice wastage are discussed in more detail in Chapter 7.

North America

As in Britain, it has been the custom in North America to distinguish between 'Earlier' and 'Later Drift' (Chamberlin 1895), the boundary between the two indicating the maximum extent of the last (Wisconsinan) Laurentide ice sheet (Fig. 1.2). For a little over half its length, the outer limit of the Wisconsinan drift sheet between the Atlantic Ocean and the Rocky Mountains is marked by a terminal moraine (Flint 1971), although there are large areas where no morphological distinction can be made between Wisconsinan and pre-Wisconsinan deposits, and mapping of the former ice limit rests largely on stratigraphic evidence. Within the Wisconsinan limits, end moraines are common, marking stillstands or readvances of the ice margin during retreat from the glacial maximum, and in many areas the identification, tracing and correlation of these moraines have tended to form the basis for the deglacial chronology of the Wisconsinan (Wright 1976).

In the Great Lakes region, where several distinct ice lobes developed, the sequence is particularly complex (Fig. 2.4). Willman and Frye (1970), for example, mapped over thirty moraines which they believed to have formed by oscillations of the Michigan lobe as it retreated across Illinois, while further west the Superior lobe wasted back from the huge St Croix moraine in a completely different manner, producing tunnel valleys instead of recessional moraines (Wright 1972), although recessional moraines beneath Lake Superior have been identified from seismic reflection profiling (Landmesser *et al.* 1982). Significant breaks in the deglacial sequence appear to have occurred around 14 500 BP with the **Cary advance** of the Michigan lobe, and readvances also of the Erie lobe in the east and the Des Moines lobe in the west; around 13 000 BP represented by the prominent Port Huron moraine of the Michigan and Huron lobes; and around 11 500 BP with the **Valders Readvance** of the Michigan lobe (Wright 1971). Further north in the Lake Superior region across northern Ontario and Michigan, a zone of end moraines formed between *ca.* 11 000 and 10 000 BP, and these reflect successive halts in general ice retreat. This **Algonquin Stage** is comparable in age with the Younger Dryas of Scandinavia (Saarnisto 1974).

Although the morphological evidence for the extent of ice in the Great Lakes area is impressive, the interpretation of the evidence is far from straightforward. In many areas, the moraine sequences are complex and younger moraines are frequently found overlapping or cross-cutting older landforms. In northern Indiana, for example, the innermost moraines of the Erie lobe are not recessional features from the last ice to invade the area, but are much older moraines veneered by the latest drift (Bleuer 1974). Correlation between lobes in different parts of the Great Lakes region has proved particularly problematical. Differences in age at the outermost moraines may reflect the westward shift of the ice dispersal centres during the Wisconsinan glaciation (Wright 1976) with the result

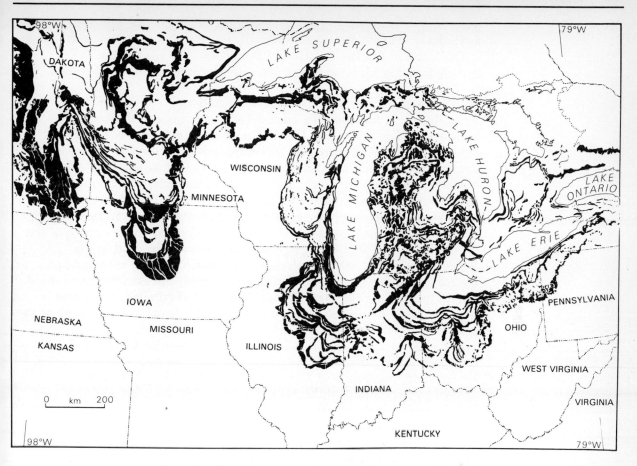

Fig. 2.4: The Wisconsinan ice limits and major recessional stages (dark shading) in the Great Lakes region, North America (based on *The Glacial Map of the United States east of the Rocky Mountains, Geol. Soc. Amer. 1959*).

that the glacial maximum was attained earlier in the east than in the west. There has been considerable discussion over the occurrence of readvances during deglaciation, and the absence in particular of independent pollen evidence for climatic deterioration (e.g. Webb and Bryson 1972) has led to speculation that some of the major readvances that have been recognised (e.g. the Valders Readvance) may not have been climatically determined, but could have resulted from glacier surges (Wright 1971, 1973, 1980). Although abundant, therefore, the morphological evidence is frequently equivocal and recourse must be made to lithostratigraphic and biostratigraphic evidence in order to obtain a proper appreciation of the extent of ice at various stages of the Wisconsinan in the Great Lakes region.

Further west on the Canadian Prairies, end moraines and the distribution of hummocky moraines and meltwater channels have been used to delimit the extent of Wisconsinan ice (Stalker 1977; Christiansen 1979), although in many parts of the region conspicuous morphological features are relatively uncommon and the limits of the last glaciation can only be established on stratigraphic grounds (Prest 1970). The large expanses of hummocky disintegration moraine suggest widespread ice sheet stagnation, however, following the Late Wisconsinan glacial maximum. On the eastern flank of the Laurentide ice sheet, the extensive lateral moraines of the Saglek system can be used to define the upper

Fig. 2.5: Late
Wisconsinan terminal
moraines, eastern Sierra
Nevada, California, USA.

limits of Wisconsinan ice in northern Labrador (Ives 1976), while
numerous end moraine sequences have been identified in the Canadian
Arctic and sub-Arctic marking recessional stages of the last ice sheet (e.g.
Falconer *et al.* 1965). These include moraines of the **Cockburn Event**,
dated to the period between *ca.* 9000 and 8000 BP, which were first
identified on Baffin Island and whose possible correlatives can be traced
across Keewatin, Ellesmere Island, northern Labrador and western
Greenland (Andrews and Ives 1978). The '**Cochrane Readvance**' in the
area between Hudson Bay and the Great Lakes also culminated a little
before 8000 BP and appears radiometrically correlative with the Cockburn
phase (Andrews 1973), although a surge has been suggested to account
for this readvance of the ice margin, probably related to the incursion of
the sea into Hudson Bay (Prest 1970). Subsequently the ice sheet split into
three residual masses, those in Keewatin and Labrador-Ungava having
virtually disappeared by 5000 BP, leaving only the Barnes Ice Cap on Baffin
Island as the last vestige of the Laurentide ice sheet (Bryson *et al.* 1969).
    Although morphological evidence for the central part of the Cordilleran
ice sheet between the Coast Range and the Rocky Mountains is limited
(Flint 1971), the valley and piedmont glaciers left a wealth of evidence in
the form of lateral and terminal moraines (Fig. 2.5), outwash terraces and
trim lines which mark both the maximal extent of Wisconsinan ice and
also recessional stages and readvances following the glacial maximum (e.g.
Wright and Frey 1965; Mahaney 1976 and papers therein). Of particular
interest are the 'fresh' and relatively unweathered moraines of the
Neoglacial period (Porter and Denton 1967) found in the mountains from

Colorado to Alaska. The evidence suggests three major phases of glacier expansion during the Holocene at 30–450 years BP, 2400–3300 years BP and 4900–5800 years BP (Denton and Karlén 1973a). In the mountains of the Yukon and Alaska, radiocarbon dates on organic material incorporated into or beneath moraines suggested a further short-lived phase of glacier activity *ca*. 1050–1300 years BP (Rampton 1970; Denton and Karlén 1977).

Glacial landforms therefore continue to act as a cornerstone in establishing the extent of former glacier ice. They are of greatest value in the investigation of recent (i.e. Lateglacial and Neoglacial) patterns of glacier activity, especially in highland regions where the features are often well-preserved and relatively easily mapped, and from which the vertical and lateral extent of glacier ice can often be reconstructed. For earlier glacial episodes, morainic landforms in particular can still be used to delimit the glacierised area, and as a basis for mapping of readvances, although it is clear that a proper appreciation of glacial and deglacial sequences rests as much (if not more) on stratigraphic verification as on morphological evidence. This is discussed more fully in Chapters 3 and 6. Meanwhile, one further line of evidence is required before ice sheets and glaciers can be reconstructed, namely the former directions of ice movement, and it is to this aspect of glacial landforms that we now turn our attention.

Fig. 2.6: Parallel, ice-moulded grooves in bedrock, reflecting dominant direction of ice movement. Western Mull, Scotland.

Direction of ice movement

In many formerly-glaciated areas, a 'grain' or streamlined sculpture is

evident in the landscape reflecting the former direction of ice movement. At a small scale, bedrock protuberances are scratched (striated), fractured, polished and grooved (Fig. 2.6), and at larger scales, whalebacks, roches moutonnées and glaciated valleys (troughs) are fashioned by overriding ice. This preferred alignment of erosional forms in a glaciated landscape is often best seen where ice has emphasised local bedrock contrasts, particularly where ice has followed the trend of geological weaknesses such as relatively incompetent strata and joint and fault lines (Sissons 1967a; Sugden and John 1976). Certain glacial depositional landforms may also be aligned in the direction of glacier flow. Careful mapping of landforms that are ice-directed therefore allows the main pattern of ice movement to be reconstructed.

Striations.   **Striations** (or **striae**) form where stones entrained within the basal layers of the ice are dragged across bedrock surfaces. The plotting of striation trends is a relatively simple field exercise and, given the availability of exposed striated bedrock in accessible areas, regional directions of ice movement can, at least in theory, be determined fairly rapidly. In practice, however, the interpretation of striation data is rather less straightforward. Not all 'scratch marks' on bedrock surfaces in formerly glaciated regions have resulted from the passage of ice. Many will simply reflect lines of weaknesses in the rock, accentuated perhaps by sub-aerial weathering, while others may have resulted from fluvial, fluvioglacial or even periglacial (e.g. solifluction) activity. Where striations are of glacial origin, they may reflect only basal ice movements determined by local bedrock irregularities. They can often be seen, for example, to follow the curvature of the face of a bedrock protuberance, and on the lee sides (with respect to ice movement) of rock obstacles are often perpendicular to the dominant or regional direction of glacier flow (Demorest 1938). In some areas, diverging sets of striae can be found reflecting perhaps more than one direction of ice movement (e.g. Virkkala 1951), while on certain rock outcrops striations with significantly different trends may be found crossing or superimposed upon one another. **Crossing striations** can arise where glaciological conditions have changed over time, or where ice has readvanced into a region. If the later ice advance has a different direction of flow and the striations resulting from the initial ice advance have not been completely erased, a second set of striations will become superimposed upon the ones that remain from the initial phase of glacier activity. In some areas, it may be possible to distinguish between different sets of cross-striations; for example, in Sweden the relative ages of the principal glacier movements have been inferred from the fineness, sharpness, relative abundance and topographic position of striations (Virkkala 1960; Stromberg 1972; Vorren 1977). In most areas, however, such distinctions are difficult to make. Overall, it would seem that striations are best used in conjunction with other lines of evidence in the reconstruction of regional patterns of ice movement.

Friction cracks.   A range of fractures or 'friction cracks' (Fig. 2.7) results from stones in basal ice being forced against underlying bedrock (Boulton 1974). Perhaps the best known and most widely reported are '**crescentic**

Fig. 2.7: Small-scale features of glacial erosion exposed on bedrock surfaces in North Wales: (a) Friction fractures (b) Crescentic scars. These features either form concave up-ice (e.g. fractures) or down-ice, and may be useful ice-directional indicators in areas where other erosional features, such as striae, are not preserved.

gouges' and 'crescentic fractures' (Embleton and King 1975a). Crescentic gouges are believed to form concave down-ice and crescentic fractures concave up-ice, and the direction of concavity has been used to interpret former direction of ice movement (e.g. Stieglitz *et al.* 1978). Consistent patterns do not always emerge, however, and the use of friction cracks in isolation as ice-directional indicators now seems a doubtful procedure (Gray and Lowe 1982). Nevertheless, where employed in conjunction with other evidence (e.g. striations) they can often lend useful support in the reconstruction of regional ice-flow patterns (e.g. Andersen and Sollid 1971; Thorp 1981).

Roches moutonnées and stoss-and-lee forms.   An irregular bedrock surface presents numerous obstacles to the passage of ice. This leads to compression of the ice and increased erosion of the upstream (stoss) sides of the obstruction, whereas plucked and frost-shattered craggy surfaces tend to characterise the downstream (lee) sides. A series of ridges running at right angles to the direction of basal ice flow will, after prolonged glaciation, be smoothed only on the up-ice side to form **stoss-and-lee** landforms, and where a consistent pattern is evident in the landscape, this can be used to infer ice-flow directions. Where a bedrock ridge runs parallel to the direction of ice flow, the bedrock will become smoothed on the up-ice sides and also along the flanks to produce the landforms known as **roches moutonnées**. These can vary considerably in size (Sugden and John 1976), and often the whole floor of a valley, or a wide expanse of lowland formerly subjected to glacial erosion, will show a consistent pattern of smoothed ice-scoured surfaces (commonly with striae) on one side, and irregular craggy faces on the other. Such ice-moulded landforms, in conjunction with striae and friction cracks, provide good evidence for former directions of ice flow, although the fact that some large scale ice-moulded features can survive more than one phase of glaciation can lead to problems of interpretation where the direction of ice movement has changed between successive glacial episodes.

Glaciated valleys.   During the build-up of an ice sheet, glacier flow is concentrated initially in pre-existing valleys and these are progressively modified to form **glacial troughs**. As ice builds up, outlets become impeded and ice from the most congested valleys overtops the lowest cols on the interfluves to escape into neighbouring valley systems. The original drainage network is therefore modified by **watershed breaching** (Fig. 2.8). In some areas, ice will exploit lines of structural weakness in which case the trend of glacial troughs may be significantly different from the original drainage pattern. Eventually, individual valley glaciers may coalesce and the mountain summits become submerged beneath an ice sheet, although the major troughs and breaches will still constitute the principal avenues of flow within the ice sheet. By plotting the trends of glacial troughs and interconnecting breaches, therefore, the principal routes taken not only by valley glaciers but also by ice streams beneath the former ice sheets can be established, and the major centres of ice accumulation and dispersal can be identified. The major glacial troughs in west-central Scotland, for example, are arranged in a radial fashion around Rannoch Moor, a pattern first noted by Linton (1957) who also drew attention to similar 'radiative dispersal systems' from the uplands of the English Lake District, southern Norway and South Island, New Zealand. In all of these regions, it is assumed that successive ice sheets developed from the same centres of ice accumulation, so that the present morphology of the major glacial troughs is a product of several phases of ice sheet and valley glacier erosion. Problems can arise, however, in such gross interpretations of former glacier movements as a result, for example, of migrating ice sheds and ice

Fig. 2.8: Watershed breach in the Kananaskis Valley area of the Rocky Mountain Front Range, Alberta, Canada.

centres during the growth of ice sheets over successive glacial phases, and from the fact that in some areas ice flowed in the opposite direction to the gradients of the pre-glacial river valleys. Examples of such 'intrusive glacial troughs' (Linton 1963) can be found in Glen Eagles in south-east Scotland where ice from the Scottish Highlands pushed southwards into the Ochil Hills, and in the Finger Lakes region of New York State where ice flowed south to intrude into the northern escarpment of the Allegheny Plateau (Rice 1977).

Streamlined glacial deposits. Subglacial debris, deposited beneath moving ice is frequently streamlined in the direction of ice movement. Streamlined glacial deposits are often found in glaciated landscapes and range in size from small-scale **fluted ground moraine** with a relief amplitude that can be measured in centimetres, through larger flutes and **drumlins**[4] to very large drift accumulations some tens of metres in thickness in **crag-and-tail** forms. These depositional landforms invariably record the basal movements of the last ice mass to have affected an area, for any subsequent glacier with a different direction of movement would have erased, or at least substantially modified, such features. Certainly almost all of the detailed work that has been published on drumlins and fluted moraine relates to forms that have developed in drift of the last glaciation.

Drumlins are undoubtedly among the most intensively studied of all glacial landforms and have been particularly widely used as ice-directional indicators. They frequently occur in 'fields' or 'swarms' in lowland areas where there was little obstruction to the passage of ice, or in piedmont

Fig. 2.9: Drumlin in the Bow Valley, near Calgary, Alberta, Canada. Ice flow was from right to left across the photograph.

zones where flow was radiative or dispersive. They are also occasionally found on the floors of glacial troughs. Many are ellipsoidal in form, some are almost circular, while at the other extreme, linear drumlins several kilometres in length have been observed (Lemke 1958). Most possess a prominent stoss end with a trailing distal slope (Fig. 2.9). It is generally agreed that the direction of the drumlin long axis reflects local direction of ice movement with the stoss end usually pointing up-glacier. The ice-moulded or streamlined form appears to be produced by variations in stress levels at the base of the ice, although the precise mode of formation of the features is far from clear (see reviews in Embleton and King 1975a; Menzies 1978).

In addition to the regional ice flow trends displayed by their long axes, the overall shape of drumlins can provide information on former glacier dynamics, such as indications of basal ice pressure (Chorley 1959), and rate and type of ice flow (Doornkamp and King 1971; Sugden and John 1976). Numerous shape indices have been developed using axial and outline ratios (e.g. Chorley 1959; Reed et al. 1962; Jauhiainen 1975) which enable comparisons that are independent of scale to be made between drumlins in different areas. Although most measurements of drumlin shapes have been made from maps and aerial photographs, detailed field mapping such as that undertaken by Barnett and Finke (1971) in the eastern United States and southern Germany is very much to be preferred, for subtleties in form are not always expressed on topographic maps, or are not always clearly identifiable on aerial photographs (Rose and Letzer 1975).

Reconstruction of former ice masses

The general trends and geographical distribution of the glacial landforms discussed above constitute a major source of evidence in the reconstruction of former ice sheets and glaciers. For example, ice-directional indicators such as drumlins, fluted moraines and striations, in association with evidence from glacial erratics and till provenance studies (Ch. 3) enable flow lines to be established, on the basis of which major ice dispersal centres, ice domes and ice divides can be located (e.g. Shilts 1980). By combining this geomorphological data with evidence for glacial isostatic rebound following deglaciation based on raised shoreline studies (see below), and glaciological theory derived from observations on present day ice sheets and glaciers, reconstruction of the morphology of Pleistocene ice sheets is possible (Andrews 1982).

A steady-state model of the last (Devensian) British ice sheet has been constructed by Boulton et al. (1977). Glacial geomorphological information (ice-directional landforms, distribution of glacial erratics, and mapped

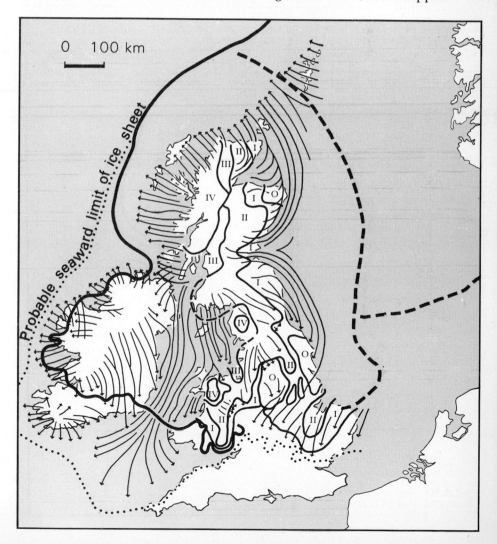

Fig. 2.10: Patterns of ice-sheet movement over Britain and patterns of glacial erosion. The dotted line shows the maximum extent of glaciation in Britain, while the bold continuous line shows the accepted limit of Late Devensian ice, dashed in the speculative area of the North Sea and omitted over central England. The arrows show generalised ice-flow directions based on geological evidence. Zones of increasing erosional intensity from 0 to IV are also shown (after Boulton et al. 1977).

glacial limits) was used to establish the probable pattern of ice sheet flow lines (Fig. 2.10) and the shape of the ice sheet surface (Fig. 2.11) was based on climatic modelling of precipitation levels in different parts of Britain during the last glacial (e.g. Williams and Barry 1974) and on analogies with present-day ice sheets. At its maximum, the ice sheet consisted of a series of domes centred on the main dispersal centres, the largest of which over central Scotland had a modelled summit height of over 1800 m. This would imply local ice thicknesses in the last British ice sheet of well in excess of 1200 m. For the Fennoscandian ice sheet, reconstructions show a major ice dome centred over the Gulf of Bothnia with a maximum ice thickness of over 2500 m (Denton and Hughes 1981; Oerlemans 1981). Numerous models have been developed for the Laurentide ice sheet in which maximum ice thicknesses range from a little over 2000 m (e.g. Peltier 1981) to over 4000 m (e.g. Sugden 1977) at the last glacial maximum. Some of these are single dome models centred over the Hudson Bay area (Denton and Hughes 1981), while others incorporate

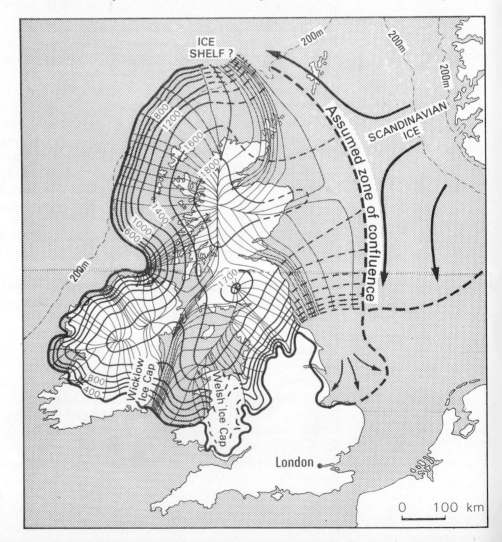

Fig. 2.11: The modelled surface topography (in metres) and flow lines of the Late Devensian ice sheet (after Boulton *et al.* 1977).

a series of ice domes over Keewatin, Labrador/Ungava and Baffin Island (Peltier 1981).

Glacier modelling exercises have also been carried out in order to assess the rate at which ice sheets develop (e.g. Andrews and Mahaffey 1976; Budd and Smith 1981). The conventional view of ice sheet build-up is of glacier development in coastal mountain ranges where precipitation is high and the gradual coalescence and expansion of these glaciers into adjacent lowland regions (e.g. Flint 1943). Time-scales of the order of tens of thousands of years are usually envisaged for the development of the great Quaternary ice sheets. More recently it has been suggested that ice sheets may develop in a more catastrophic manner by the coalescence of snow banks spread over wide areas, both coastal and inland (Ives *et al.* 1975). The term **instantaneous glacierisation** has been introduced to refer to this kind of glacier development. Dynamic models of the Laurentide ice sheet incorporating this concept suggest that under favourable mass balance conditions, very large ice sheets can develop in the space of 10 000 years (Andrews and Mahaffey 1976).

These glaciological reconstructions are a major recent development and they clearly have profound implications for many aspects of Quaternary study. They provide new insights into global sea-level variations, plant and animal migrations and glacial stratigraphy and chronology. They may also help explain many of the observed morphological characteristics of glaciated landscapes. On the basis of the inferred thermal régime of the British ice sheet model, for example, Boulton *et al.* (1977) suggest that the conditions under the centres of ice outflow, the Scottish Highlands, the Lake District, North Wales and the Southern Uplands, would not be conducive to high rates of erosion at the glacier maximum. Hence the landscapes of glacial erosion in those areas may have been fashioned during periods of more limited glacierisation when local ice caps existed in these areas. In North America, Sugden (1978) identified a zone of maximum glacial erosion which formed a ring between the centre of the Laurentide ice sheet and its periphery. This coincided with a zone where, on the basis of the ice sheet model (Sugden 1977), meltwater from the ice sheet centre is calculated to have frozen on to the bottom of the ice sheet. It was therefore suggested that this regelation incorporated basal debris into the ice and therefore afforded a highly efficient means of debris evacuation.

Clearly, however, reconstructing Quaternary ice sheets is a difficult and complex field, and many of the models produced so far must be regarded as speculative (although they can act as hypotheses against which to test field evidence). The geomorphological evidence upon which some of the reconstructions have been based is equivocal and, in particular, evidence for the maximal extent of many parts of the last great ice sheets is lacking. There is, for example, no clear evidence for the maximum position of the last British ice sheet (Fig. 2.3) on either the continental shelf to the north-west of the British Isles, nor for the eastern margin of the ice sheet in the North Sea. Moreover, throughout the life of an ice sheet, ice thickness, ice flow directions and ice marginal positions will vary, and this clearly poses problems for both dynamic and steady-state modelling. Further, most models of ice sheet development are based on observations of present-day glacier distributions and climatic patterns and thus may not provide valid analogues for the very much more extensive Quaternary ice sheets. These and other aspects of reconstructing former ice sheets are discussed in detail by Andrews (1982).

Less speculative reconstructions are possible for smaller ice caps (Fig. 2.12) and for glaciers that were restricted to cirque and valley situations (Fig. 2.13). These include the Loch Lomond Readvance glaciers of Britain (Sissons 1974a, 1979b) and the Neoglacial glaciers that developed in the mountains of both the northern and southern hemispheres (e.g. Porter 1975). The shape of the former glacier can be traced by joining those points or areas where clear ice-marginal evidence (e.g. terminal or lateral moraines; valley-side limits or downvalley termination hummocky moraine) is preserved (Fig. 2.14). A problem here is that glaciers often leave abundant depositional evidence in the lower ablation zones, but little in the higher accumulation zones. If trim-line evidence is unavailable for these upper areas, then extrapolation becomes necessary throughout often large sections of the outline of the reconstructed glacier. When the glacier outline has been determined, ice-surface contours can be estimated by analogy with typical contour patterns on present-day glaciers. Ice-surface contours are commonly perpendicular to valley walls near the median altitude of a valley glacier, and they become progressively more convex towards the glacier terminus and more concave towards valley headwalls.

Fig. 2.12: Surface form and thickness of the Gaick ice cap, Grampian Highlands, Scotland. 1: Ice margin. 2: Ice shed. 3: Ice-surface contour (50 m interval). 4: Land contour, 100 m interval. 5: Ice thickness in hundreds of metres. Ice-free areas stippled (after Sissons 1974a).

Once the ice-surface contours have been drawn, it is possible to estimate the volume of ice within the former glacier, and also to calculate the altitude of the **equilibrium line**. The equilibrium line is the line on a glacier separating the **accumulation area** (the area where a glacier gains in mass) from the **ablation area** where a net loss of mass occurs. A term that is often used synonymously with equilibrium line is the **firn line**, which is

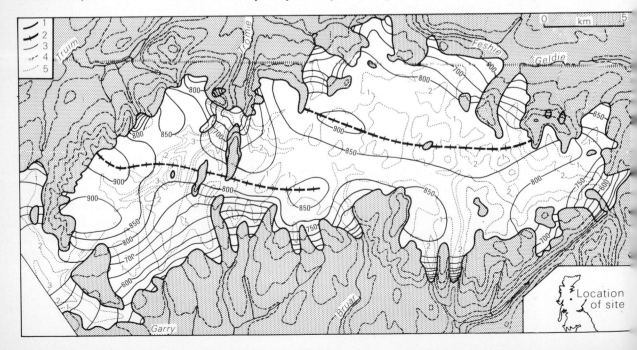

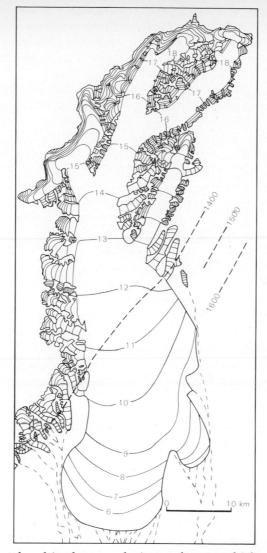

Fig. 2.13: Reconstructed topography of the Tasman glacier and smaller nearby glaciers, South Island, New Zealand, during the Balmoral (Late Pleistocene) Advance. Contour intervals on glaciers – 100 m. Isolines of equal ELA shown by dashed lines (in metres) (after Porter 1975).

the altitude on a glacier surface to which consolidated granular snow (**firn**) recedes on surviving a full summer season's melt (Fig. 2.15). In fact the two are not quite the same,[5] but for the purposes of the present discussion the **equilibrium line altitude (ELA)** and the **firn line altitude (FLA)** can be taken to be synonymous. Once the ELA/FLA has been established for individual glaciers, the regional firn line (or snowline) can then be reconstructed.

Two basic methods are employed to estimate the approximate altitude of the ELA or FLA on reconstructed glacier surfaces. The most easily accomplished is the calculation of an **accumulation area ratio (AAR)** for the glacier, which is the ratio between the accumulation area and the total area of the glacier. It has been found that the AAR for present-day glaciers in a steady-state (stable ELA) commonly lies between 0.6 and 0.7 (Porter 1970). Assuming that former glaciers had AARs approximately equal to this value, the altitude of the ELA can be rapidly computed from maps or aerial photographs where the altitudinal distribution of the

Fig. 2.14: Hummocky moraine in the Pass of Drumochter, Grampian Highlands, Scotland. The clearly-defined altitudinal limit to the moraines may reflect the maximum surface altitude of the last glacier in the area.

Fig. 2.15: Firn line on the Scud Glacier, Northern Coast Mountains, British Columbia, Canada, showing the contrast between fresh snow (upper) and bare ice (lower).

former glacier surface can be measured, for the ELA will be approximately equivalent to the lower limit of the accumulation area.

The second method is more elaborate and involves calculations of the variation in altitudinal distribution of the former ice masses. It rests on two assumtions:

1. That the reconstructed glaciers were at their maximum extent and therefore in equilibrium, so that the firn line at the end of the ablation season marks the line where total accumulation and ablation were exactly balanced.

2. That a linear relationship exists between the **ablation gradient** (the rate of decrease in the ablation rate with altitude) and **accumulation gradient** (the rate of decrease of accumulation with decrease in altitude).

The equilibrium firn line altitude can then be calculated from the following equation:

$$x = \frac{\sum\limits_{i=0}^{n} A_i h_i}{\sum\limits_{i=0}^{n} A_i}$$

where x = the altitude of the firn line in metres; A_i = the area of the glacier surface at contour interval i in km^2; h_i = the altitude of the mid-point of contour interval i; and n + 1 = the number of contour intervals (Sissons 1974a). The technique was used to reconstruct the firn line altitude of the Gaick ice cap that formed during the Loch Lomond Stadial in the Grampian Highlands of Scotland (Fig. 2.12), the data for which are shown in Table 2.1, and equilibrium FLAs have subsequently been calculated on this basis for a large number of Loch Lomond Readvance glaciers in northern Britain (e.g. Sissons 1977a, 1979b, 1980; Cornish 1981). From these data regional equilibrium firn lines or snowlines have been obtained (Fig. 7.20).

Contour interval (m)	h_i (m)	A_i (km^2)	$A_i h_i$
900–950	955	12.5	11 562.5
850–900	875	71.4	62 475.0
800–850	825	90.1	74 332.5
750–800	775	40.2	31 155.0
700–750	725	27.2	19 720.0
650–700	675	19.2	12 960.0
600–650	625	13.8	8 625.0
550–600	575	9.5	5 462.5
500–550	525	6.2	3 255.0
450–500	475	2.9	1 377.5
400–450	425	1.1	467.5
350–400	375	0.1	37.5
		294.2	231 430.0

Table 2.1: Calculation of firn-line altitude based on morphological evidence for the former Gaick ice cap in the Scottish Highlands (after Sissons 1974a).

The firn line is calculated at $\frac{231\ 430.0}{294.2}$ m = 787 m, or in general terms at 780–790 m.

Palaeotemperature estimates from glacial geomorphological evidence

The expansion of ice masses in the mid-latitude regions resulted from lower global temperatures which, in turn, caused widespread lowering of firn-line altitudes. If FLAs can be estimated for times in the past, then by studying the relationship between climatic factors and corresponding present-day firn-line altitudes, former temperature and, in certain cases, precipitation values can be inferred.

Cirque-floor altitudes.   The precise relationship between the ELA of a small cirque glacier (Fig. 2.16) and the altitude of the cirque floor is difficult to define, but a close correspondence is generally assumed. Since the regional snowline approximates the ELA of cirque glaciers (the exact relationship can be measured for any particular area) the latter can be derived for times

Fig. 2.16: A present-day cirque glacier – the Angel Glacier, Mt Edith Cavell, Alberta, Canada.

in the past from the altitude of cirque floors. Where the average annual temperature is known for the altitude of the present snowline, the temperature reduction required to depress the snowline to the altitude of cirques presently devoid of glaciers can be calculated from regional temperature lapse rates. A further index used in the same way is the **glaciation limit**, the altitude above which glaciers exist on mountains. Estimates can be derived for former glaciation limits by comparing mountain summits with cirques with mountain summits showing no morphological evidence of cirque glaciation (Sugden and John 1976). From various estimates of past snowlines, Andrews (1975) has suggested that average annual temperatures during the last glaciation in the United States were most probably about 6 °C to 10 °C lower than at present.

The use of cirque-floor altitudes in palaeoclimatic reconstruction is limited by a number of factors. These include the fact that (a) the snowline associated with cirque and valley glaciers is normally somewhat lower in altitude that the average altitude on exposed summits and slopes, since wind-drifting and low insolation protects accumulated snow and ice within cirque basins; (b) it can only provide a means of estimating the snowline, and hence palaeotemperatures, for time when the cirques were occupied by small cirque or valley glaciers, and it is not suitable for calculating temperatures during periods of more extensive ice cover; and (c) not all cirques in the landscape were occupied by ice contemporaneously, and since some variation in the altitude of cirque glaciers can be expected through the influence of factor (a), contemporaneity of cirque glacier development may be very difficult to establish.

ELA/FLA method.   More precise palaeotemperature measurements are possible using this technique. Since the ELA on a glacier is governed by both temperature and precipitation, graphs can be produced showing the relationship on a steady-state glacier between mean temperature during the ablation season and the accumulation-ablation balance at the firn line which is a reflection of precipitation (Fig. 2.17). On reconstructed glaciers, therefore, for a given precipitation value, the summer temperature can be obtained from such graphs directly. This approach has been widely used by J. B. Sissons to derive July–September temperatures at the firn line on reconstructed glaciers of Loch Lomond Readvance age in northern Britain.

Two problems are encountered in the application of this method, however. First, precipitation/temperature graphs have to be employed which are applicable to the area under investigation. In the case of the British Isles, for example, where no glaciers exist at the present day, Sissons has assumed that the curves calculated for Norwegian glaciers (e.g. Ahlmann 1948; Liestøl 1967) provide the closest analogues for Loch Lomond Readvance glaciers in northern Britain. Secondly, in order to derive a temperature value from the graphs, precipitation levels at the firn line must be known. In his work on the Gaick ice cap, Sissons (1974a) estimated precipitation at the equilibrium firn line to have been around 80 per cent of present-day levels. On that basis, summer temperatures at the firn line (780–790 m) were calculated to have been about 1.5 °C. This is equivalent to a sea-level temperature of 7.6 °C for July, about 7 °C lower

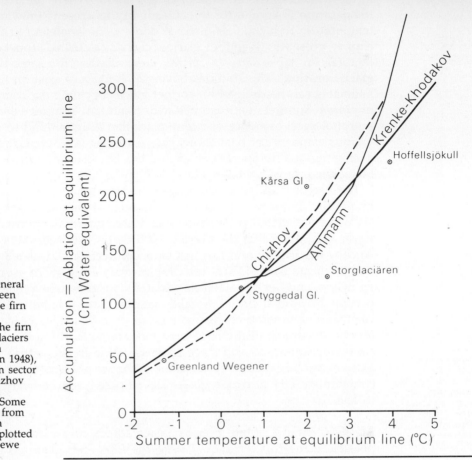

Fig. 2.17: Curves illustrating the general relationship between temperature at the firn line (x axis) and accumulation at the firn line (y axis) for glaciers around the North Atlantic (Ahlmann 1948), and in the eastern sector of the Arctic (Chizhov 1964; Krenke and Khodakov 1966). Some additional points from around the North Atlantic are also plotted (adapted after Loewe 1971).

than at the present day. In the English Lake District, similar reasoning produced a July mean temperature at sea-level of *ca.* 8.0 °C (Sissons 1980). These palaeotemperature estimates are in good agreement with those based on independent biological evidence.

A refinement of this method, in which both temperature and precipitation values could be estimated for former glaciers was described by Sissons and Sutherland (1976). Equilibrium firn-line altitudes were calculated for twenty-seven Loch Lomond Readvance glaciers in the southeast Grampian Highlands of Scotland, and for each glacier, the average mass balance was derived from an equation incorporating glacier altitude, regional ablation gradient, direct radiation, the influence of avalanching and blowing of snow, and final glacier volume. The influence of direct radiation on the glaciers was calculated taking into account the transmissivity of the atmosphere, glacier aspect and surface gradient and the albedos of ice and snow. From the mass balance equation for twenty-five of the glaciers, the distribution of annual precipitation was obtained (Fig. 2.18). Average July sea-level temperatures of *ca.* 6 °C were calculated on the basis of the Norwegian graphs, and a January sea-level temperature of *ca.* −8 °C was also inferred. The slightly lower summer temperature in the eastern Grampians during the Loch Lomond Stadial,

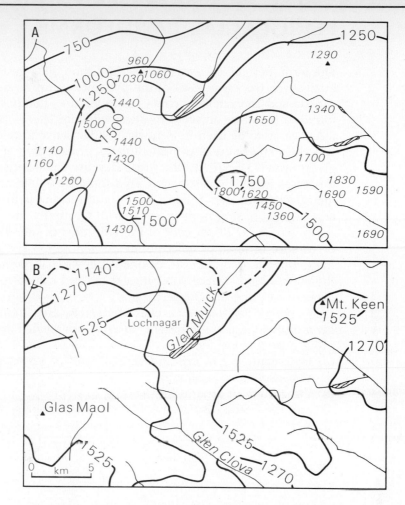

Fig. 2.18: A: Inferred precipitation during the Loch Lomond Stadial in part of the south-east Grampian Highlands, Scotland. Calculated precipitation (mm) for 25 glaciers is indicated. B: Present precipitation for the same area. Isohyets for both maps are at 250 mm intervals (after Sissons and Sutherland 1976).

by comparison with the Gaick region to the west (see above) was attributed to the heavier summer cloud cover over the former area reflecting the movement of snow-bearing winds from the south-east.

In addition to uncertainties inherent in the assumptions outlined above, a further problem with this technique arises from the fact that the relationship between the ELA/FLA and temperature may be imprecisely known for present-day glaciers. The FLA for an individual glacier can only be satisfactorily determined after several years of detailed glaciological investigations, and this information has so far been obtained for comparatively few glaciers. In addition, computed FLAs may not be in phase with present-day climatic conditions. Miller *et al.* (1975), for example, have pointed out that various time lags, ranging from a few to over 100 years, may characterise the response of glaciers to climatic changes, and some glaciers may still be reacting to budget changes initiated during the 'Little Ice Age'. On the other hand, the broad measure of agreement that has so far emerged between palaeotemperature estimates based on firn-line altitudes and those derived from independent lines of evidence suggests that this method is reasonably reliable, and constitutes a valuable new approach to Quaternary palaeoclimatic reconstruction.

PERIGLACIAL LANDFORMS

The term 'periglacial' was first used by the Polish geologist Walery von Lozinski in the early years of the present century to describe both the climate and characteristic cold-climate features (landforms and sediments) found in the areas adjacent to Pleistocene ice sheets. Since then, however, it has come to be used in a much broader sense to refer to non-glacial processes and features of cold climates, irrespective of age or proximity to glacier ice (Washburn 1979). Areas that are characterised by cold-climate processes in which frost action predominates together constitute the **'periglacial domaine'** (French 1976), and include both the high-altitude and high-latitude regions of the world. Currently the periglacial domaine extends over *ca*. 20 per cent of the land area of the globe, although the occurrence of fossil or relict periglacial landforms and deposits throughout the temperate mid-latitude regions suggests that perhaps a further fifth of the earth's land surface was affected by cold climate processes on occasions during the Quaternary.

The periglacial domaine is characterised by an extremely active geomorphological environment, in which processes operating on the ground surface include:

(a) frost-shattering of exposed bedrock and of particles within unconsolidated sediments;

(b) the growth of ground-ice, leading to upheaval of the ground surface, and the lateral displacement of surface materials;

Fig. 2.19: Tor forms on Dartmoor, south-west England.

(c) accelerated wind erosion in an environment where vegetation cover is sporadic and unconsolidated materials are often exposed;

(d) thermal erosion by fluvial activity which can be important in late summer when water temperatures have risen;

(e) accelerated solifluction (or **gelifluction**) where surface thawing results in a saturated surface layer overlying a still-frozen substrate which can induce mass flow on slopes with angles as low as 2°.

Some of these processes are unique to the periglacial environment, most notably those associated with the growth of ground ice, while others, including fluvial, aeolian and solifluction processes are particularly effective in the high-latitude and high-altitude regions of the world. Collectively, they give rise to a suite of landforms that is highly distinctive and that is characteristic of a periglacial landscape (see Péwé 1969; Embleton and King 1975b; French 1976; Washburn 1979).

Within the periglacial domain, three broad morphogenetic landscape units can be recognised. On upper slopes, exposed bedrock is highly fractured, angular and craggy in appearance as a result of frost action. **Tors** (upstanding masses of bedrock) are common on summits (Fig. 2.19) and on hillsides (Derbyshire 1972) and a characteristic step-like profile typically evolves in which a process of excavation of bedrock by frost action and the transport of frost-riven material away by solifluction produce **altiplanation** or **cryoplanation benches** (Demek 1969). Footslopes tend to develop smooth low-angled profiles with the accumulation of **talus** and **solifluction sheets** and **terraces**. These are often lobate in form (Fig. 2.20), and on occasions, different phases of solifluction can be

Fig. 2.20: Inactive solifluction lobe, Glen Nevis area, Scotland.

detected with one generation of lobes overlying another (e.g. Benedict 1970).

In valley bottoms, on hillside benches and on plateau surfaces a number of features occur in easily-recognised regular geometric patterns, referred to generally as **patterned ground**. This is an 'umbrella' term which covers landform assemblages that have been produced by a variety of processes. The most ubiquitous form, and perhaps the best known, is that of the **surface polygon** (Fig. 2.21), which can develop either through ground cracking at very low temperatures, or by sorting processes, where coarser materials are selectively separated from finer particles. **Striped ground** tends to form where sorting processes operate on steeper slopes (Fig. 2.22). Patterned ground may also be found, however, on mountain summits and on solifluction terraces and altiplanation benches. Finally, regular depressions reflect the original locations of ground ice masses. Following thawing, ground collapse occurs and in present-day arctic-alpine environments the resulting hollows are normally filled with water (**thaw lakes**). The landscape is commonly described as **thermokarst** (see e.g. Czudek and Demek 1970), from a visual analogy with karst regions, which are often marked by numerous sink-holes. Thermokarst is normally associated with continuously-frozen (**permafrost**[6]) or seasonally-frozen ground.

Many of these distinctive features of the periglacial landscape have been recognised in the fossil form in Eurasia and in North America, either through field mapping or through the use of remote sensing techniques. The patterned ground forms, for example, can frequently be identified on aerial photographs, for these geometrical patterns tend to be emphasised by differences in crop growth resulting from drainage variations (e.g. Christensen 1974). These provide good evidence for the former existence of periglacial conditions within an area although in most cases they only allow the most generalised of climatic inferences to be made. Some

Fig. 2.21: Active polygonal patterned ground, Mt Edziza, British Columbia, Canada.

Fig. 2.22: Active striped patterned ground, Mt Ediziza, British Columbia, Canada.

landforms, however, may form the basis for more detailed palaeoclimatic reconstructions and these are considered in the next section.

Palaeoclimatic inferences based on periglacial landforms

Certain periglacial landforms are unique to present-day arctic and alpine environments and aspects of the prevailing climate under which they are known to have evolved can sometimes be quantified. Where such features can be identified in the fossil form, therefore, they can be used as a basis for estimating climatic parameters for times during the Quaternary. Two examples of periglacial landforms that have been used in this way are rock glaciers and pingos.

Rock glaciers are active tongue-shaped or lobate accumulations of rock debris (Fig. 2.23), containing an ice core or interstitial ice, which are often found spreading downvalley from a cirque or true glacier (see e.g. White 1971, 1976). On the basis of the present-day distribution of rock glaciers, Kerschner (1978) has suggested that the large fossil rock glaciers in the west Tyrol, Austria, which lie 500–600 m below their active counterparts imply a depression of mean annual temperature below the present of 3–4 °C. The fossil features were last in motion during the Younger Dryas Stadial (*ca.* 11 000–10 000 years ago), and from the above evidence it has been inferred that a more continental climatic régime persisted during that period in parts of the eastern Alps, with summers cooler (*ca.* 2 °C below present) and winters much colder (5–6 °C lower).

Pingos, on the other hand, are dome-shaped hills that occur in permafrost regions as a result of uplift of frozen ground by the growth of a large convex mass of ground ice in the substratum. Melting of the ice lens leads to ground collapse, forming a central depression or crater, with a characteristic rampart of displaced substrate around the depression (see

Fig. 2.23: Active lobate rock glaciers near Lytton, Southern Coast Mountains, British Columbia, Canada.

e.g. Mackay 1962; Holmes *et al.* 1968; Muller 1968; Watson and Watson 1974). Using the prevailing climatic conditions in the contemporary permafrost environment of western Alaska where pingos are currently forming as an analogue, Watson (1977) has suggested that pingos that developed during the Younger Dryas Stadial in England and Wales reflect a former mean air temperature of −4 to −5 °C, representing a fall of 13–14 °C compared with the present day. On the basis of similar reasoning, Maarleveld (1976) has suggested that fossil pingo remains in the Netherlands imply a mean annual air temperature of not more than −2 °C during the last full glacial stage. Maarleveld calculated that an approximate lowering of mean annual temperature of 15 °C would be required to account for the southern limit of former permafrost in Europe. This figure also agreed with an estimate based on the amount of lowering of mean annual air temperature required to reduce the lower altitudinal limit of patterned ground in the Pyrenees to those levels where fossil evidence of patterned ground exists.

The distribution of fossil periglacial landforms therefore not only provides evidence for the former extent of the periglacial domaine, but in certain cases their occurrence can be used to derive palaeotemperature estimates for times during the Quaternary. This approach to palaeoenvironmental reconstruction is not without its problems, however, and some of these are considered in the section on periglacial sediments in Chapter 3.

SEA-LEVEL CHANGES

There is abundant morphological evidence from many parts of the world for variations in sea-level during the Quaternary. This includes former coastal landforms now standing above present sea-level, such as rock platforms sometimes with well-preserved backing cliffs, 'raised beaches'[7] consisting of sand and gravel deposits (Fig. 2.24), deltas, spits, shingle ridges, stacks, caves and coral reefs. Evidence for sea-level change can also be found offshore in the form of submerged landforms, including caves, platforms, beaches, reefs and river valleys.

In many areas there is also clear stratigraphic evidence for changing levels of land and sea, and although this chapter is concerned principally with landforms, it is more convenient to consider the morphological and stratigraphic evidence together. Stratigraphic evidence for sea-level change is usually particularly well-preserved in estuarine situations where large volumes of sediment have accumulated and where oscillating sea-levels have resulted in complex sequences of marine, estuarine, freshwater and terrestrial deposits. Detailed histories of sea-level changes can often be reconstructed in such localities. For example, a succession in which marine deposits give way to terrestrial peats suggests that a **regression** of the sea has occurred, whereas if a peat layer is succeeded by salt-water deposits, then a marine **transgression** can be inferred (Fig. 2.25). Where sufficient borehole evidence is available, the stratigraphic evidence can sometimes be resolved into buried morphological units. Such a situation

Fig. 2.24: Late
Pleistocene beach gravels
in south-east Sicily which
have been elevated to *ca.*
30 m above present sea-
level.

Fig. 2.25: A river bank section in lower Strathearn, eastern Scotland, showing marine sands at the base, overlain by compressed peat (dark layer), and about 6 metres of marine clays. All the deposits lie above present sea-level. Radiocarbon dating of the base and top of the peat layer indicates a marine regression at *ca.* 9600 BP and a transgression at *ca.* 7500 BP (Cullingford *et al.* 1980).

occurs in the Forth Valley in Scotland where raised beaches which formed at the close of the Lateglacial and the early part of the Flandrian have been buried by the deposits of the subsequent Flandrian marine transgression (Sissons 1966). By combining morphological evidence with sedimentary data from boreholes or exposures therefore, a sequence of sea-level changes in a particular locality can be established, often in considerable detail.

Relative and absolute sea-level

The position of the sea relative to the land can change through the movement of *either* the sea *or* the land surfaces or both of these. Where subsidence of the land takes place at a time of stable ocean levels, there will be a rise in sea-level; conversely, land uplift will lead to the elevation of littoral features, and an apparent fall in sea-level along a particular stretch of coastline. Where changes take place in sea-level either through land or through sea-level movements, they are referred to as **relative sea-level** changes, that is, a change in the position of the sea relative to the land. Such changes are essentially local in effect. World-wide sea-level changes, on the other hand, which result from fluctuations in the amount of water in the ocean basins, are termed **eustatic**. However, while the degree of eustatic change is the same for all coasts, this need not result in the same geomorphological response. Where a eustatic rise of 50 m is experienced over a 1000-year period for example, this will result in a relative rise in sea-level of 50 m in tectonically stable areas, but a 50 m fall in relative sea-level where the land is being uplifted at a rate of 10 cm per year.

Some land movements are long-term and result from tectonic activity associated with the migration of the great continental plates across the surface of the globe. Others may be of shorter duration and are generally more localised in their effects; these are termed **isostatic** movements. The term **isostasy** refers to the state of balance that exists within the earth's crust so that a depression of the crust by the addition of a load (sediment, ice, water, etc) in one locality will be compensated for by a rise in the crust elsewhere. The state of isostatic equilibrium is maintained by the flowage of material deep within the mantle of the earth.

In order to understand Quaternary sea-level variations, therefore, it is first necessary to establish how the separate effects of isostatic and eustatic changes have affected a region. In tectonically-stable areas, where the eustatic effect has been a major factor influencing sea-levels, a sequence of **absolute** as opposed to relative sea-level changes can be reconstructed. Absolute sea-levels cannot easily be discerned in areas where crustal movements have occurred as it is difficult to distinguish between the isostatic and eustatic effect in shoreline sequences. One way in which the extent of isostatic movement can be established is by applying absolute sea-level curves obtained from 'stable' coastlines, and this approach is discussed more fully in the next section.

Eustatic changes in sea-level

There has been a long-term trend of falling sea-levels throughout the Tertiary and the Quaternary (Colquhoun and Johnson 1968; Mercer 1968). The precise reasons for this are not known, but the two most recently-advanced theories suggest changes in the capacity of the ocean basin due to the progressive accumulation of sediment and tectonic activity resulting primarily from the movement of the continents. Certainly, sea-floor spreading must be an important factor, and Bloom (1971) has estimated that over the past 100 000 years or so, the oceans have grown by some 2.6×10^6 km^3, an expansion in ocean area that could accommodate about 6 per cent of the returned meltwater resulting from the wastage of the last ice sheets. If correct, these figures suggest that Holocene shorelines would be almost 8 m lower than those that formed during the interglacial of a little over 100 000 years ago. This, in turn, implies that Early and Middle Quaternary shorelines may be found at relatively high altitudes, although in the absence of further information on rates of sea-floor spreading and former extent of land ice, such a suggestion must remain speculative.

Superimposed on these long-term trends are a number of major sea-level oscillations that result directly from expansions and contractions of the Quaternary ice sheets. During periods of glacier build-up water from the world's oceans was abstracted and stored in the form of ice. It has been estimated that during the growth of the last ice sheet, sufficient water was removed from the world's oceans to produce a lowering of eustatic sea-level by over 100 m. The complete melting of the present Greenland and Antarctic ice sheets would, it is believed, raise world sea-levels by some 65–70 m. Sea-level changes which are controlled by the growth and contraction of the world's ice sheets are termed **glacio-eustatic.**

Glacio-eustatic changes dominate the sea-level history of the tectonically stable areas of the world, and in such regions, lengthy sequences of sea-level changes can sometimes be reconstructed. For example, on the island of Bermuda, ^{230}Th/^{234}U dating (see Ch. 5) of corals and speleothems from fossil coral reefs and beach deposits lying both below and above present sea-level revealed a chronology of eustatic sea-level fluctuations over a vertical range of +15 to –11 m which spanned almost 200 000 years (Table 2.2). Periods of high sea-level correspond with interglacial intervals while low sea-levels could be related to glacial phases (Harmon *et al.* 1978). The sequence was found to be very similar to that derived from the analysis of shore platforms in the Bahamas (Neumann and Moore 1975) and Barbados (Mesolella *et al.* 1969). The Bermudan evidence suggests that eustatic sea-level changes can occur very rapidly, with transgressions and regressions of 5–10 m/1000 years being inferred.

Glacio-eustatic sea-level changes are perhaps known in greatest detail for the last 15 000 years, during which the Laurentide, Fennoscandian and British ice sheets disappeared completely,[8] and the Antarctic and Greenland ice sheets contracted markedly. Evidence for the rise in sea-level that followed the wastage of these ice masses is available from many

Table 2.2: Sequence of sea-level changes inferred from ²³⁰Th/²³⁴U dates on corals and speleothems from both emergent and submerged shorelines around the coasts of Bermuda (data from Harmon *et al.* 1978).

Years BP	Inferred sea level (m) relative to present
195 000	Approximately –8 m
195 000–150 000	At least –7 m
ca. 125 000	+4 to +6 m for a short time
120 000	Below –6.5 m
ca. 114 000	Stabilised at *ca.* –8 m for a short time
ca. 97 000	Short-lived episode of platform submergence
38 000–10 000	At least –8 m
6000	–5.5 m

parts of the world (e.g. Fairbridge 1961; Shepard 1963; Jelgersma 1966), and some of these data are shown in Fig. 2.26. Such evidence is not only available from stable areas, however. In the British Isles, for example, the general rise in sea-level over the past 10 000 years has been calculated at numerous points around the coastline (e.g. Kidson and Heyworth 1973; Devoy 1977; Tooley 1978), largely on the basis of the radiocarbon dating of peat layers which are often found below present-day sea-level intercalated with marine and estuarine sediments. Some of these curves are shown in Fig. 2.27. Unlike the curve shown in Fig. 2.26, however, these are *relative* sea-level curves as they are derived from localities where some degree of crustal warping has occurred (see below).

It was suggested above that one method of deriving *absolute* sea-level curves from the data shown in Fig. 2.27 would be to utilise the evidence on eustatic sea-level rise obtained from stable coastlines. One problem here, however, is in the recognition of a 'stable' coastline. Few areas of the world are absolutely stable tectonically, and it is now recognised that many areas formerly considered to be stable are, in fact, not so. For

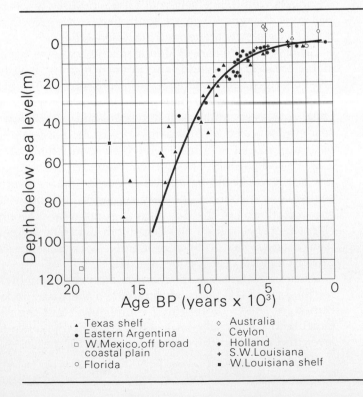

Fig. 2.26: Data points contributing to a eustatic sea-level curve taken from a number of supposedly stable areas (after Jelgersma 1966).

▲ Texas shelf
● Eastern Argentina
□ W.Mexico,off broad coastal plain
○ Florida

◇ Australia
△ Ceylon
✳ Holland
+ S.W.Louisiana
■ W.Louisiana shelf

Fig. 2.27: Relative sea-level curves indicating subsidence trends in England and Wales. Transgression (upward arrows) and regression (downward arrows) contacts from intercalated peat and clay deposits in north-west and eastern England. Relative sea-level curve 1 from north-west England; curve 2 from south-west England; curve 3 from the lower Thames estuary; curve 4 represents the movement of MHWST (mean high water spring tide) at Tilbury, lower Thames estuary (after Devoy 1977). Line a — a_1 is explained in text.

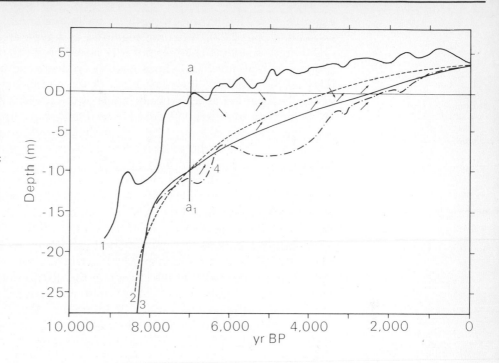

Fig. 2.28: Raised rock platforms, eastern Mallorca, Spain.

example, the Mediterranean, with its well-developed horizontal raised shorelines (Fig. 2.28), was long regarded as a type area for the study of Quaternary sea-level change, but it is now known that many of the prominent high-level shorelines have been displaced by earth movements (Hey 1978). Similarly, the coast plain of the south-east United States is generally regarded as tectonically stable at the present day, but indications are beginning to emerge of post-Lower Pleistocene uplift of certain shoreline features (Cronin 1980). A further complication arises out of the phenomenon of geoidal eustasy. The earth is not spherical, but is flattened at the poles and bulging at the equator, and it has generally been assumed that the free ocean-surface (the **geoid**) parallels that of the earth. That this is not so is indicated by maps of the geoid which show startling variations including a 180 m difference in altitude between a high near New Guinea and a trough off the south of India (Mörner 1976). These irregularities are caused by gravitational variations, which are determined by the earth's pattern of rotation and by its structure and gravity. Of particular significance is the fact that the change in the distribution of ice during glacial and interglacial phases results in gravitational changes which, in turn, led to variation in the geoidal surface. This suggests that the geoidal ocean surface can intersect different land masses simultaneously at different absolute altitudes (Tooley 1978). Although this does not affect the construction of sea-level curves for individual localities, it does raise questions about the applicability of eustatic sea-level curves to areas other than those for which they were originally derived.

Effects of tectonic activity

Evidence for shoreline displacement resulting from long-term earth movements is available from many parts of the world. Tectonically-displaced shorelines in the Mediterranean basin have already been noted, but others include the high-level Holocene shorelines around the coasts of the north-west Pacific (Pirazzoli 1978), the tilted Late Quaternary marine terraces of Baja California (Woods 1980), the raised coral terraces of the Loyalty Islands near New Caledonia in the western Pacific (Marshall and Launay 1978), and the spectacular flights of raised shorelines on the emergent coastline of the Huon Peninsula in New Guinea (Bloom *et al.* 1974; Chappell 1974a). All of these localities are adjacent to the boundaries of major continental plates, and the various raised marine features are clearly a reflection of the long-term process of continental drift (Fig. 2.29).

The effects of isostatic earth movements are also widespread and the most important for the study of sea-level changes are those that resulted from the expansion and contraction of the Quaternary ice sheets. Melting of the ice sheets and the subsequent release of large quantities of meltwater into the ocean basins would, it is believed, have led to crustal warping through the process of **hydro-isostasy** (Walcott 1972). It has been estimated that following deglaciation, the ocean basins may have been depressed by about 8 m on average, and that hydrostatic warping could have introduced differences of up to 30 per cent in estimates of marine transgression rates between ocean islands and continental crusts (Chappell 1974b). Important though the effects of hydro-isostasy may have been,

Fig. 2.29: Early Pleistocene deep-water marine calcarenites, southern Italy. In localities around the Straits of Messina, these 'Calabrian' sediments have been raised by tectonic activity to positions up to 1400 m above present sea-level.

they in no way compare to the crustal deformation that resulted from the build-up of the great ice sheets, a phenomenon known as **glacio-isostasy**.

The effects of glacio-isostasy near an ice sheet margin are shown in Fig. 2.30. The consequences of glacial loading will vary with the rigidity of the crust, but it is clear that the earth's crust does not behave as a solid block, but is subject to warping and differential subsidence according to local ice loads. In general, maximum loading occurs near the centre of an ice sheet and there is a gradual rise in the level of the crust towards the ice sheet margins (Walcott 1970). However, crustal depression at one point must be compensated elsewhere and it is thought that marginal

Fig. 2.30: Schematic diagram of the effect of an ice mass on a land surface. In general terms, the amount of isostatic depression of the land surface A–B increases with local ice load, so that greater depression occurs towards the centre of an ice mass than at the margin (cf. h_1 and h_2).

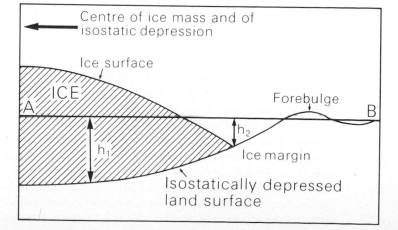

displacement of the crust involving a degree of upward bulging (**forebulge**–Fig. 2.30) may be one aspect of this compensation. If so, then the effects of glacial loading by the large Quaternary ice caps may have extended for tens or even hundreds of kilometres beyond the former ice margin (Andrews 1975).

Andrews (1970) has considered the process of isostatic recovery under three headings. The rapid crustal adjustment that occurs at a site between the ice sheet beginning to lose mass and deglaciation is referred to as **restrained rebound**. Following ice wastage more gradual **postglacial rebound** takes place and this continues up to the present day. The amount of uplift still required to establish the pre-glacial crustal equilibrium is termed **residual rebound**. Isostatic recovery can therefore be seen as a process that accelerates rapidly at first, but which then slows down gradually as the pre-glacial state of crustal equilibrium is approached. In many areas, isostatic uplift is not complete. Highland Britain, for example, is still rising relative to the more stable southern parts of the country, while in the Hudson Bay region of northern Canada, it has been estimated that around 150 m of residual rebound remain before isostatic recovery following the wastage of the Laurentide ice sheet is complete (Andrews 1968).

Shoreline sequences in areas affected by glacio-isostasy

During glacial episodes, eustatic sea levels were low, but because the crust was also depressed beneath the weight of ice, shorelines appear to have formed in a number of places close to the margins of the Quaternary ice sheets. In parts of Scotland, for example, this is indicated by the fact that many raised shorelines terminate inland in glacial outwash and related deposits (e.g. Cullingford and Smith 1966, 1980; Sissons and Smith 1965), and similar relationships have been noted in Scandinavia and Canada. Following deglaciation, both land uplift and sea-level rise occurred and hence sequences of raised shorelines have developed that reflect the often complex interplay of isostatic and eustatic factors.

Fig. 2.31 shows the type of shoreline sequence that might be found in an area undergoing isostatic recovery. As the ice front recedes from A to B, uplift occurs so that the shoreline formed when sea-level stood at SL-1 is raised above the new sea-level, SL-2. Because isostatic recovery decreases with distance from the centre of an ice sheet, the shoreline RS-1 will be tilted away from the ice-sheet centre. During glacier retreat from B to C the shoreline that developed while the sea stood at SL-2 is raised and tilted (RS-2), but it will be less steeply inclined than RS-1 which has now been even further deformed. However, because the rate of isostatic recovery at that time was accelerating due to rapid ice wastage, the **marine limit**[9] (ML-2) of RS-2 has been raised to a higher altitude than the marine limit ML-1 of shoreline RS-1. Subsequently, a third raised shoreline (RS-3) develops, but by this time isostatic recovery has slowed down, so that the marine limit ML-3 is found at a lower altitude than either ML-1 or ML-2. Moreover, a combination of decreased uplift and an increase in the rate of eustatic sea-level rise means that the *relative sea-level*

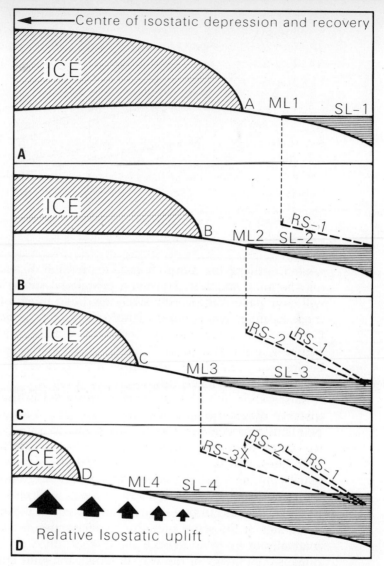

Fig. 2.31: The development of a raised shoreline sequence reflecting isostatic recovery and eustatic rise following deglaciation. ML – marine limit; SL – sea-level; RS – raised shoreline. The effects of the forebulge (Fig. 2.30) have been omitted from the diagram. For explanation see text.

in the area is rising and therefore later shorelines will progressively truncate the older and more steeply-inclined features (Fig. 2.31D). In general therefore, the oldest and most steeply tilted shorelines will form at the greatest distance from the ice centre, younger shorelines will be less steeply inclined and more extensively developed, and older features will either have been destroyed during the formation of younger ones or, in certain cases, will be buried beneath later sediments or will be found below present sea-level. The vertical interval between individual shorelines shows the amount of isostatic uplift that has occurred between the times of shoreline formation (X–Fig. 2.31).

Raised shorelines can be both depositional and erosional in form, and the extent to which a clear morphological feature develops depends on a range of factors including the length of time that relative sea-level remained stable, and on the operation of local glacial, fluvial and marine processes. Continuous shoreline features will have evolved in some areas

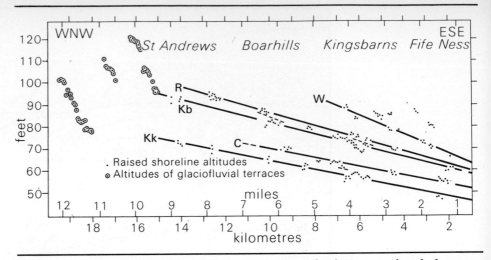

Fig. 2.32: Height-distance diagram for shoreline fragments in part of eastern Scotland (after Cullingford and Smith 1966).

while in others, the morphological expression of relative sea-level change may be more sporadic. However, postglacial subaerial and marine activity will have destroyed or extensively modified much of the evidence, even in those localities where coastal landforms were originally well developed. Consequently, only shoreline fragments remain in most areas, and careful mapping and instrumental levelling of each shoreline remnant is necessary before individual shorelines can be reconstructed and inferences made about former sea-levels (Sissons 1967a; Gray 1975; Rose 1981).

Raised shoreline data are usually presented in the form of a **height-distance diagram** (Fig. 2.32), which is a plot of all the individual data points in a vertical plane running parallel to a line towards the assumed ice centre. Shoreline fragments are resolved into a series of inferred shorelines, and the gradients of the features can then be calculated, usually by means of regression analysis. Where prominent shorelines of the same age are found in different areas, **isobases** can be constructed for these shorelines. Isobases join points of equal altitude (or uplift) on shorelines of the same age. The reconstruction of isobases, either manually or by trend surface analysis (Smith *et al.* 1969), gives a three-dimensional image of the way in which the land surface was deformed by ice (Fig. 2.33).

Where isobases of different shorelines are parallel to one another, *i.e.* the centre of uplift has not changed, and there have been no perturbations in the pattern of uplift, alternative methods of representing shoreline data are available. These include **shoreline relation diagrams** and **shoreline displacement curves**. Where extensively developed and reliably-dated shorelines are found in an area, these can form the basis for a shoreline relation diagram, in which the dated shorelines are used as reference levels. Since the gradients of some of these shorelines are known, the ages of shorelines of intermediate gradient can be calculated by extrapolation in areas where the rate of isostatic uplift can be assumed to have been constant (Andrews 1969). The shoreline displacement curve, on the other hand, plots the altitude of shorelines of known age against time and, therefore, shows the changing altitude of relative sea-level at any one locality. Curves for different areas can then be compared to

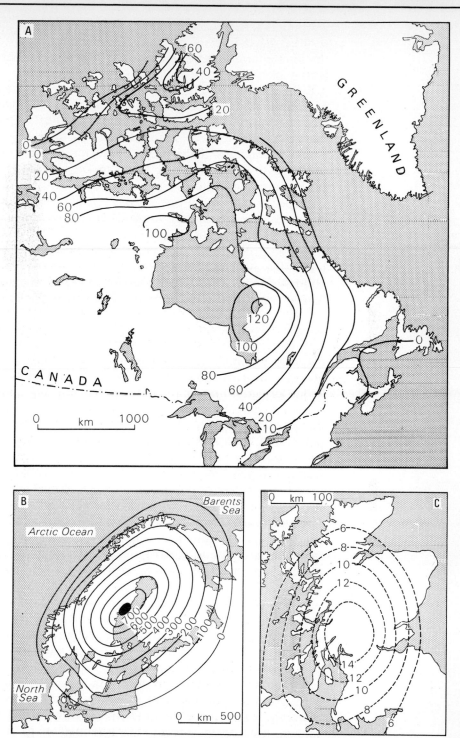

Fig. 2.33: A: Isobase map for northern and eastern North America showing shoreline emergence (in metres) since *ca.* 6000 BP (after Andrews 1970). B: 100 m contours (Isobases) showing absolute uplift of Scandinavia during the Holocene, based on evidence from a number of sources (after Mörner 1980a). C: Generalised isobases (in metres) for the Main Postglacial Shoreline (dated *ca.* 7000–6000 BP) in Scotland (after Sissons 1976).

isolate differences in degree of crustal warping over time. On Fig. 2.27, for example, the line a-a[1] crosses points on the curves that show sea-level in different parts of England at 7000 BP. These localities will all have experienced the same *relative* sea-level at that time, so that differences in altitude between the dated levels indicates the relative amounts of warping that have occurred over the past 7000 years. This partly reflects residual glacio-isostatic recovery in northern Britain (curve 1) and partly the longer-term trend of subsidence in areas adjacent to the North Sea (curve 3).

If the altitude and age of a particular shoreline are known, the amount of uplift that has taken place at a locality can be obtained by adding the elevation of the shoreline to the amount of eustatic sea-level rise that has been estimated to have occurred since the shoreline was formed. A shoreline fragment that lies at +30 m, for example, and which has been dated at 10 000 BP, will have been isostatically raised by 65 m, for reference to Fig. 2.26 shows that sea-level at that time was approximately 35 m below that of the present day. In this way, shoreline displacement curves can be converted into **uplift curves** which show the total isostatic or tectonic uplift over time (Andrews 1968). Such curves are largely restricted to the last 10 000 years or so, however, for this is the period for which the most reliable data are available. However, uncertainties may still be introduced as a consequence of crustal movements in supposedly 'stable' areas, and also through the effects of geoidal eustasy (see above).

Palaeoenvironmental significance of sea-level changes

Changing levels of land and sea impinge on many aspects of Quaternary research, and a knowledge of the causes and effects of sea-level change is of fundamental importance to all those involved in the analysis of Quaternary environments. Information on the sequence of Late Quaternary climatic change can be derived from those areas where an absolute chronology of sea-level fluctuations can be established (e.g. Mesolella *et al*. 1969), while high interglacial sea-level stands provide indirect evidence of global ice volumes at those times. On tectonic coasts, dated raised shoreline sequences show the rates at which uplift along plate margins has taken place during the Late Quaternary, and this clearly has implications for assessments of, for example, the rates of mountain building over the past two million years or so. In those areas affected by glacio-isostasy, gradients of tilted shorelines allow estimates to be made of the amount of crustal warping that has resulted from glacial loading, and this information is now being used as an additional data input in the reconstruction of former ice sheets (e.g. Peltier *et al*. 1978; Clark 1980). Raised shoreline sequences in areas affected by glacio-isostasy are also useful in the establishment of deglacial chronologies (e.g. Sugden and John 1973). Finally, a knowledge of the position of sea-level is extremely important in understanding the movements of flora and fauna during the Quaternary. In the British Isles, for example, the contrasts between the Flandrian flora and fauna of Ireland on the one hand, and mainland Britain and adjacent parts of north-west Europe on the other, are very largely explicable in terms of the fact that, at the close of the last glacial, the Irish Sea was flooded at least 5000 years before the Straits of Dover were formed, and Britain was finally isolated from the European mainland. Of even greater significance were the relative sea-level changes that occurred between the coasts of Alaska and the eastern parts of the USSR, and which led to the development of a land bridge across the

Bering Straits on numerous occasions during the Late Quaternary (Hopkins 1967). Not only were the migration routes of plants and animals affected in this case, but the peopling of the New World was controlled principally by the changing levels of land and sea.

RIVER TERRACES

In most river valleys, terraces are found either flanking the valley sides, or alongside the present river channel on the floodplain (Fig. 2.34). Sometimes they occur as single features, but on occasions they are arranged in vertical successions forming a flight or 'staircase'. Terraces may be erosional, with bedrock being planated to form a low-gradient **strath** which is often covered by a thin veneer of gravel, or they may represent the upper level of aggradation before subsequent downcutting *i.e.* the surfaces of former floodplains. The gradient of a river is determined by the position of based-level, the lower limit of potential erosion by river incision. Ultimately this is sea-level, but local base-levels may temporarily control downcutting where, for example, a resistant rock bar crosses a river valley (Fig. 2.35). A change in base-level will automatically lead to a change in river bed (and floodplain) gradient, as the altitudinal difference between river source and mouth has been altered (Fig. 2.35). Thus the gradients of abandoned floodplains (**terraces**) can be used to reconstruct former long profiles of rivers, and to infer past

Fig. 2.34: A suite of terraces formed by river incision into glaciofluvial gravels. Roy Valley, near Fort William, Scotland.

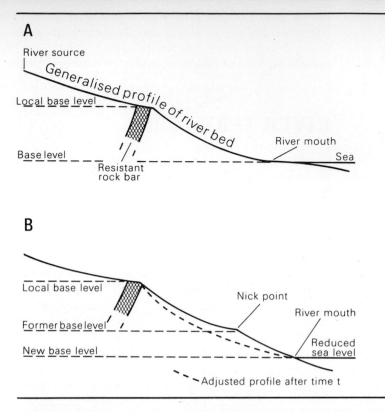

Fig. 2.35: The effect of changing sea level on river long profiles. For explanation see text.

changes in base-level. However, other factors may also influence the evolution of river long profiles, for gradients can change between contemporaneous terrace surfaces due to the influence of local base-levels (Fig. 2.35), as a result of local variations in sediment supply or water volumes, or through changes in run-off patterns (see below).

River terraces occur in all geomorphological and climatic environments and are a reflection of the operation of universal fluvial processes. They may be preserved as either 'paired' or 'unpaired' terraces. Where there have been significant and comparatively sudden changes in the river's behaviour, due for example to aggradation as the result of an increase in sediment load, or incision by the river into the valley floor, then paired terraces may form on both sides of the valley. Where the river begins to meander, however, lateral migration of the stream channel leads to erosion of the floodplain gravels on the outer edge of the meander and a single or unpaired terrace develops. Terrace sequences, therefore, reflect both lateral migrations of the river channel and also vertical displacements through a sequence of downcutting and aggradational phases, a process that is usually referred to as 'cut and fill'.

The extent to which terraces will be preserved depends on the nature of fluvial activity and the amount of time over which processes operate for any particular base-level position. In Fig. 2.36A, the river is normally confined to channel C-C[1] except during seasonal or occasional floods, when a floodplain F-F[1] of alluvium is formed. In Fig. 2–36B, heavy flooding has resulted in the widening of the channel, which may also have been deepened by incision following a change in base-level.

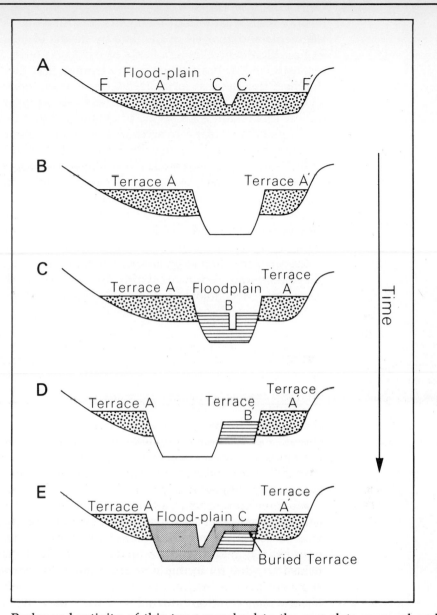

Fig. 2.36: The development of fluvial terraces by lateral displacement of the channel and by 'cut and fill' processes.

Prolonged activity of this type may lead to the complete removal and subsequent redeposition of former floodplain sediments within some valley sections. In Fig. 2.36C a renewed phase of aggradation has produced a second but lower floodplain (B), and renewed channel widening (Fig. 2.36D) gradually leads to partial removal of the deposits of both floodplains A and B. In this instance, evidence of floodplain level B has been completely removed from one side of the valley. Older river terraces may even be buried by subsequent aggradation (Fig. 2.36E) producing morphological and stratigraphic sequences of great complexity in floodplain and valley-bottom situations.

Because river terraces are most frequently developed in unconsolidated sediments, they are easily destroyed by subsequent fluvial action, and hence a previous floodplain surface will usually only be preserved in the form of individual terrace fragments. Instrumental levelling of the fragments and analysis of the data by means of height-distance diagrams will enable downvalley gradients of particular terrace fragments to be reconstructed. It has usually been assumed that the highest (and generally the most fragmented) forms in a terrace series represent the oldest river levels, and lower terraces reflect successively younger stages. In broad outline this relationship seems to obtain, but it is now appreciated that the sequences in many river valleys are more complicated than this and stratigraphic evidence is usually required if the fluvial history of a river valley is to be reconstructed. For example, older terrace surfaces may be buried beneath younger alluvial fills (Fig. 2.36E) and other important palaeoenvironmental indicators such as buried soils, peats, fossils and datable materials are usually found only in sections or in boreholes. Increasingly, therefore, in the study of river terrace sequences recourse is being made to sedimentary, biological and archaeological evidence, both in the reconstruction of palaeoenvironments associated with alluvial deposition, and also in the dating and correlation of individual terrace fragments. Morphological evidence, therefore, represents only one aspect of the more comprehensive study of alluviation.

Origin of river terraces

River incision into a valley floor leading to the abandonment of floodplain levels, or renewed aggradation along the river channel may result from a range of factors. These include:

Eustatic changes in sea-level. Until comparatively recently, all river terraces were interpreted as reflecting changes in sea-level. Eustatic sea-levels have fluctuated through a range of perhaps 150 m (see above) during the glacial and interglacial cycles of the Quaternary. With falling sea-levels, active downcutting and river incision would be expected to have occurred, while higher sea-levels would produce periods of aggradation, particularly in the lower reaches of river valleys, and long profiles would be progressively graded to these interglacial sea-level positions. Terraces related to sea-level in this way are termed **thalassostatic** terraces.

Climatic change.   The influence of climate on river valley evolution has long been recognised on the Continent (Zeuner 1945), and is still regarded as a major influence on river terrace development in non-glaciated areas such as the Mediterranean (Vita-Finzi 1969; Bintliff 1975; Gladfelter 1972). The relationship between river behaviour and climate depends on the nature of climatic change and the effects of such changes on discharge and sediment load. In semi-arid and arid regions, river incision may occur during the pluvial intervals (see below), when discharge is more constant and runoff is reduced by an increase in vegetation cover within the drainage basin catchment. Phases of aggradation, on the other hand, may reflect more arid periods, when discharge was erratic due to seasonal flooding and to increased sediment yields at times of reduced vegetation cover. In temperate areas increased aggradation may occur in colder periods when rivers are heavily debris-laden. High discharges resulting from snowmelt floods characterise periglacial regions today (McCann *et al* 1972), and areas experiencing periglacial conditions in the past would have witnessed seasonal flooding due to snowmelt, and increases in sediment yield due to a relatively sparse vegetation cover and to ground disturbance by periglacial processes. During warmer periods sediment yield would be reduced and discharge variations less marked. Terraces related primarily to climatic events have been referred to as '**climatic terraces**', but clearly some geographical variation in the relationship of terrace formation to climatic factors is to be expected. In addition, climate is only one of a number of factors affecting discharge, sediment load, and sedimentation.

Effects of glaciation.   In areas covered by glacier ice, terrace development in river valleys will clearly be limited to the period following deglaciation. Following ice wastage, numerous relatively small-scale terraces develop reflecting the modification of glacial and glaciofluvial drift as the postglacial rivers adjusted to local changes in base-level or sediment supply, or migrated during downcutting into drift deposits. Such terrace forms are common on outwash plains and in proglacial areas and are usually described as **outwash terraces.** Throughout the mid-latitudes, however, a number of major drainage basins were not directly affected by glacier ice, but lay sufficiently close to the ice margin to be indirectly influenced by glaciofluvial activity during deglaciation. As a consequence the geomorphology of these drainage basins may have been affected by the increased discharge into the drainage systems during deglaciation and also by the large amounts of debris liberated from the ice which would have added significantly to stream loads. At times of ice melt, therefore, considerable aggradation might be expected in the upper reaches of such rivers and terraces would remain as evidence of these aggraded surfaces following river incision in the subsequent warmer phase. However, although it is likely that glacier activity in peripheral and adjacent parts of drainage basins would find some local expression in terrace formation, the role of proglacial aggradation in the development of river terrace sequences on the regional scale may have been over-estimated (Clayton 1977).

Tectonic changes.   Tectonic uplift leads to rejuvenation of rivers and therefore to increased gradients. The effects are best seen in the Alpine orogenic areas of, for example, New Guinea and New Zealand, where relatively rapid uplift has produced sequences of widely-separated terrace levels in many valleys (e.g. Pullar 1965; Pullar *et al*. 1967). Uplift of part of the Danube terrace sequence during the Quaternary has resulted in some terraces lying over 200 m above the present river bed (Ronai 1965). Those areas affected by glacio-isostasy would also have experienced gradient changes as a result of differential land uplift, in addition to base-level changes caused by relative sea-level changes. In large river valleys it is possible for only part of the basin to be affected by uplift or tectonic downwarping, so that the original valley long-profiles indicated by contemporaneous terrace fragments become warped. This particular factor complicates the investigation of terraces of the River Rhine, since downwarping characterises that part of the drainage basin adjacent to the North Sea, whereas significant uplift has occurred in the upper parts of the drainage basin (Brunnacker 1975). Clearly crustal warping must be taken into consideration when interpreting height-distance diagrams of river terrace sequences and when correlations are being made between terrace fragments on the basis of morphological evidence.

Effects of man.   It is apparent that in many river basins, discharge variations and changes in floodplain levels have been affected by anthropogenic activity (Nelson 1966; Gooding 1971). Throughout the Mediterranean region, for example, archaeological evidence suggests that major alluvial fills post-date the main phase of Roman agricultural development (Butzer 1980), and it now appears that many terraces formerly interpreted as 'climatic' can be related to the activities of man (Wagstaff 1981). Forest clearance and cultivation practices reduce the vegetation cover, which in turn reduces the water retention capacity of the soil and leads, therefore, to increased run-off. Over-grazing and over-cultivation also result in substantially increased erosion and therefore higher stream loads. A very considerable number of late Holocene river terraces in areas like the Mediterranean basin could, it seems, be the product of anthropogenic influences.

River terraces and palaeoenvironments

River terraces have long attracted the interest of those concerned with Quaternary environments. Not only do they provide evidence, through their different gradients, of former positions of sea-level and of changes in climate, but they often contain sediments rich in floral and faunal remains. Fossils are preserved, for example, in abandoned meander scrolls, in oxbow lakes or in backswamp deposits, and from these the nature of the environment under which the terrace levels evolved can often be established. Indeed, in both western Europe and North America, some of the most important fossiliferous sites of both cold and warm phases are found in deposits associated with river terraces. In Britain, the terraces of the major rivers of central and southern England, particularly the Severn-Avon, the Trent and the Thames (see below) have been resolved into full glacial, interstadial and interglacial sequences largely on the basis of contained fossil assemblages. Moreover, a combination of morphological, sedimentological, and biological evidence and, in some instances, radiometric dates have allowed these individual terrace systems to be correlated over wide areas (Clayton 1977). In providing links between sequences in different areas therefore, river terraces become valuable datum surfaces and provide a potentially useful framework for Quaternary chronology.

The Thames terraces

One of the most intensively studied river terrace sequences is that of the River Thames in southern England. The Thames basin contains a long record of terrace evolution that reflects the influence of a number of the factors discussed above. Moreover, the Thames terrace system also exemplifies some of the difficulties that are encountered in the analysis and interpretation of river terrace sequences.

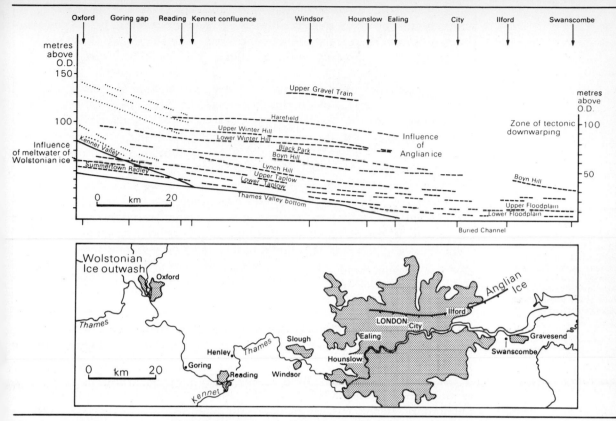

Fig. 2.37: Reconstructed long profiles of the terraces of the River Thames (dashed lines), and of one of its tributaries, the River Kennet (dotted lines), along an approximate axis from Oxford to Swanscombe. Based on data from Evans (1971), Clayton (1977) and Jones (1981).

A height-distance diagram of the Thames terrace fragments is shown in Fig. 2.37. This generalised scheme is largely based on one introduced by Wooldridge and Linton (1955) which drew upon work that had been published over a number of years (e.g. King and Oakley 1936; Wooldridge 1938; Hare 1947). Although partially modified by the results of more recent research, the general reconstruction of the terrace sequence has changed little since then (Clayton 1977).

The interpretation of the sequence, however, has been less straightforward, and correlation between the individual terrace fragments (Fig. 2.38) remains speculative in many parts of the basin. Throughout the valley, a number of terraces are locally poorly-defined or extremely fragmented. A particularly problematic area is the Goring Gap through the Chiltern escarpment. In that part of the valley, terrace remnants are scarce and, as a consequence, it is difficult to link the terraces of the middle Thames with those of the Oxford region further up-stream. A further complication is that the river has changed its course during the Quaternary, partly as a result of glacier ice moving in from the north. This is apparent from the distribution and gradient of certain terraces, from the occurrence of buried valleys and from the spread of fluvial gravels (Hare 1947; Gibbard 1977; McGregor and Green 1978). Finally, there is evidence to show that tectonic deformation has occurred in and around the Thames basin during the Quaternary. This includes local displacements due to flexuring and unloading (Jones 1974), and long-term regional subsidence of the southern North Sea and adjacent areas

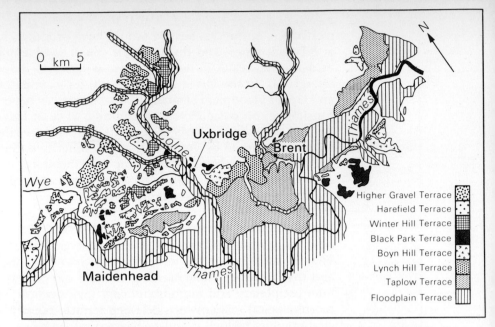

Fig. 2.38: The distribution of Thames terraces in the Middle Thames area and west and central London (after Hare 1947; Wooldridge and Linton 1955).

(McCave *et al.* 1977). The terraces may, therefore, have been locally deformed, and this further complicates the correlation of individual terrace remnants.

The traditional interpretation of the Thames terrace system is that the sequence is primarily a reflection of sea-level changes throughout the Quaternary. In the Thames estuary and adjacent parts of the North Sea bed, for example, boring and seismic investigations have revealed a 'palaeo-drainage' channel which appears to be related to glacial low sea-levels (D'Olier and Madrell 1970; D'Olier 1975), and similar buried channels have been found in the middle reaches of the Thames Valley. Terraces above present sea-level are generally regarded as thalassostatic, but it is now apparent that this interpretation may not always apply. For example, biostratigraphic evidence suggests that a number of terraces aggraded under cold conditions (e.g. Briggs and Gilbertson 1980), when low sea-levels presumably prevailed. This has prompted speculation that at least some of the Thames terraces, particularly those in the lower reaches of the valley, may have formed following surges of the world's ice sheets (Hollin 1977). In the upper parts of the drainage system, on the other hand, many terrace forms may relate to increased aggradation during deglaciation. When the tectonic effects described above are also taken into account, it is clear that a straightforward thalassostatic interpretation of the Thames terrace sequence can no longer be sustained.

Little is known about the absolute ages of the Thames terraces, but relative ages have been assigned to some of the terrace fragments on the basis of the contained biological and archaeological evidence. The Boyn Hill terrace, for example, which can be traced intermittently from Reading to Clacton (Fig. 2.37) has been assigned to the Hoxnian Interglacial (Waechter *et al.* 1970), while the sedimentology, fauna and periglacial evidence from the Handborough terrace suggest a Wolstonian age for that particular feature (Briggs and Gilbertson 1973). The problem with this means of dating, however, is that the repeated glacial/interglacial oscillations of the Quaternary would have produced similar environments at different times, and therefore the use of sedimentological and biological evidence can result in terraces being assigned to the wrong glacial or interglacial. In the lower Thames basin, for example, the Ilford, Aveley and Upper Floodplain (Trafalgar Square) terraces have usually been considered, on palaeobotanical grounds, to be of Ipswichian Interglacial age, whereas on the basis of mammalian evidence, the Trafalgar Square terrace appears to be younger than the other two (Sutcliffe 1975). Terraces once regarded as being of the same age, therefore, may in fact belong to different interglacials. Similar inconsistencies are beginning to emerge in the dating of other terrace remnants.

Despite over a century of research, therefore, many aspects of the Thames terraces remain enigmatic, and several parts of the sequence are now being revised (Jones 1981). It is apparent that the river terraces of the Thames basin reflect the influence of factors other than sea-level change, and that a proper understanding of these can only be achieved from interdisciplinary and multidisciplinary investigations involving not only morphological relationships between terrace fragments, but also the use of stratigraphic, biological and archaeological evidence, supported throughout by radiometric dates. A growing appreciation of the value of such integrated studies is leading to a more comprehensive approach to the study of Quaternary fluvial landforms and deposits, not only in the Thames basin, but in river basins throughout the world.

Quaternary landforms in low latitudes

There is abundant geomorphological and lithological evidence to show that major climatic changes have affected the tropical, subtropical and warm temperate regions of the world during the Quaternary. The periodic expansions and contractions of the great ice sheets in the high and mid-latitudes were accompanied by shifts in the main climatic zones of the low latitudes, and these produced marked spatial and temporal variations in regional rainfall totals, seasonal distribution of rainfall, annual temperatures, and wind directions and strengths. Although these climatic changes were experienced throughout the tropics and subtropics, their effects were most pronounced in the desert and savannah margins of, for example, the Sahara, north-west India and parts of Australia. In many of these areas, fossil landforms and deposits are preserved that can be used to infer climatic changes during the Late Quaternary. Of particular importance are lacustrine features that provide evidence for wetter conditions at times in the past, and sand-dune complexes indicating former episodes of increased aridity.

Pluvial lakes

In many arid and semi-arid regions of the world, there are indications that saline lakes and playas have experienced phases of expansion and contraction during the Quaternary. Marked fluctuations have also been detected in the former levels of a number of present-day lakes in tropical and sub-tropical regions. These lakes and fossil lake features are common in those areas where geological structures have produced large basins and depressions in which drainage is predominantly internal and where throughflow has been minimal. Lakes that show evidence of expansion and contraction unrelated to world-wide changes in base-level have been termed **pluvial lakes**, as their high-water stages have been attributed to wetter climatic phases known as **pluvials** or **pluvial intervals** (Morrison 1968a). Low water levels or periods of complete desiccation, in turn, reflect **interpluvials** or **interpluvial intervals**. Because of the very close

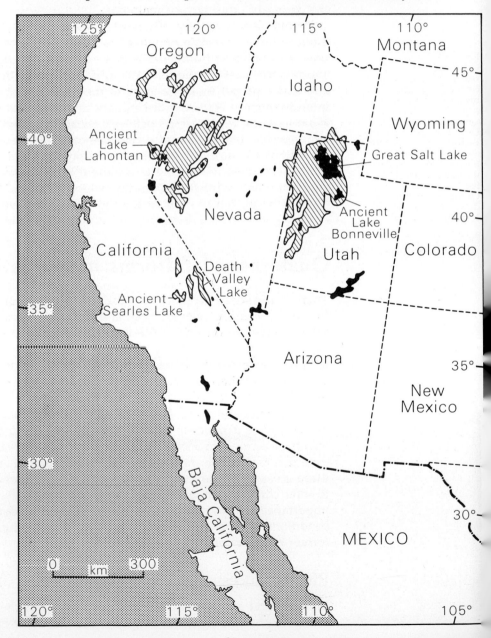

Fig. 2.39: The principal 'pluvial' lakes of parts of the western United States. Existing lakes in Nevada, Utah and California are shaded, and the maximum extents of the 'pluvial' lakes are shown by diagonal lines (adapted from Flint 1971).

relationships that appear to exist between pluvial lakes and precipitation and evaporation, fluctuations in water level in closed-basin lakes are potentially useful indicators of continental palaeoclimates during the latter part of Quaternary time.

Morphological evidence for the existence of pluvial lakes and for oscillations in water level includes abandoned clifflines and shorelines, beaches, bars and deltas, and abandoned watercourses that acted as overflows at times of high lake-level. These features can be mapped in the field, with altitudes and height correlations of individual forms being established by instrumental levelling (e.g. Washbourn-Kamau 1971). Although in the majority of cases the shorelines are horizontal features, care is needed in their correlation for the differential effects of loading and unloading can, in certain situations, lead to shorelines of a similar age being found at different altitudes (Morrison 1965). In heavily-vegetated areas and regions of difficult terrain, increasing use is being made of remote sensing techniques. In the Rift Valley of southern Ethiopia, for example, Grove *et al.* (1975) were able to detect recessional strandlines on aerial photographs by contrasting tonal bands that reflected alternating dense and sparse vegetation marking deep and shallow soils developed on calcreted beach sediments. Lake shoreline features can also be recognised on satellite images produced by LANDSAT where vegetational contrasts around abandoned shorelines show up well on bands 6 and 7 of the electromagnetic spectrum (Ebert and Hitchcock 1978). Strandline evidence is particularly useful in establishing the former extent of pluvial lakes, some of which occupied considerable areas. In the south-west

Fig. 2.40: Death Valley, California, much of which is a dry, salt-covered lake bed underlain by a considerable thickness of Quaternary deposits. During 'pluvial' episodes, Death Valley lake formed as one of a chain of five pluvial lakes (which included Searles Lake) fed by waters from the Owens River flowing down from the Sierra Nevada (Fig. 2.39).

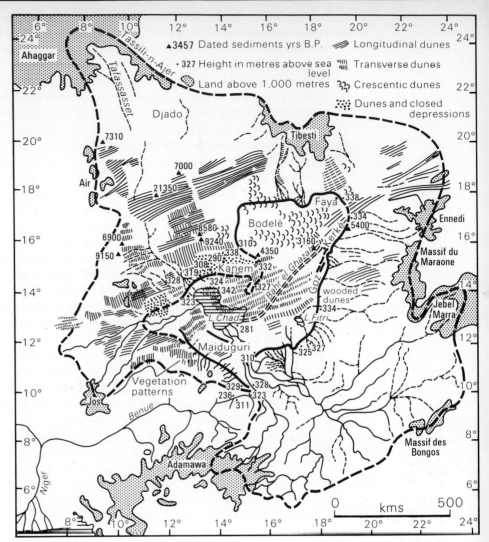

Fig. 2.41: The Chad basin, showing dune systems and the shoreline of 'Mega-Chad' (bold dashed line) at about 320 m (after Grove and Warren 1968).

United States, where one of the greatest concentrations of pluvial lake features in the world is to be found (Figs 2.39 and 2.40), Lake Bonneville, which was a former extension of the Great Salt Lake, occupied some 50 000 km² at its maximum, and was over 330 m deep (the present-day Great Salt Lake is *ca*. 4000 km²), while nearby Lake Lahontan was approximately half that size and two-thirds the depth (Morrison 1965). In North Africa, Lake Chad (Fig. 2.41) covered over 300 000 km² at its maximal extent during the Late Quaternary (its present area fluctuates between 10 000 and 25 000 km²), yet was relatively shallow, being no more than 50 m deep (Grove and Pullan 1963). The most extensive former lake so far recorded, however, is the Aral and Caspian Sea systems which together formed a water body of over 1.1 million km² (Goudie 1977), although this enormous lake was pluvial only in part as it was fed by glacial meltwaters via the Volga, Ural and Oxus drainage systems.

Abandoned lake shorelines in the low latitude regions of the world therefore provide clear evidence of both the former extent of pluvial lakes, and of fluctuations in lake-water level, reflecting periods of increased precipitation during the Late Quaternary. More detailed climatic reconstructions can be made on the basis of stratigraphic evidence, however, while the dating of sedimentary sequences within some lake basins enables a chronology of climatic changes to be established. These aspects of pluvial lake sequences, along with some of the problems that are encountered in the interpretation of evidence from pluvial lakes, are discussed in Chapter 3.

Dunefields

Increased aridity at times in the past can be demonstrated by the existence of desert landforms, especially sand dunes, in areas where such features are no longer evolving at the present day. Even when heavily vegetated, dunefields often stand out clearly on aerial photographs and satellite images (Fryberger 1980), so that rapid mapping is possible of inactive dune systems beyond the margins of the present desert regions. Stabilised dunes have been found along the margins of many of the world's deserts. They are particularly well-developed along the southern margins of the Sahara and can be traced in a discontinuous band spanning some 5° of latitude from Senegal in the west to the upper reaches of the Nile Valley (Grove and Warren 1968), while further south stabilised sand dunes flank the northern margins of the Kalahari Desert (Grove, 1969). The Australian arid zone is ringed on three sides by fossil dunes (Mabbutt 1977), they occur in extensive belts along the north-eastern margins of the Indian subcontinent (Goudie *et al.* 1973), and they have also been mapped in belts in parts of the western United States (Smith 1965). The geomorphological evidence in all of these regions points to phases of increased aridity at times during the Late Quaternary.

In some areas, it may be difficult to make a clear distinction between active and fossil dunes, particularly where sand has been remobilised and secondary dune patterns have become superimposed on an older primary set (*e.g.* Smith 1965). However, most fossil dunes can be distinguished by features indicative of a period of stability under more humid climatic conditions. These include the presence of an established vegetation cover; evidence of gullying and truncation by fluvial erosion; features produced by pedogenesis and chemical weathering including chemical alteration of clay minerals, decalcification and staining by iron oxide; and discordance with currently-prevailing wind directions. In addition, archaeological evidence of prolonged history of cultivation in dunefields can provide a further indication that the landforms are inactive (Goudie 1977).

Some fossil dunefields are found on the beds of former pluvial lakes (Fig. 2.41) where the presence of lacustrine deposits blanketing the inter-dunal depressions, and lake shorelines etched into dune flanks provide convincing evidence for the alternation of arid and pluvial phases. In the Seistan Basin of Afghanistan, for example, the presence of sand dunes on the bed of a vast playa lake that formerly covered some 65 000 km² (Fig. 2.42) indicates clearly that wetter conditions in the past have given way to a more arid climatic régime which still prevails at the present day (Smith 1974). In the Lake Chad basin of northern Nigeria, morphological evidence from both lake shorelines and fossil sand dunes points to an

Fig. 2.42: The western half of the Seistan Basin, Afghanistan, showing the margin of the palaeolake and directions of strong winds inferred from dune orientations (after Smith, 1974).

even more complex sequence of arid and pluvial phases, with the major dune-building episodes dated to between *ca.* 3200–5400 BP and 12 000–21 000 BP (Table 2.3) separated by wetter intervals during which lake levels rose markedly (Grove and Warren 1968; Pias 1970). Multiple dune systems have also been found in the Sudan. The most recent of these appear to be Holocene in age and indicate a southward movement of the wind belts by about 200 km, while an earlier period of aridity represents a shift of some 450 km. The fact that the dune belts cross the Nile valley implies that the river may have been dry at the time of their formation (Warren 1970). Extensive dune systems of two different ages have also been traced across Mali, Niger and Mauritania (Fryberger 1980). Collectively, the evidence from the southern margins of the Sahara suggests that during the arid phases of the Late Quaternary, the major wind and precipitation belts may have shifted southwards through some 5° of latitude.

In addition to providing evidence on the extent of arid zones at times in the past, palaeo-wind directions can also be inferred from fossil sand dunes. In the Lake Chad basin, for example, systems of dunes formed during one or more Late Quaternary arid phases under the influence of

Table 2.3: Sequence of lake stages and aeolian episodes in the Lake Chad basin, Nigeria. TR = Transgression; R = Regression (from Mabbutt 1977, after Pias 1970).

Stage	Lake level (m)	Lake area (km^2 × 10^3)	Limit of dunes	Dune Formations	Age (BP)
Present	282	24 (24–28)	16°N	Barkhans	
4th TR	287–290	180		Drowned archipelago of lake	1800 to 3200
Arid R			12°N (3rd erg)	Barkhans and intersecting dunes with enclosed depressions	
3rd TR	320	350 'Megachad'			5400 to 12 000
Arid R			12°N (2nd erg)	Transverse dunes NNW-SSE	
2nd TR	350–400				21 350 to 30 000
Arid R			10°N (1st erg)	Alab ridges NE-SW	
1st TR	400				

winds from between north east and east (Fig. 2.41) while a wind direction from the north east is also implied by the orientation of dune fields in the 'empty quarter' of the western Sahara (Grove and Warren 1968). Fossil longitudinal dunes in Australia show a clear counter-clockwise swirl about the arid interior and the evidence suggests a summer anticyclonic system during the major dune-building episodes at least 5° further north than at present (Bowler 1976; Mabbutt 1977).

Although fossil dunefields are therefore potentially extremely valuable indicators of the former extent of arid environments, care must be exercised in their use as palaeoclimatic indicators. Desert conditions result from a variety of climatic factors, the most important clearly being temperature, precipitation and wind. A change in the balance between these factors may be sufficient to alter the balance between adequate groundwater retention for plant growth and excessive evaporation leading to severe drought. Hence, a change towards a phase of dune-construction may result from higher temperatures, lower precipitation levels, increased wind strength, or a combination of all three. For transitional periods therefore, additional lithological and biological evidence may be required if correct palaeoclimatic inferences are to be made. A second problem arises out of the activities of man on the desert and savannah margins. It is clear that throughout the Late Holocene, man has played an increasingly important role in the expansion of the desert regions (Glantz 1977) and man-induced desertification is now regarded by the United Nations Organisation as one of the major global problems.

The extent of anthropogenic influence on the landscape of the lower latitudes during earlier Quaternary periods is not known, but clearly this is a factor that must be borne in mind when palaeoenvironmental inferences are being made for the Holocene on the basis of dunefield evidence. Finally, where fossil sand dunes are being employed as indicators of palaeo-wind directions, the problems encountered in the interpretation of fossil dunes found at high latitudes, which are discussed in Chapter 3, must always be considered.

Additional morphological evidence

Other geomorphological evidence that has been used to infer changing climatic conditions in low latitudes includes fluvial landforms and weathering crusts. Fossil fluvial features are found throughout the tropics and subtropics. In the Sahel of West Africa LANDSAT imagery has revealed the presence of integrated drainage networks of the River Niger extending far northwards into the present arid zone (Talbot 1980). These drainage channels were believed to be active in the early Holocene but are relict features at the present day and provide further evidence of a change from wetter to more arid conditions during the Holocene. Similar relict drainage networks have also been detected by LANDSAT and SLAR imagery in the Amazon lowlands (Tricart 1975). In the arid and semi-arid regions of Australia formation and subsequent dissection of alluvial fans has been taken as evidence of climatic change (e.g. Williams 1973).

In reconstructing palaeoenvironments of the arid and semi-arid zone from relict fluvial landforms, non-climatic factors must be taken into account before drawing palaeoenvironmental inferences (Rognon 1980). Fluvial processes are governed by a range of environmental variables which include, in addition to climate, geology, relief, vegetation cover and the influence of man. All of these, in combination, will affect the rate at which fluvial activity proceeds as well as the geomorphological response, and climate may not always be the dominant factor. For example, in highly arid regions, stream loading may result from increased rainfall, whereas in the semi-arid zone, reduced rainfall may lead to increased stream loading as a result of reduced vegetation cover (Mabbutt 1977).

Weathering crusts, or **duricrusts**, are resistant surface mantles or cappings commonly found as protective layers at the surface of eroded bedrock or sediments in low latitudes. They originate through the concentration in soils, sediments or permeable rocks of certain chemical constituents displaced through solution or translocation. These concentrations develop as hardpans and when exposed form cemented layers which are more durable than adjacent layers. There is a wide range of weathering crusts of different chemical constituents, but in general they can be divided into those that originate as weathering layers in humid tropical environments, and those that develop under arid or semi-arid conditions. **Laterites**, for example, evolve through the accumulation of hydroxides of aluminium and iron in humid tropical soils and when these layers are exposed they form an extremely hard cemented horizon. Their occurrence as caprocks in some desert and savannah regions indicates a major change in climatic conditions, for they are believed to form where both temperature and precipitation are high. Precise climatic parameters governing their formation are difficult to establish, however (McFarlane 1977) and, moreover, as the formation and exposure of laterites normally operates over a time-scale much longer than the Quaternary time-span (Thomas 1974), they are of limited value in inferring climatic change over the last million years or so.

Weathering crusts that are believed to have developed under relatively arid conditions include **calcrete**, also known as **caliche**, which is composed of cemented calcareous horizons and which often forms the hard rim of exposed escarpments in deserts and savannah regions, **gypcrete** or gypsum cement (Watson 1979), and **silcrete** (siliceous crusts). All of these crusts are found throughout the landscapes of the low latitude regions, and the degraded nature of many of the crusts suggests that they are relict forms from previous arid phases. Again the climatic conditions governing their formation are difficult to quantify, although Goudie (1973) has suggested that in many parts of the world, the 500 mm isohyet appears to be significant as a division between calcrete and other duricrust types. The age of these weathering crusts is uncertain, but it appears that many may be pre-Quaternary in age. Duricrusts and related forms are discussed in detail by Goudie (1973) and Mabbutt (1977).

CONCLUSIONS

Morphological evidence, therefore, provides a useful starting point in the investigation of Quaternary environments. By using modern landforms as analogues, aspects of former glacial, periglacial, fluvial, marine and aeolian environments can be inferred. In many cases, only rather generalised conclusions are possible, but in certain instances specific climatic parameters can be deduced. Of particular significance in this respect has been the recent research on the reconstruction of Late Quaternary ice sheets and glaciers for, in addition to providing information on such aspects as ice thickness, extent of glacial loading, and rates of ice sheet growth and decay, this offers an ingenious new method of estimating former climatic conditions from, in the first instance, geomorphological evidence. Throughout this chapter, however, it has been stressed that morphology is but one of the lines of evidence used in the reconstruction of Quaternary environments, and that wherever possible, landform evidence should be used in conjunction with other types of information. A second major data source lies immediately beneath the earth's surface, and it is to the stratigraphic record that we now turn our attention.

NOTES

1. The use of aircraft requires survey by wide-angle camera lenses in order to cover the maximum possible area. This results in increasing distortion towards the margins of each photograph. Satellite images, on

the other hand, are based on narrow-angle 'lenses'; distortion is therefore minimised and a single exposure from these higher altitudes can also cover a much larger area.

2. **Sandar** is an Icelandic term used to describe the alluvial plains that form in lowland areas in front of valley glaciers and ice sheets. The singular is **sandur**.

3. As there are indications that glacier ice may have disappeared completely from highland Britain prior to the Loch Lomond Stadial, this renewed glacier activity should, perhaps, be referred to as an 'advance' rather than a 'readvance' (Sissons 1977a).

4. **Drumlins** are low hills with an oval outline and are usually less than 60 m in height. They are formed mainly of till, although some contain stratified material or a bedrock core.

5. In some glaciers, particularly in the High Arctic, there may be a complex zone of transition between the accumulation and ablation areas. This is usually termed the **superimposed ice zone** in which ice will have formed as a result, for example, of the refreezing of meltwater runoff or avalanche material from upglacier. The lower boundary of the superimposed ice zone is the equilibrium line (above that the glacier is gaining in mass) whereas the firn line, *i.e.* the line dividing fresh snow from ice, forms the upper boundary. On most maritime temperate glaciers where snowfall is heavy, there is frequently no superimposed ice zone and therefore the firn line and the equilibrium line coincide (see Embleton and King 1975a, and Paterson 1981 for further discussion).

6. **Permafrost** has been defined by Muller (1947 p. 3) as 'a thickness of soil or other superficial deposit, or even of bedrock, at a variable depth beneath the surface of the earth in which a temperature below freezing has existed continually for a long time' (from two to tens of thousands of years).

7. Strictly speaking, the term 'raised beach' relates to beach deposits that have been raised by land uplift above the level at which they were formed. However, the term is conventionally used to describe all beach features found above present sea-level, irrespective of whether their position results from actual uplift of the land or a fall in sea-level.

8. In fact a small remnant of the Laurentide ice sheet still exists in the form of the Barnes Ice Cap on Baffin Island.

9. The **marine limit** simply refers to the farthest point of marine influence inland.

CHAPTER 3

Lithological evidence

INTRODUCTION

Although geomorphological evidence can provide useful insights into former climatic régimes and environmental conditions, a more detailed impression of events during the Quaternary can often be gained from the sedimentary record. Not only can valuable data on Quaternary environments be obtained from the sediments themselves by relating observations on present depositional environments to features preserved in the recent stratigraphic record (Reineck and Singh 1973), but as many deposits are fossiliferous, inferences based on lithological changes can often be supported directly by those based on fossil evidence. Further, because sedimentary sequences frequently reflect sediment accumulation over an extended time-period, some appreciation can be gained of both spatial and temporal aspects of environmental change. Finally, while morphological evidence is restricted largely (although not wholly) to the terrestrial environment, sedimentary data can be obtained from beneath the waters of present-day lakes, from the world's ice caps and, perhaps most important of all, from the deep-ocean floors where lengthy sequences of virtually undisturbed deposits are preserved.

Quaternary sediments are generally unconsolidated and are of two principal types: inorganic deposits, consisting of mineral particles ranging in size from large boulders to very fine clays, and biogenic sediments, consisting of the remains of plants and animals. Biogenic sediments can, in turn, be divided into an organic component of humus and the decayed remains of plants and animals, and an inorganic component of such elements as mollusc shells and diatom frustules (West 1977a). In this chapter we are concerned primarily with inorganic sediments such as tills, aeolian and cave sediments, and fossil soils, although some of the properties of biogenic sediments are also considered. A full discussion of the fossil record contained in Quaternary sediments can be found in Chapter 4.

METHODS

(1) Field procedures

(a) Sections

Wherever possible lithological investigations should be carried out on open sections so that variations in stratigraphy, both vertically and horizontally, can be carefully recorded. Before commencing fieldwork the section should be cleaned of slumped material, and a 'fresh' face revealed by cutting back into the exposure. On large sections steps can be cut in the slumped material or the face to provide temporary working surfaces, but a ladder may be necessary to reach the less accessible parts. Careful drawing of the exposure is the first stage in analysis, and this should be supported wherever possible by a photographic record. It may be useful to grid the face with measuring tapes as this will provide an accurate scale for section drawing. Detailed notes should be taken on all aspects of the exposed stratigraphy, using Munsell colour charts to obtain a relatively precise description of colour changes between and within lithostratigraphic units. Where necessary, instrumental levelling from a benchmark will enable the various stratigraphic features to be related to a common datum, for purposes of altitudinal comparison.

The type of sampling framework employed will depend on the nature and purpose of the investigation. For certain types of study, for example the analysis of soil profiles, a sequence of samples may be required from a representative vertical section of the exposure. Sometimes a set of monoliths measuring perhaps 25 cm square may be cut from the face for subsequent laboratory analysis, using either a sharp spade or a metal box specially designed for the purpose that can be hammered into the face. Alternatively, bulk samples may be taken at a set vertical interval or, depending on the aims of the investigation, small samples of only a few cubic cm may be all that are required. In all of these cases, however, the sampling horizons should be carefully related to a measuring tape attached to the free face at the side of the sampling line; the trowels, spades, knives and spatulas should be cleaned between the extraction of each sample, and care must be taken over the packaging, sealing and labelling of the sediment samples. Detailed notes should be made throughout the sampling process. In other types of investigation, it may be necessary to take a number of samples from points scattered across the face of the section. These can be selected subjectively, but if the face has been gridded and the squares numbered, random number tables can be used to achieve a more objective sampling framework (Rose 1974). The same method can be applied where measurement is required of the orientation and dip of the pebbles or other clasts (fabric — see pp. 100–101).

(b) Coring

Although section work is preferable to coring, relatively few natural sections are to be found, and there are many situations where it is impossible to excavate exposures due, for example, to problems of time, expense, sediment thickness and the likelihood of waterlogging. Hence, recourse must frequently be made to coring.

Equipment.   A range of coring equipment is now available for the sampling of Quaternary deposits. For the most cohesive sediments, such as gravels and tills, heavy motor-driven drilling equipment is needed. Typical samplers include the **Percussion system** which is capable of taking 10 cm diameter cores of up to 2 m in length, and the **Rotary corer**, although the latter has been less widely used in Quaternary research. In the majority of cases, however, hand-operated corers are employed, particularly for the sampling of peat and lake sediments. All of these work on a similar principle, namely that a closed metal chamber is pushed down through a sequence of sediments to a predetermined depth at which point the chamber is opened and sampling begins. There are two basic types of corer: one group, which includes the **Hiller** and **Russian** designs (Fig. 3.1), samples from the side of the chamber, while the other group including the **Dachnowski** and **Livingstone** models, are filled from the end (Fig. 3.2). The side-filled corers are relatively light and easy to use, but suffer from problems of contamination since the chamber must pass *through* the sediments from which samples are to be taken. In addition, the more delicate sedimentary structures are often disturbed during the sampling process. The piston corers, on the other hand, are more robust and are less prone to contamination, but they are usually more cumbersome to use and the cores obtained may be compressed. Most hand-operated samplers take cores of either 50 cm or 1 m length and from depths of up to 20 m or so, depending on the nature of the sediments. Hand-operated corers can also be used to sample sediments from modern lakes and ponds, either from boats or from the surfaces of frozen lakes (Cushing and Wright 1965). Indeed, the Livingstone sampler

Fig. 3.1: A core obtained from a lake bed in north Wales, using the Russian sampler (length 50 cm). White diatomite at the base grades upwards into dark brown organic lake muds.

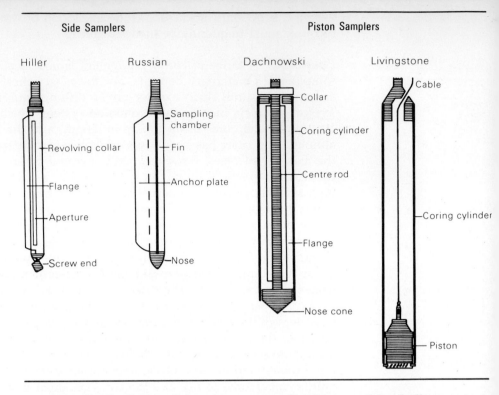

Side Samplers   Piston Samplers

Hiller   Russian   Dachnowski   Livingstone

Fig. 3.2: Four types of hand-operated coring equipment for unconsolidated sediments.

was specifically designed for use on a small raft moored in shallow water (Livingstone 1955). The major problems with this form of coring are contamination if the equipment is not watertight, difficulties involved in providing a stable platform for sampling, restrictions imposed by water depth, and the problems involved in maintaining stratigraphic control during the removal of successive cores. Longer cores can be obtained using more sophisticated equipment, such as the piston sampler designed by Kullenberg (1947) for the coring of sediments on the ocean bed, and which is capable of raising relatively undisturbed cores up to 15 m in length from beneath extremely deep water, and the **Mackereth corer** (Mackereth 1958) which can take cores up to 6 m in length using compressed air (Fig. 3.3). Further details on coring equipment can be found in Faegri and Iversen (1975), Wright *et al.* (1965), West (1977a), and Moore and Webb (1978).

Sampling procedures. Because of the considerable amount of palaeoenvironmental information that has been obtained from peat bog and lake sediments using hand-operated coring devices, the procedures involved in this form of fieldwork will now be considered in some detail. The location of the actual sampling point in a bog or lake site will be determined by a number of factors, including the depth of sediment infill, the type of sampling equipment available, and the questions that the investigator is attempting to answer. In a typical site, the most complete stratigraphic record will usually be preserved in the deepest part of the basin, and wherever possible, samples should be taken from that point.

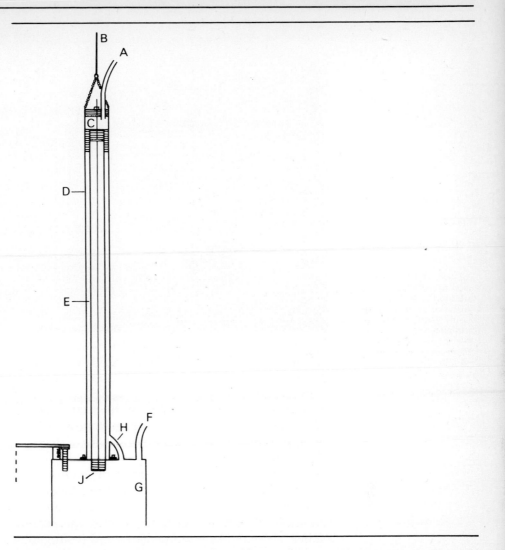

Fig. 3.3: The Mackereth Piston Corer for sampling present-day lake sediments. The apparatus is transported by small boat to the sampling point and lowered to the lake floor by cable (B). As water is pumped out of the anchor drum (G) by pipe (F), it sinks into the mud and provides a solid base for sampling. Compressed air is admitted to chamber (C) by pipe (A) and the core tube (E) is driven down into the sediment. When the sampling chamber has almost passed out of the outer tube (D), air escapes through pipe (H) and lifts the whole apparatus to the surface. The core itself is extruded from the sampling chamber by means of the piston (J).

This is particularly important where lake sediments are being investigated for it is not uncommon to find that, where former lake levels have fluctuated, marginal areas will have been eroded, and therefore only in the deeper parts of lakes will a complete sedimentary record be preserved. Although redeposited sediments from the lake margins may pose problems, these can often be recognised in the cores either by an abrupt change in lithological characteristics or by a marked increase in the numbers of deteriorated fossil remains (Ch. 4). In any event, it is preferable to be confronted by this particular problem than by a gap in the sediment sequence.

In an infilled basin, the deepest point can be located by random test bores, or by taking a transect across the site. However, the shape of many infilled basins is complex, and therefore the most reliable means of finding the deepest point is by sounding the infill along a rectilinear grid, using a level mounted at the side of the site to provide a common datum for individual test bores (e.g. Walker and Lowe 1977). An alternative

Fig. 3.4: A level being used to establish a common datum for hand-coring of an infilled lake basin. Torness, Mull, Scotland.

approach is to use electrical resistivity surveys (Thompson 1978a), although this technique has not so far been widely employed in the morphometric analysis of infilled basins.

From the above discussion on sampling equipment, it will be appreciated that the sampling of most lake and bog sediments involves the extraction of successive cores of 50 cm or 1 m lengths. However, because most coring devices disturb the horizons immediately beneath those from which the samples are being taken, either by the penetration of part of the coring equipment (nose cone, auger head), or by compression and contamination of the sediments by material carried down from above, it is common practice to take alternate cores from adjacent boreholes. During sampling the level of the ground surface and local water table are likely to fluctuate, and stratigraphic control is best maintained by using an instrumental level mounted on a tripod near the sampling point (Fig. 3.4). This provides an independent datum to which the two boreholes can be related. In this way, alternate cores can be removed from the two boreholes and a continuous stratigraphic sequence recovered.

A further problem in the coring of peat bogs and infilled lake basins arises from the fact that, at depth, lithostratigraphic boundaries are seldom horizontal, and in fact can be highly variable (e.g. Lowe and Walker 1981). Moreover, records of important palaeoecological changes are often contained within a few cm of sediment. In order to ensure that a complete stratigraphic record has been obtained, therefore, overlapping cores should be taken from the adjacent boreholes. It is also important to ensure that the corer is driven vertically into the sediment. If these procedures are not followed, an incomplete and potentially misleading stratigraphic record may be obtained (Fig. 3.5). Finally throughout the sampling process careful recording, labelling and storing of samples are essential. Equipment must be carefully cleaned and contamination

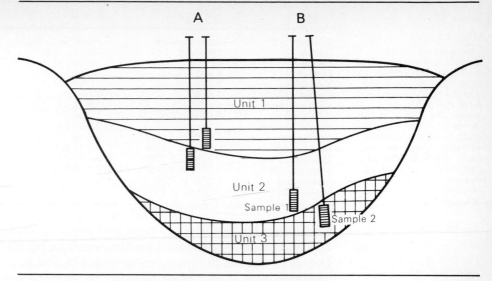

Fig. 3.5: Diagram illustrating the problems that can arise in coring. If overlapping cores are not taken, important lithostratigraphic boundaries can be missed (A). If the corer is not inserted vertically boundaries can still be missed even where overlaps have been allowed for (sample 2-B).

avoided. Hours of painstaking laboratory work may be wasted by careless field sampling. As Birks and Birks (1980) have observed: 'Field work is an extremely important part of any palaeoecological investigation. If the site selected is unsatisfactory, or the samples taken badly, no amount of sophisticated laboratory techniques will compensate for the loss or confusion of information' (p. 38).

(2) Laboratory methods

Laboratory analysis of Quaternary sediments is an integral part of environmental reconstruction. Both the physical and chemical properties of sediments can provide valuable data on the nature of former depositional environments, and changing sedimentological characteristics are often useful indices of climatic and other environmental changes. Set out briefly below are some of the methods commonly employed in the analysis of Quaternary environments. It should be pointed out, however, that this list is by no means exhaustive, for a very wide range of sedimentological techniques have been used by Quaternary scientists, and the reader is directed to the various textbooks and manuals that cover these in more detail (e.g. Krumbein and Pettijohn 1938; Carver 1971a; Folk 1974; Avery and Bascombe 1974; Allen 1975; Briggs 1977; Goudie 1981).

Particle size.   Measurement of particle size is a most important sedimentary technique, for it is a valuable tool in the description and classification of sediments, as well as aiding in the understanding of modes of sediment transport and deposition (Folk and Ward 1957). The **particle size distribution** of coarser grades of sediment (sand size and above – Table 3.1) can be established by sieve analysis, but sedimentation methods are required for finer materials. Different grades of sediment suspended in a liquid will settle out at different rates and these can be established by using either a hydrometer to record changes in the density of the suspension over time, or by extracting subsamples from the

Table 3.1: The Wentworth Scale of particle-size fractions and the equivalent (phi) units. The phi units are obtained by conversion from the mm scale, where phi is $-\log_2$ of the diameter in mm. The phi scale has the advantage of using integer numbers only, and also makes the statistical description of sediments more straightforward. See also Fig. 3.6.

Name	mm scale	ϕ units
Boulder	more than 256	more than -8.0
Cobble	256 to 64	-8.0 to -6.0
Pebble	64 to 4	-6.0 to -2.0
Granule	4 to 2	-2.0 to -1.0
Very coarse sand	2 to 1	-1.0 to 0
Coarse sand	1 to 0.5	0 to 1.0
Medium sand	0.5 to 0.25	1.0 to 2.0
Fine sand	0.25 to 0.125	2.0 to 3.0
Very fine sand	0.125 to 0.0625	3.0 to 4.0
Coarse silt	0.0625 to 0.0312	4.0 to 5.0
Medium silt	0.0312 to 0.0156	5.0 to 6.0
Fine silt	0.0156 to 0.0078	6.0 to 7.0
Very fine silt	0.0078 to 0.0039	7.0 to 8.0
Coarse clay	0.0039 to 0.001 95	8.0 to 9.0
Medium clay	0.001 95 to 0.000 98	9.0 to 10.0

suspension by means of a pipette and then measuring the changing concentrations of suspended matter over time by successive weighings. Details of these techniques can be found in Galehouse (1971) and Avery and Bascombe (1974). More recently, an electrical sensory method has been developed for particle size work by using a **Coulter Counter** (McCave and Jarvis 1973). Sediments are suspended in an electrolyte and passed through an electrode-flanked aperture. Voltage pulses proportional to the volumetric size of the particles can then be counted and the particle size distribution established. Particle size data are usually presented in the form of sigmoidal curves on probability graph paper (Fig. 3.6) or by means of ternary diagrams (triangular graphs) or histograms.

Particle shape.   Particle shape has been used to distinguish between sediments that have accumulated in different depositional environments, and has been used particularly effectively in the analysis of glacial and glaciofluvial sediments (e.g. King and Buckley 1968; Gregory and Cullingford 1974). Techniques include the purely visual assessment of particle shape, the use of prepared charts (e.g. Powers 1953) as a basis for the division of pebbles into classes ranging from angular to rounded, and the direct measurement of particles themselves. Shape measurements are then based on the axial ratios of the particles and include the Zingg (1935) method of plotting b/a against c/b[1] to derive classes of blades, rods, spheres and discs; indices of sphericity, for example, $(c^2/ab)^{0.333}$ (Sneed and Folk 1958); flatness indices, for example $(a + b/2c)$ (Cailleux 1947); and roundness indices, for example $(2r/a \times 1000)$ (Cailleux 1945). More sophisticated automated methods of grain counting and index measurement are discussed in Orford (1981).

Surface textures of quartz particles.   Different sedimentary environments (e.g. glacial, marine, aeolian) give rise to particular textural features on the surfaces of quartz and sand grains and these can be analysed using an **electron microscope**.[2] Characteristic textural features on quartz grains that can be detected by scanning electron microscopy (SEM) include fracture patterns, scratches, grooves, chatter marks and cleavage flakes (Krinsley and Doornkamp 1973). Moreover, it is often possible to identify

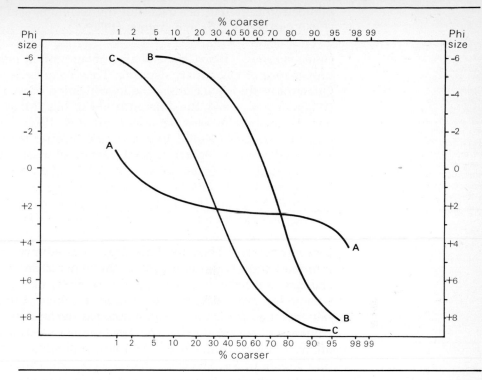

Fig. 3.6: Particle size distributions shown on probability graph paper. A: aeolian sand; B: glaciofluvial sediments; C: till.

superimposed features and hence more than one palaeoenvironment of modification can sometimes be inferred.

Organic carbon content.   The determination of organic carbon content is of considerable importance in palaeolimnology where it provides an index of biological productivity in former lake basins. It is also useful in establishing the amount of organic material that is likely to be required for the radiocarbon dating of a sample (Ch. 5). The most widely used method is loss on ignition (e.g. Belcher and Ingram 1950), in which the amount of carbon in a sample is indicated by the weight loss following combustion in a furnace. More accurate results can usually be obtained using standard titration methods (e.g. Tinsley 1950) or colorimetry techniques (e.g. Walkley and Black 1934; Metson 1961).

Metallic elements.   The variations in proportions of metallic ions of, for example, calcium, potassium, sodium or magnesium in late Quaternary lake sediments are now regarded as important indicators of the changing erosional history of lake catchments (Mackereth 1965; and explained below). The concentration of such elements in a sediment sample can be determined using either a flame photometer or an atomic absorption spectrophotometer. The former operates on the principle that a metallic salt drawn into a non-luminous flame ionises and emits light of a characteristic wavelength, while the AAS measures the concentration of an element by its capacity to absorb light of its characteristic resonance while in an atomic state. In both cases, the light emissions are recorded photoelectrically. Further details of the technique can be found in Dean (1960) and Reynoulds and Aldous (1970).

Heavy minerals. Heavy mineral assemblages often reflect the derivation or provenance of Quaternary deposits. They have been used in a number of Quaternary studies including the investigation of weathering profiles (Willman *et al.* 1966), the differentiation of tills (Madgett and Catt 1978), and the analysis of loess deposits (Catt *et al.* 1974; Rose *et al.* 1976). Heavy minerals are those with a specific gravity greater than 2.85 and are usually separated from the lighter mineral fraction in a sample by settling in a heavy liquid such as bromoform (SG 2.89). The heavy mineral assemblage is then dried and mounted on a slide, and the percentage of individual types can be determined using a petrological microscope. Full details of the method can be found in Carver (1971b) and Hutchinson (1974).

Clay mineralogy. The clay mineralogy of a sediment can provide information on both the origins of the material and on any chemical changes that have occurred due, for example, to the effects of different weathering processes since deposition. In Quaternary research, clay mineral analysis has also been employed in the differentiation of tills (Willman *et al.* 1963, 1966). The most widely used method in clay mineral analysis is X-ray diffraction (XRD) which involves the rotation of a sample in a stream of directed electrons. The clay minerals present (e.g. illite, chlorite, montmorillonite) are identified by observing and comparing the spacing and intensity of peaks on diffractometer traces (Griffin 1971). A short account of the method is provided by Yatsu and Shimoda (1981), but a more detailed coverage can be found in Andrews *et al.* (1971) and Carroll (1970).

GLACIAL SEDIMENTS

Glacial sediments of Quaternary age cover large areas of the earth's surface, particularly in the mid-latitude regions. In Europe, for example, glacially-derived **diamictons**[3] form an intermittent blanket over at least one third of the land area, while in North America, such deposits are spread over half the continent. It has already been shown (Ch. 2) that the moulding of these deposits into characteristic glacial landforms can be used to establish former glacier extent and direction of ice movement, and can form the basis for glacier modelling and climatic reconstruction. However, equally important palaeoenvironmental data can be derived from an analysis of the deposits themselves, for the distinctive properties of many glacial sediments allow inferences to be made about former glacier types, about the mode of deposition and about ice-flow directions. Indeed, in view of the widespread nature of glacial deposits by contrast with well-defined landforms, such as moraines and drumlins, it could be argued that the lithological evidence has a more important role to play in the reconstruction of Quaternary environments, although without doubt the most secure palaeoenvironmental inferences are drawn from situations where morphological and sedimentological evidence are employed together.

The nature of glacial sediments

The nomenclature of glacial deposits is often confusing. The superficial sediments seen to blanket the landscape in many parts of Europe were originally believed to have been derived from a great flood and were termed **diluvium**, although Sir Charles Lyell referred to them as **drift** because, along with many of his contemporaries in the early years of the nineteenth century, he believed that the deposits had been derived primarily from the melting of icebergs that had drifted in during a marine inundation. Curiously, the latter term has survived in the literature to the present day and is still used to refer to '. . . all deposits which are formed by, or in association with glacier ice, and to all deposits which owe their origin predominantly to glacier ice' (Francis 1975, p. 44). The fundamental division of glacial drift into **stratified** and **unstratified drift** was first proposed by Chamberlin (1894) and has formed the basis for the classification of glacial deposits until relatively recently. Unstratified drift is usually referred to as **till** or **boulder clay**. The former term was first used by Geikie in 1863 to describe the coarse stony soil found on the bouldery drifts of northern Britain. It is preferable to, and now more widely used than, the term 'boulder clay' as till frequently contains neither boulders nor clay. The term is also more satisfactory than '**ground moraine**' which has been employed to describe spreads of glacial deposits. We would follow Francis (1975) in suggesting that the term 'ground moraine' be no longer used in this context and that the word moraine be applied only to topographical features. All deposits formed directly by ice or from glacier ice without the intervention of water should be referred to as **till**.

Till is one of the most variable sediments to be found on the earth's surface. According to Goldthwait (1971), the principal characteristics that collectively distinguish till from other diamictons are:

(a) a lack of complete sorting which usually means the presence of some pebbles or boulders much larger than the dominant clay, silt or sand;
(b) a homogeneous mix lacking any smooth laminations or regular graded bedding, combined with
(c) a mixture of mineral and rock types, some of which are of distant provenance and are not represented in the local strata.

A number of additional identifying features frequently found associated with till are:

(d) at least a small proportion of striated stones and microstriated grains;
(e) common orientation of the long axes of elongated grains and pebbles;
(f) relative compactness or close packing of sediment;
(g) presence of a striated surface on the underlying rock in certain cases;
(h) subangularity in clasts of all sizes due to frequent breakage in transport coupled with particle smoothing by abrasion.

Stratified drift is characterised by the sorting of material by the action of meltwater. Such deposits are usually termed **glaciofluvial** and include sediments formed adjacent to, or in contact with, glacier ice (**ice contact**) and deposits accumulating in front of a glacier terminus (**proglacial** or **outwash**). The sediments are typically of gravel and sand size and frequently contain large boulders and cobbles. Deposits accumulating in glacial lakes (**glaciolacustrine**) are usually also included under the general heading of stratified drift.

The interpretation of Quaternary glacigenic sequences

Till is often subdivided into **lodgement** and **ablation** types, the former considered to be the result of deposition from beneath moving ice while the latter accumulated either from above or within the ice as the glacier melted. For many years, glacial sequences in both Europe and North America were frequently interpreted in terms of this simple depositional model, and there has been a general assumption that most of the Quaternary tills in these formerly glaciated regions are of the subglacial lodgement type. As a consequence, where individual till units have been found which are separated by stratified sediments, it has been customary to interpret such a sequence as indicating two distinct glacial phases (represented by the tills) and an intervening 'non-glacial' interval as shown by the glaciofluvial deposits (e.g. Poole and Whiteman 1961).

One of the first workers to challenge the validity of this depositional model was R. G. Carruthers (1947–48, 1953). In many parts of the north-east of England, Carruthers found glacigenic sequences which, he believed, suggested that glacial sedimentation was perhaps a more complex process than had hitherto been considered, a fact that was conclusively demonstrated some twenty years later by the research of G. S. Boulton. In a series of papers (Boulton 1967, 1968, 1970a, 1970b) Boulton described the results of fieldwork carried out around the glaciers of Spitsbergen which necessitated a fundamental reappraisal of previous theories of glacial sedimentation. Three main types of till could be identified: **flow till, melt-out till** and **lodgement till** (Table 3.2.). Flow till is released as a water-saturated fluid mass from the downwasting glacier surface. Flowage of successive generations of this material leads to the interbedding not only with outwash and lacustrine sediments deposited on the ice surface, but also with spreads of outwash beyond the glacier terminus. As a consequence, highly complex stratigraphic sequences of till, outwash and lacustrine sediments can result from a single phase of ice wastage. Lodgement till is released from basal ice and accumulates on the subglacial floor either through pressure against bedrock protuberances or against patches of stagnant ice underneath the moving glacier body. Melt-out till consists of englacial debris released from melting ice either above the glacier sole or at the glacier surface. In the former situation it is

Traditional classification	New classification		
Ablation till	Supraglacial till	{ Flow till	Melt-out till
Lodgement till	Subglacial till	{ Melt-out till	Lodgement till

Table 3.2: Classification of tills (after Embleton and King 1975a).

confined beneath the overlying ice body and the glacier bed, while at the surface it is trapped beneath the overburden of flow till. Hence, melt-out till can be either subglacial or supraglacial in origin (Boulton 1980).

A fundamental factor in the determination of the type of till that will be deposited at any one locality is the position at which debris is transported within the ice and this, in turn, is governed largely by the thermal régime of the glacier (Boulton 1972b). The thermal régime is determined by ice thickness, mass balance and, above all, climate, and can be used to define four boundary conditions at the glacier sole. These are:

A. a zone of net basal melting where more heat is provided to the glacier sole than can be conducted through the glacier;

B. a zone in which a balance exists between melting and freezing where the heat provided at the glacier sole is approximately equal to the amount that can be conducted through the glacier per unit time;

C. a zone of net basal freezing but where sufficient meltwater may still be present to raise the temperature and maintain parts of the sole at the melting point;

D. a zone in which the amount of heat provided at the sole is insufficient to prevent freezing throughout.

In zone A the glacier slips over its bed and material entrained in the basal ice layers will subsequently be deposited where frictional retardation against the bed is high. Similar processes operate in zone B although lodgement of material will tend to be greater with lower amounts of meltwater present. In zone C plucking of subglacial material occurs as the glacier slides over its bed, little lodgement till is deposited and material tends to be carried up into the ice through shearing action. Subsequent deposition therefore tends to be in the form of melt-out and flow tills. In zone D, the bed is frozen, no basal sliding occurs and glacier movement is entirely a result of internal shearing. Again, melt-out and flow tills are the dominant depositional types. Zones A and B tend to be associated with 'wet-based' or temperate glaciers, while zones C and D are found principally in 'cold-based' or polar glaciers. Hence in Spitsbergen the great thicknesses of till which have been released during retreat of the primarily cold-based glaciers are dominated by flow and melt-out tills, while in areas such as Iceland, Norway and the Alps where temperate glaciers predominate, most of the till deposited is of the lodgement type (Boulton 1972a).

There are several aspects of this work that have profound implications for the interpretation of Quaternary glacigenic sequences. First, if the relationships between glacier types and the form of deposition are applicable to older Quaternary glaciers, then it should be possible to make inferences about the thermal régime of former glaciers on the basis of their deposits. Boulton (1970a) has suggested that many of the late Quaternary tills that have been described at localities in both Europe and North America, and which have been interpreted as subglacial in origin, bear striking resemblance to the supraglacial tills of Spitsbergen, and are therefore more likely to be flow tills than lodgement tills. This has led to speculation that the last British ice sheet, for example, was largely cold-based, at least in its outer parts where it adjoined the permafrost zone beyond the ice margins (Boulton 1972a; Boulton *et al.* 1977). Secondly, it is

Fig. 3.7: Multiple till sequence at Glanllynnau, North Wales interpreted by Boulton (1977) as showing basal lodgement till and overlying melt-out and flow tills interbedded with glaciofluvial gravels.

becoming increasingly apparent that multiple till sequences that have previously been interpreted in terms of glacier advance, retreat and readvance, can now be viewed as the product of a single retreating ice mass releasing flow tills at its margin. For example, Boulton (1977) has described how such an interpretation can be applied to the complex sequence of till and outwash deposits at Glanllynnau on the Cardigan Bay coast of Wales (Fig. 3.7), while Eyles and Slatt (1977) reported a Pleistocene drift sequence on the coast of the Avalon Peninsula in Newfoundland in which complexly-related lodgement till, melt-out till, flow till, supraglacial and proglacial outwash and supraglacial rhythmites can be attributed to a single ice melt phase. In both of these cases previous interpretations involved two glacial advances separated by a non-glacial interval. Thirdly, the fact that many Quaternary tills may not be of the lodgement type has important implications for the interpretation of till fabric data, discussed in the following section.

Ice-directional indicators

It has already been shown how certain landforms such as drumlins and roches moutonnées can be used to infer the direction of ice movement across a glaciated area. However, some characteristics of glacial sediments can also yield valuable ice-directional information. The most widely used are indicator erratics, till fabrics and certain properties of the till matrix.

Erratics.   The term 'erratic' is derived from the phrase *terrain erratique* used initially by the French geologist de Saussure in the late eighteenth century to describe areas where material of foreign origin overlay local bedrock. The shortened term is now widely applied to a particle of any size that is not indigenous to the area in which it is currently found. Although erratics occur in a variety of deposits, including beach sediment

Fig. 3.8: The giant quartzite erratic near Okotoks, Alberta, Canada.

and river gravels, they have been most widely employed in glacial geological studies as a means of reconstructing patterns of ice movement. In this respect, the most valuable types of erratic are those for which the sources are known, which are resistant to erosion, and which have distinctive appearances, unique mineral assemblages or unique fossil contents thereby allowing unequivocal identification. These are usually termed **indicator erratics** (Fairbridge 1968). Indicator erratics may be contained within glacial deposits or they may lie on the surface. They range in size from finely-comminuted fragments to large blocks weighing several hundred tons. The famous Okotoks erratic (Fig. 3.8) in south-west Alberta, for example, is a block of quartzite weighing over 18 000 tonnes and is considered to have travelled over 100 km from its source in the Rocky Mountains to the north-west of Calgary.

Erratics are useful in two particular respects: in the reconstruction of local glacier flow patterns, and as indicators of the direction of regional ice movement. With regard to the former, the localised nature of flow lines within glacier ice can be demonstrated by the often narrow debris trails which lead away from particular rock outcrops. One such example is the fan of erratics spreading eastwards from the outcrop of Essexite near Lennoxtown to the north-east of Glasgow (Fig. 3.9), and which can be traced eastwards in a narrow band no more than a few kilometres in

Fig. 3.9: The Lennoxtown boulder train in the Forth lowlands of central Scotland, based on A. M. Peach (1909) (after Sissons 1967a).

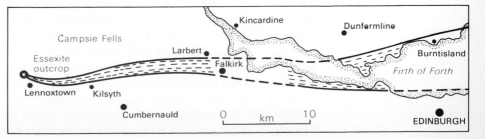

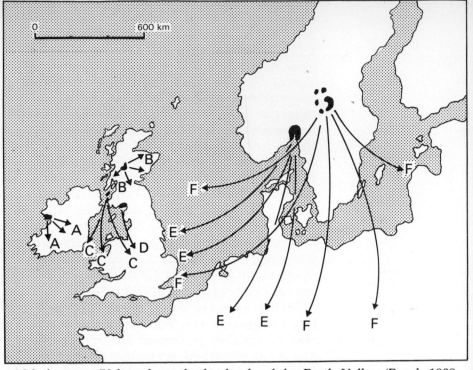

Fig. 3.10: Distribution of some indicator erratics by ice in Britain and north-west Europe.
A: Galway granite
B: Rannoch granite
C: Ailsa Craig riebeckite-eurite
D: Criffell granite
E: Oslo rhomb porphyry
F: Dala porphyries
(after Sissons 1967a; Fairbridge 1968; West 1977a).

width for over 50 km along the lowlands of the Forth Valley (Peach 1909; Shakesby 1978).

On the regional scale important data about former ice movement can be derived from indicator erratics. For example, the occurrence in the tills of eastern England of the distinctive rhomb porphyry from the Oslo region of Norway and the Dala porphyry from southern Sweden indicate the former passage of Scandinavian ice across the North Sea basin. Similarly, the presence of the Ailsa Craig riebeckite-eurite in glacial deposits around Cardigan Bay is evidence of ice movement from Scotland southwards down the Irish Sea basin (Fig. 3.10). In North America, erratics from outcrops to the north of Lake Superior can be found as far south as the state of Missouri, while some from even further afield on the Canadian Shield have been traced south over distances of up to 1200 km (Flint 1971). In northern Canada, flow patterns and ice divides of the Laurentide ice sheet have been reconstructed on the basis of erratic trains (Andrews and Miller 1979; Shilts 1980).

One problem with using erratics in this way, however, is that their presence in a glacial deposit may not always reflect primary derivation (*i.e.* an erratic could have been removed during a previous glacial episode and reincorporated into a younger till) and this can lead to erroneous interpretations of former patterns of ice movement. Hence, although erratics can sometimes provide useful information on directions of ice flow, they are perhaps best used in conjunction with information obtained from other independent sources.

Till fabrics. The arrangement of particles of any size in a till is termed the **till fabric**. It was noted at a very early stage in the development of glacial geological studies that pebbles within a till often displayed a preferred orientation (Miller 1884), although it was somewhat later that a quantitative relationship was established between pebble orientations in till and patterns of regional ice movement (Richter 1936; Holmes 1941). Since then, till fabric analysis, namely the study of the orientations and dips of particles within a till matrix, has become one of the most important means of reconstructing former ice-flow directions. The technique rests on the assumption that within the constantly deforming regelation layer between the glacier sole and the underlying bedrock, stones will become orientated to adopt the line of minimal resistance with their long axes parallel to the direction of glacier movement (Glen *et al*. 1957). Subsequent deposition of the subglacial debris as lodgement till therefore preserves a record of the former direction of ice movement, and this can be discerned by measuring the orientation of pebbles in till exposures using a compass. Measurement of the dip by means of a clinometer or similar instrument may also provide useful ice-directional information as a tendancy has been observed for pebbles to dip up-glacier (e.g. Harrison 1957a). Methods of collecting till fabric data in the field have been described by Andrews (1971a).

Till fabric data are usually presented in the form of polar graphs. Where two-dimensional data only have been obtained, a rose diagram is constructed showing the number or proportion of pebbles in different azimuthal classes. However, because each measured stone is represented by two opposite azimuthal values (e.g. 30° and 210°), the rose diagram actually consists of two reflected halves or mirror images. The data can be shown by a line through the middle of each sector (Fig. 3.11a), by the linking of such lines to form the typical rose diagram (Fig. 3.11b), or by the shading of each azimuthal class to the extent of the line marking the outer limit of each class (Fig. 3.11c). Where the dip of the pebbles has been recorded, orientation measurements are taken in the *down-dip* direction, and therefore each pebble is represented by a single dip and orientation value. In this case the diagram will show a full 360° distribution (Fig. 3.11d). Orientation and dip can be plotted together in the form of a scattergram with the radius divided into degrees showing the angle of dip and the circumference divided into degrees showing the orientation of the pebble (Fig. 3.11e). This is one of the most commonly used methods for depicting till fabric data, and in some cases the visual effect is enhanced by contouring the diagram as shown in Fig. 3.11f.

The technique of till fabric analysis and its application in Quaternary research has attracted much critical discussion. The accuracy of till fabric data collection has been questioned (e.g. Hill 1968; Drake 1977), problems connected with the graphical presentation of the data have been considered (e.g. Andrews and Smith 1970; Andrews 1971a), and a lengthy debate has ensued over the most appropriate method of statistical analysis of orientation data. Two-dimensional till fabric data have been analysed using a variety of procedures ranging from simple calculations of mean and modal azimuths (Briggs 1977), to more advanced methods including linear transformations (e.g. Krumbein 1939), vector analyses (e.g. Reiche 1938; Curray 1956), Chi-square tests (e.g. Harrison 1957b; Andrews and Smithson 1966) and the Tukey Chi-square test (Harrison 1957b). Three-

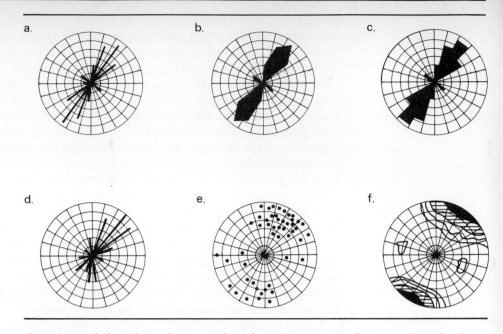

Fig. 3.11: Different methods for representing till fabric data. For explanation see text.

dimensional data have been analysed using more sophisticated methods including vectorial techniques (e.g. Andrews and Shimizu 1966) and eigenvalue procedures (e.g. Mark 1973). There is, however, no universally accepted method for the statistical analysis of till fabric data, and it has even been argued that many of the more complex methods are in fact either inappropriate or of little practical value in the analysis of this type of orientation data (Cornish 1979).

The reconstruction of regional ice flow patterns from till fabric data must be based on fabrics derived from lodgement till. Studies of till fabrics in the deposits around the Spitsbergen glaciers (Boulton 1971) have shown that flow till is often characterised by a fabric which simply reflects local flowage of saturated debris on a downwasting glacier surface. No reliable regional component of ice movement can therefore be derived. Melt-out tills may retain their englacial fabric, but as this is easily lost during the final melting phase, these too are unreliable indicators of regional ice movement. Even with lodgement till there is often between-site variation, irrespective of any change in direction of glacier movement, reflecting the mode of deposition and the form of the surface onto which the till is deposited. Similar between-site variability has also been detected in Quaternary lodgement tills within a very limited area (Andrews 1971b) and even within a single section (Rose 1974). A further complicating factor is that subsequent ice advances can cause reorientation of the fabric in previously-deposited tills, as has been shown in the Edmonton region of western Canada (Ramsden and Westgate 1971).

In order to derive meaningful ice-flow patterns from till fabric data therefore, a number of basic rules must be observed:
1. Particular care must be taken with field measurement in order to minimise operator errors.
2. Samples should only be taken from lodgement till. This, however, is often easier said than done, for flow tills sometimes resemble

lodgement tills and it may not always be possible to differentiate between the two, particularly in limited exposures. In order to infer a systematic trend in till fabrics, therefore, and also to eliminate results from fabrics reflecting local depositional variations, a large number of till fabrics will always be required from an area. Moreover, stratigraphic and sedimentological characteristics of the tills should always be considered in the interpretation of fabric data.

3. In view of the above, it is, of course, axiomatic that fabrics should always be taken from what is known to be the same till.

4. Fabrics should be used, as far as possible, with other indicators of ice-flow direction.

Properties of the till matrix.   Certain physical and chemical properties of glacial tills can be used to make inferences about regional ice movement, and also to differentiate between tills from different source areas. This approach rests on the assumption that glacial till will inherit certain textural and chemical characteristics from the bedrock over which the glacier has travelled. In certain cases, where a particularly distinctive lithological type lies up-glacier, it may be possible to detect the influence of that rock outcrop on the till matrix. In a sense, therefore, certain properties of glacial tills can be used in the same way as indicator erratics. For example, in Arctic Canada Andrews and Sim (1964) used variations in carbonate content in tills in relation to the known distribution of limestone bedrock to infer the radial passage of ice from the Foxe Basin westwards onto Melville Island and eastwards onto Baffin Island. In Norway, Jørgensen (1977) showed how particular size contrasts between tills derived from Pre-Cambrian bedrock differed from those derived from sedimentary Cambro-Silurian rocks, and then augmented these data with the results of chemical and mineralogical analyses to determine transport distance and direction of the till material. Finally, in the Rocky Mountain foothills of western Alberta, Boydell (1970) used particle size and heavy mineral data in addition to erratic counts and till fabric analyses to differentiate between glacial tills derived from glaciers flowing eastwards from the mountains, and those derived from the Laurentide ice sheet moving south-westwards from the Canadian Shield. Properties of the till matrix can therefore provide useful insights into patterns of Quaternary ice movements, although in the majority of cases, tills tend to exhibit characteristics derived from the bedrock immediately up-glacier from the point of deposition. This suggests that the use of till matrix properties in inferring ice flow patterns may be more of local than regional application.

Careful study of glacial sediments, particularly tills, can provide a considerable amount of information about Quaternary glacial environments. A range of techniques of both a morphological and sedimentological nature is available for the establishment of ice flow direction, and it is the judicious application of these methods, particularly in combination, which makes possible the type of reconstruction shown in Fig. 3.11. In addition, the considerable advances that have been made in recent years in the understanding of processes and patterns of glacial deposition are beginning to offer the tantalising prospect of being able to infer the type of glacier responsible for the deposition of a particular sediment body.

PERIGLACIAL SEDIMENTS

In the 'periglacial domaine' (p. 48), freeze-thaw activity causes fracture of the country rock and the accumulation of coarse, angular debris. This material moves downslope through the combined processes of flowage (solifluction) and creep induced by the growth and melt of interstitial ice and a landscape of low-angled slopes with smooth profiles results. The deposits are known by a variety of names, including '**head**', '**coombe rock**', '**tjaele gravel**' and, where coarse stratification has developed, '**stratified screes**' or '**grêzes litées**' (Fig. 3.12).

Sediments that have been affected by periglacial action during the Quaternary are widespread throughout the mid- and high-latitude regions of the world. Frequently their presence reflects not only the breakdown of bedrock by cold-climate processes, but also the reworking and redistribution of pre-existing drift deposits. They can be recognised by a number of distinctive characteristics including the occurrence of predominantly angular material within the sediment matrix as a result of frost-riving, the vertical alignment of many stones reflecting the upward movement of particles with the expansion and melting of ice lenses (e.g. Watson and Watson 1971), the presence of structures associated with ground ice (see below) and, where the deposits have been moved by solifluction, the preferred alignment of the larger particles downslope. Fabric analysis in particular may be used to distinguish between, for example, soliflucted and undisturbed glacial drift, for in periglacial deposits fabrics taken over a wide area should exhibit a constant relationship between stone orientation and local slope (Watson 1969).

Structures associated with permafrost

Although the presence of frost-shattered bedrock and extensive spreads of solifluction deposits are indicative of a former periglacial climatic régime within a particular area, they can only provide the most generalised of information about former environmental conditions. Certain structures within these deposits, however, offer a basis for more detailed palaeoclimatic reconstructions, and those which appear, at present, to offer the greatest potential in this direction are **ice-wedge casts** and, to a lesser extent, **involutions**.

Ice wedges are considered to be diagnostic structures of permafrost (Péwé 1973; Black 1976) and form where thermal contraction of the active layer in winter opens vertical cracks into which water seeps and subsequently freezes. The addition of successive increments of ice leads to the growth of the wedge but upon melting the ice is replaced by material falling into the crack from above and from the sides. In this way a cast or **pseudomorph** of the original form of the ice wedge is preserved (Fig. 3.13). Ice wedges typically form as part of a network of thermal contraction cracks which appear as interconnected polygons on the ground surface, although in small sections they are often found as single features. They develop in a variety of sediments and occur only in the zone of continuous permafrost (Péwé 1966); in the discontinuous

Fig. 3.12: Stratified scree
(foreground) and fossil
protalus rampart (middle
distance), near Cader
Idris, Wales.

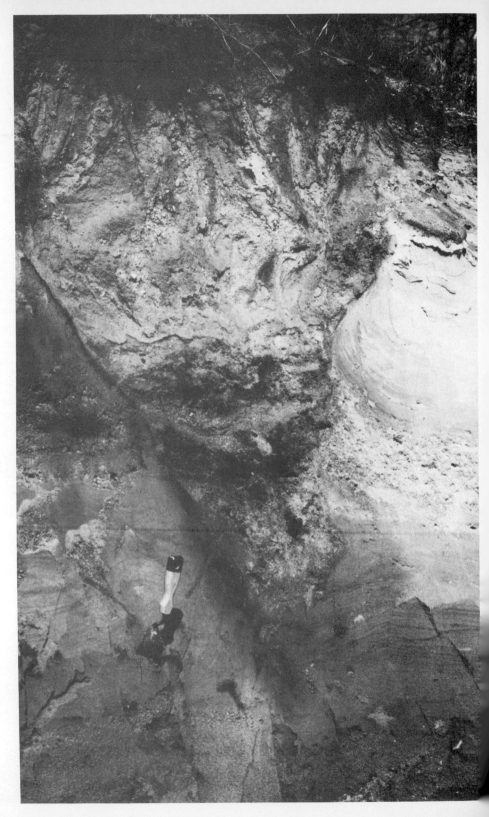

Fig. 3.13: Ice-wedge pseudomorph, Westdorp, Netherlands.

Fig. 3.14: Sand wedge,
Westdorp, Netherlands.

permafrost zone, most ice wedges appear to be inactive (Brown and Péwé 1973). In arid and semi-arid periglacial regions and in some localities that are free-draining, frost fissures that develop from thermal contraction of the ground are frequently filled with aeolian sediment and are termed **sand wedges** (Fig. 3.14) or **tesselations** (Péwé 1959). Sand-wedge pseudomorphs can sometimes be distinguished from ice-wedge casts by the characteristics of the sediment infill (e.g. Worsley 1967; Morgan 1973), but a clear differentiation between the two is not always possible in the field. A considerable body of data is now available on ice wedges, and further discussion can be found in Embleton and King (1975b), French (1976) and Washburn (1979).

In many areas where periglacial conditions prevailed, unconsolidated sediments in open sections frequently display contortions in bedding, the interpenetration of one layer by another, and pockets which resemble load structures.[4] Such features are termed **involutions** or **festoons** (Fig. 3.15) and are widely interpreted as reflecting differential pressures induced by freezing within the active layer above the permafrost table (Washburn 1956). Curiously, involutions appear to be relatively uncommon within present periglacial environments, and it is clear that not all involutions are a product of the growth of ground ice. Nevertheless there is often a close field relationship between involutions and other undoubted periglacial phenomena, such as ice wedges and upturned stones, and thus where the depositional context is clearly periglacial it is generally accepted that the involutions can also be safely interpreted as such. Their mode of origin is not completely understood, although two hypotheses, one involving the

Fig. 3.15: Section in head deposits at Pitstone, near Aylesbury, Buckinghamshire. The upper layers display involution structures.

squeezing of a moist plastic layer between rigid permafrost beneath and a newly frozen crust at the ground surface, and the other involving differential ice segregation in materials of varying frost susceptibility, have gained widespread acceptance (Embleton and King 1975b). On the other hand, it is also clear that moisture-controlled density differences following the thawing of ice-rich sediments can produce involution structures, similar to the load-cast structures described by geologists (Kostyaev 1969). Such an interpretation would not necessarily involve periglacial activity and French (1976) has therefore cautioned against the interpretation of involutions as periglacial in origin in the absence of other unequivocal indicators of either frost action or permafrost.

Palaeoclimatic significance of periglacial evidence

As ice wedges and, in some cases, involutions are associated with permafrost, the former extent of permanently frozen ground can be deduced from the distribution of ice-wedge casts and involution structures. The present permafrost zone of the northern hemisphere is characterised by daytime temperatures of less than 0 °C for three-quarters of the year, less than −10 °C for half of the year and rarely more than 20 °C; precipitation is typically less than 100 mm in winter and over 300 mm in the summer months (Velichko 1975). These figures are, of course, highly generalised and significant differences are noted with, for example, altitude and proximity to coastlines. Although permafrost may begin to form where the mean annual air temperature falls to 0 °C or slightly colder (Péwé 1973), many studies in the Arctic have shown that discontinuous to continuous permafrost develops where mean annual air temperatures have been in the range −4 to −6 °C or lower for decades or centuries (Black 1976). This is confirmed by the fact that in North America the southern boundary of the continuous permafrost zone lies in the neighbourhood of the −7 to −8 °C mean annual air isotherm, which corresponds to a mean annual ground temperature at sea level of −5 °C (Gold and Lachenbruch 1973).

Ice wedges, however, require markedly colder conditions than permafrost in order to develop. Lachenbruch (1966), for example, has suggested that mean winter temperatures must fall below −15 or even −20 °C before cracking of the ground will occur. Studies in Alaska have shown that active ice wedges are almost entirely confined to the area where mean annual air temperatures range from −6 to −8 °C (Péwé 1966), although slightly lower values have been reported from Siberia. Williams (1975), however, has argued that in areas prone to seasonal flooding and where the deposits are largely of coarse sands and gravels, mean annual air temperatures may need to fall below −8 to −10 °C in order for ice wedges to form. If these figures are representative, they imply that where ice-wedge casts occur, mean annual air temperatures at the time of their formation must have been around −8 °C or even lower.

Ice-wedge casts, therefore, provide useful information on former mean annual air temperatures, and can also give an indication of the levels of freezing that occurred during the winter months. They do not, however, give any indication of former summer temperatures, nor of the range in annual air temperatures. One way in which such data may be derived has been outlined by Williams (1975) and involves the use of involutions. As outlined above, involution structures develop in the active layer overlying permafrost, the depth of which is controlled by the warmth of the summers (*i.e.* the accumulated temperatures above zero). In many areas of Britain where involutions are found, thaw depths revealed by thicknesses of involutions are comparable with depths of thaw recorded in the interior of Alaska and Siberia. The accumulated temperatures above zero in those regions (**thawing index**) for the summer months ranges between 900 and 1500 degree days, and most of the areas experience July temperatures of 10 °C or more. At many sites in central and southern England, ice wedge casts have been found associated with the involution structures, and a mean annual air temperature of less than −8 °C is thus implied. Assuming therefore (a) a mean annual air temperature of −8 °C, (b) a thawing index of 900 degree days, and (c) average temperatures for the warmest month in excess of 10 °C, a curve can be constructed showing annual temperature fluctuations at the time the involutions were forming (Fig. 3.16). The curve implies a high degree of continentality, a mean annual temperature range of *ca.* 30 °C and, although at present conjectural, agrees well with temperature curves for the last cold phase in central England based on coleopteran evidence (Ch. 7).

In addition to palaeotemperature values, some indication may also be gained from periglacial evidence of former precipitation levels. It was noted above, for example, that sand wedges are often characteristic features of arid periglacial environments. Active sand wedges have been observed in Antarctic regions where mean annual precipitation values are believed to be less than 160 mm (Berg 1969), while French (1976) has stated that sand wedges occur in arid regions with less than 100 mm of

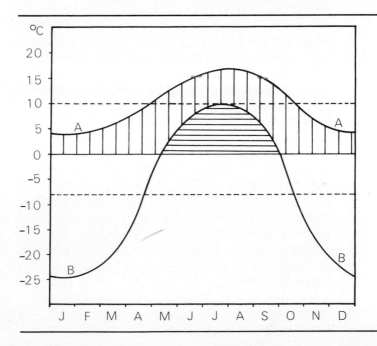

Fig. 3.16: Present mean monthly temperatures in central England (curve A) and suggested values during the coldest part of the Devensian cold stage (curve B) (after Williams 1975).

precipitation per year. Fossil sand wedges may, therefore, be indicative of arid periglacial environments. Ice-wedge casts, on the other hand, are indicative of more humid conditions although it would seem that true ice wedges also are unlikely to form where precipitation levels are high. Evidence from present-day Arctic areas suggests that low winter snowfall is as vital a condition for the growth of ice wedges as severe cold, for as snow is an excellent insulator the ground will not crack if there is a snow cover in excess of 15–25 cm (Williams 1975). A 25 cm snow cover is equivalent to a little over 100 mm of winter rainfall. Since about 60 per cent of total precipitation in many Arctic regions falls in the summer months, an annual precipitation of 250 mm (in rainfall equivalent) would appear to be the maximum possible if ice wedges are to form. An annual precipitation of 250 mm is markedly less than that found in many mid-latitude regions where ice-wedge pseudomorphs occur implying significantly drier conditions in these areas at the time that the ice wedges were forming. Although at present speculative, this line of reasoning is supported by the fact that in those areas of Alaska where ice wedges are best developed, mean annual precipitation is around 200 mm (Péwé 1966).

Fossil periglacial phenomena, therefore, appear to be a useful data source for the reconstruction of Quaternary environments. However, a number of workers have urged caution over the use of this type of evidence in palaeoclimatic reconstruction. A major problem arises from the fact that present day Arctic areas are not necessarily good analogues for Quaternary mid-latitude periglacial environments, where a higher-angled sun would have produced a solar radiation régime markedly different from that which prevails at the poles. As a consequence, many areas of the mid-latitudes would have experienced considerably more diurnal freeze-thaw cycles than the present High Arctic where the seasonal periodicity of daylight and darkness favours longer, more severe cycles and deeper ground freezing (Washburn 1979). Certainly, mid-latitude areas did not experience the extremes of freezing that now characterise, for example, interior Siberia and Canada and, in view of the proximity of large glacier masses, probably had a climatic régime which differed in a number of respects from that of the present-day high latitudes, particularly in terms of precipitation, wind direction and wind intensity (French 1976). Difficulties may also arise from the sometimes ambiguous nature of periglacial evidence, and from the fact that, in periglacial regions, geomorphological processes are affected as much by local site factors as by prevailing climatic conditions. Finally, there is the problem of dating. Relatively few sites have been found where periglacial phenomena can be dated with any degree of precision, and it seems likely that in both Europe and North America, relict periglacial phenomena belong to more than one cold phase. On the other hand, the close similarities that are beginning to emerge between climatic estimates based on different types of periglacial evidence (Karte and Liedtke 1981) and those derived from other, quite different sources (Ch. 7), are encouraging. This suggests that relict periglacial features are potentially useful data sources in the analysis of Quaternary environments, although they are perhaps best employed in a supplementary rather than in a primary capacity.

AEOLIAN DEPOSITS

During the cold phases of the Quaternary, wind was a particularly effective geomorphological agent, and the results are seen today in the extensive spreads of aeolian sands and silts that are found in many parts of the world. Much of this material was derived from the greatly-expanded periglacial regions where a combination of a sparse vegetation and seasonally-dry ground meant that large areas of unconsolidated and friable sediment were left exposed to wind action. Particularly vulnerable were the unvegetated surfaces of till and outwash sediments which were rapidly stripped of their finer components by strong winds blowing off the nearby glacier surfaces. The periglacial domaine was not the only source of wind-blown sediment, however, for considerable quantities of sand and fine silt were removed from the continental shelves which had been exposed by a eustatic fall in sea-level of over 100 m, while great spreads of sand, which originated in the warm temperate and subtropical deserts, were deposited in the desert margins and savannah regions of the world (Saarnthein 1978).

Fine-grained Quaternary aeolian deposits are often referred to as **loess.** True loess consists largely of silt with a grain size ranging from 2–64 μm, although there may also be a significant proportion of material in the clay size category (less than 2 μm). Loess often has a high carbonate content, sometimes exceeding 40 per cent by weight, and frequently possesses a distinctive heavy mineral and clay mineral suite which may be useful in identification and correlation between individual deposits (Frye *et al.* 1962). Blown sand on the other hand is characterised by a size range from 64 μm to 2 mm. The sand grains are mainly rounded and possess a surface texture which has a matt or 'frosted' appearance as a result of wind abrasion, by contrast with those which have been deposited in water which are usually more angular and shiny (Cailleux 1942). Typical fracture patterns on sand grains which have been moved by wind action can be detected under the electron microscope (Krinsley and Funnell 1965). Where the origin of blown sand is offshore, the deposits often contain fragments of marine shells and other calcareous matter and may have a carbonate content as high as 65 per cent. In general, blown sand moves by deflation and is less far-travelled than loess, the lighter particles of which may be carried aloft by air currents and deposited hundreds or even thousands of kilometres from their source. Loess deposits characteristically blanket the landscape covering hilltops, plateau surfaces and valley floors alike. They often grade into, and are interbedded with outwash and till, and in areas such as central Europe, the buff/yellow loess deposits (Fig. 3.17) are found intercalated with the darker humic horizons of buried soils (Kukla 1975). Loess exposures often display no apparent stratification, but careful examination of the sections may reveal faint, well-developed beds (Washburn 1973). In eastern Europe, the Ukraine and parts of Asia, loess sheets attain thicknesses of over 100 m, while in the New World loess exposures up to 30 m thick have been reported from Kansas and Argentina. The major spreads of loess are found in northern China, in the pampas of Argentina and Uruguay, on the Great Plains of North America,

Fig. 3.17: Loess deposits with well-developed palaeosols, southern Germany. The buried soils are the darker bands near the surface and in the middle of the section. A third, less well-developed palaeosol lies between these two.

and in a discontinuous belt stretching across central Europe from northern France to the Ukraine. In Britain, the major concentrations are found in eastern and southern England (Fig. 3.18).

Wind-blown sands occur either as an irregular veneer over the land surface or in the form of dunes. In Holland, Belgium and north-east France, a mantle of yellow sand several metres in thickness is widespread, and is known as **coversand**. These deposits usually have flat surfaces, are often stratified with alternating loamy and sandy layers, and may be at least partly niveo-aeolian in origin, reflecting the contemporaneous transport and deposition of snow and sand grains. The source of this material was most probably the adjacent Rhine delta or the exposed floor of the North Sea basin (Maarleveld 1960). In many parts of Europe and North America, the aeolian sands have formed dunes. The most spectacular are found in North America, particularly in Nebraska where dunefields occupying an area of over 50 000 km^2 have been described (Smith 1965), around the Great Lakes, and in a belt across Europe stretching through north Germany, Poland and into the USSR. In Britain, coversands are thin and patchy (Fig. 3.18), and dunes have only been found in a few localities (e.g. Straw 1963; Matthews 1970).

Aeolian phenomena and Quaternary wind directions

The principal contribution that the study of wind-blown deposits can make to the understanding of Quaternary environments is in the evidence that they provide of former wind directions. This information can be derived from the particle size and chemical composition of the aeolian material as well as from sand dune morphology. It has long been recognised, for example, that wind-blown sands and loess become progressively thicker and more extensive in an easterly direction across Europe (Williams 1975), thus implying a predominantly easterly airstream

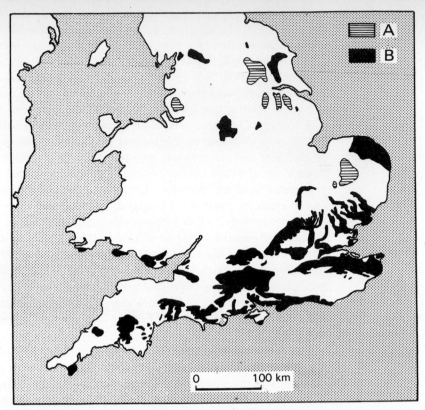

Fig. 3.18: Distribution of coversands (A) and loess (B) in England and Wales (after Catt 1977, 1978).

during the phases of loess deposition. In Britain, the analysis of aeolian deposits at sites from the Kent coast to south Devon (Pitcher *et al.* 1954; Harrod *et al.* 1973) revealed a westward decrease in coarser particles and a complementary increase in light flaky minerals (mica and chlorite) which can most easily be interpreted as reflecting transport of the loess by northeast or easterly winds (Catt 1977). An easterly wind direction was also indicated by the close similarities between the mineralogy of the silt-size fraction of Late Devensian glacial deposits and loesses in eastern England which suggested that the source for the loessic sediments was glacial outwash in the North Sea basin (Catt 1978). An easterly wind direction implies the development of a major anticyclonic circulatory system centred over north-west Europe (Lill and Smalley 1978). Aeolian materials have also been found in ocean sediments. For example, aeolian deposits in deep-sea cores taken from the vicinity of the Cape Verde Islands were interpreted as reflecting the transmission of dust particles by easterly winds from the Sahara Desert (Bowles 1975), and are evidence of greater aridity in North Africa during the glacial phases of the Quaternary (Parmenter and Folger 1974).

Dune morphology has been used in both Europe and North America to infer former wind directions. In the Sandhills of Nebraska and South Dakota dune systems of different ages suggest the dominance of northerly winds during the Wisconsinan glacial, by contrast to the predominantly north-westerly winds of the present day (Smith 1965). In Britain, studies of aeolian dunes in the Vale of York which appear to have formed during the Loch Lomond Stadial, suggest predominantly north-westerly winds around 11 000 to 10 000 years ago (Matthews 1970), while dune orientations in the Netherlands indicate a west or south-westerly wind direction during the same time period (Maarleveld 1960, 1964). The Dutch evidence also shows, however, that during the earlier cold phase of the Older Dryas, winds were dominantly north-westerly and this therefore implies a significant shift of wind direction during the Lateglacial part of the last cold stage.

Although Quaternary aeolian deposits can be useful indicators of wind direction and hence of the former positions of major pressure cells, the evidence is not always easy to interpret. Many former wind-blown deposits have been affected by the action of ground ice, and by fluvial and alluvial processes, and are therefore frequently intermixed with other deposits. In many areas of south-east England, for example, loessic sediments have been incorporated with slope-wash deposits (Gruhn et al. 1974), fluvial deposits (Kennard 1944) or soliflurcted chalk (Evans 1972) to form **brickearths**, so called because of their value to the brick-making industry. In many of the brickearths, the diagnostic physical and chemical properties will have been lost and hence recognition of the original wind-blown nature of the sediments may be difficult. Post-depositional changes induced by weathering can further change the character of loessic deposits.

The interpretation of palaeo-wind directions from fossil sand dunes may also pose problems (Smith 1949). The orientation of sand dunes may more often reflect the direction of movement of storms, rather than the direction of the prevailing winds which may blow from a different quarter. Moreover, the morphology of dunes may be an unreliable indicator of former wind directions as this may have been affected by weathering prior to the stabilisation by vegetation. Where sections are available, it may be possible to use the dip of foreset beds providing that these are steeply inclined (30–35°), for the true dip in both U-shaped dunes and barchans is always to the leeward (Embleton and King 1975b). The method is often less satisfactory in practice, however, as many dunes possess beds whose dips are shallow and variable. Finally, there is the recurrent problem of dating of dunes, coversands and loess. The contemporaneity of Quaternary dune systems can seldom be established, and relatively few radiometric dates are yet available on loess spreads and coversands. It is this lack of a secure dating framework that is perhaps the most serious obstacle to the more widespread use of fossil aeolian phenomena in palaeoclimatic reconstruction.

Aeolian deposits and Quaternary environments

Because aeolian deposits are often rich in carbonate they frequently contain molluscan remains (Ch. 4) and these can be used to obtain information on environmental conditions prevailing at the time of deposition. In the Peoria Loess of the Great Plains of North America, Leonard and Frye (1954) found a molluscan fauna indicative of grassland interspersed with patches of woodland. No species occurred that were tolerant either of drought or of high temperatures. The evidence suggested that during the time of loess deposition in the Late

Wisconsinan period (between *ca.* 22 000 and 12 000 BP) conditions were more favourable to animal life on the Great Plains than at the present day, and that although wind-blown sediment was being deposited, the landscape was far from barren or arid. In Britain, a brickearth in Kent of Late Devensian age yielded a gastropod fauna which included species indicative of cold conditions such as *Columella columella*, which today is widespread in the Alps and Scandinavia (Kerney 1971a). In central Europe, the calcareous loesses of Czechoslovakia and Austria are rich in snail shells, many of which are intact and whose occupants were clearly alive at the time of loess deposition. Some of the loesses are characterised by the *Pupilla* group of land snails and are interpreted as reflecting a cold loess steppe environment with a highly continental climate and a large daily seasonal temperature range. Other loesses are dominated by the *Columella* group of gastropods and are considered to be indicative of loess tundra conditions, perhaps with permafrost, and with very low winter temperatures. A third group, dominated by the *Striata* snail fauna, resembles that found today in central Europe and reflects warm loess steppe conditions with warm dry summers and cold winters (Lozeck 1964, 1972). Elsewhere, in the loess deposits of Austria and Moravia, Frenzel (1964) has recorded pollen grains of Gramineae and Cyperaceae, but other palynological evidence from loess deposits is rare.

PALAEOSOLS

A soil that has developed on a land surface of the past is termed a palaeosol (Ruhe 1965). Such soils will frequently have formed under environmental conditions that differ markedly from those currently prevailing at a site, and therefore by relating the fossil horizons to those of present-day soils, deductions can be made about their conditions of formation. In particular, inferences can often be made about two of the principal soil-forming factors, namely climate and vegetation. Additional palaeoenvironmental information may be obtained from the fossil content of the palaeosols, for acid soils often contain pollen (Dimbleby 1957), while molluscan remains are often found in soils in calcareous areas (Evans 1972). Palaeosols are also important in Quaternary stratigraphy as they form time-parallel marker horizons, and where widely developed, have been used for regional and inter-regional correlation (Ch. 6).

The nature of palaeosols

Three types of palaeosols are usually recognised: **buried palaeosols, relict palaeosols** and **exhumed palaeosols**. In practice, however, one type may grade into or merge with another, and a clear differentiation may not always be possible.

Buried palaeosols.   Buried palaeosols occur where the land surface upon which they developed has been completely covered by younger sediments. They can be subdivided into those that are deeply-buried and therefore lie well below the present zone of biological action, and those that lie nearer the surface and which are therefore affected by present-day soil forming processes (Ruellan 1971). The latter group may be indistinguishable from relict palaeosols unless the old ground surface can be seen. The recognition of buried palaeosols is not always straightforward, for a wide range of weathered materials will have been buried in Quaternary landscapes which may, or may not, be soils (Valentine and Dalrymple 1976). The most important property of a soil, and that which distinguishes it from other sediments, is that it has developed distinctive, vertically-differentiated layers or horizons in response to variations in levels of physical, chemical and biological weathering, and the subsequent movement of these weathering products up and down the profile. The often abrupt nature of the horizon boundaries, the truncation of underlying geological structure, and the areal or lateral continuity of soil are further characteristics that aid in the recognition of buried palaeosols. The soil profile forms the upper part of the **weathering profile** (Fig. 3.19), although some confusion has arisen, particularly in the American literature where the two terms have sometimes been used interchangeably. In some cases, however, the weathering and soil profile may be indistinguisable where, for example, erosion has removed the A and B horizons leaving only the weathered subsurface materials exposed in sections. In most palaeosols the organic content of the A horizon is not retained after burial, although the mineral part of the A horizon may still be present and may be recognised by a slightly lower clay content than in the underlying B horizon (Birkeland 1974). Generally however, it is the B horizon which is of greatest importance in the identification of buried palaeosols. Some important diagnostic properties of the B horizon include colour changes, texture variations, weathering characteristics on both non-silicate and clay minerals, levels of mineral depletion and carbonate

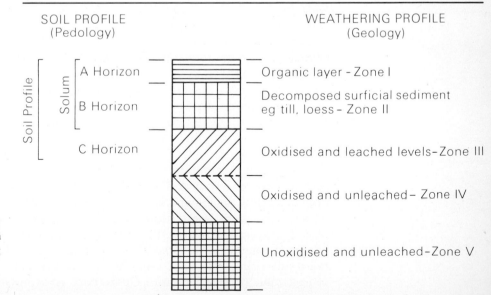

SOIL PROFILE
(Pedology)

WEATHERING PROFILE
(Geology)

Soil Profile

Solum

A Horizon — Organic layer - Zone I

B Horizon — Decomposed surficial sediment eg till, loess – Zone II

C Horizon — Oxidised and leached levels–Zone III

Oxidised and unleached– Zone IV

Unoxidised and unleached–Zone V

Fig. 3.19: Comparison of the weathering profile and soil profile. The soil profile is the upper part of the weathering profile and may be composed of different kinds of horizon in different combinations. All of these soil horizons are considered in the weathering profile as simple zones (after Ruhe 1965).

content. These properties can be used either singly or, more preferably, in combination, to demonstrate evidence of pedogenesis, and also to establish the type of environmental conditions under which the soil evolved. More recently, increasing emphasis has been placed on **soil micromorphology** in palaeopedology (Brewer 1964). Soil micromorphology is the term used to describe the distinctive arrangement of particles and voids comprising a **soil fabric**, and this can be established by an examination of soil thin sections under a microscope. It is widely regarded as one of the most reliable methods for detecting evidence of pedogenesis, and also as a means of interpreting the environment of soil formation (Dumanski 1969). Care is needed, however, in the interpretation of all soil properties. Changes in colour may indeed reflect soil development under different climatic régimes, but they could also be due to changes in parent material, or to subsurface weathering associated with groundwater flow (Valentine and Dalrymple 1975). Those buried palaeosols that lie near to the surface may experience post-burial alteration and this may take the form either of the addition of new features or the destruction of existing ones. In each case, this can lead to further problems of identification. At depth, sediments are subject to **diagenesis**[5] resulting in physical and chemical characteristics which, in the buried soil, appear to be little different from those produced by pedogenesis, and which are one of the major sources of confusion in the recognition of buried palaeosols (Valentine and Dalrymple 1976). Because of the equivocal nature of the evidence of soil properties, these authors suggest that, in order to prove beyond any reasonable doubt the existence of a buried palaeosol, it is necessary to show that its morphology changes laterally as well as vertically. It should vary logically across a landscape and should form a **palaeocatena**, for this is the one ubiquitous soil characteristic that sedimentation and diagenesis will not produce (Valentine and Dalrymple 1975).

Relict palaeosols.   Palaeosols developed on exposed surfaces and not subsequently buried by younger sediments are termed relict. Most soils are, in fact, polygenetic and contain characteristics inherited from a former climatic régime or from previous topographic or drainage conditions. These have survived because more recent soil-forming processes have been insufficiently powerful to change the characteristics acquired during the earlier phase (or phases) of pedogenesis. These include the results of such essentially irreversible processes as clay accumulation under humid conditions, and carbonate accumulation in arid or semi-arid environments (Morrison 1978). Equally, the colour of a soil may have been inherited from a former weathering régime. The distinctive red colouring of the soils in parts of South Wales (Ball 1960), for example, or the Reddish Prairie Soils of the American Midwest may be indicative of pedogenesis under warmer and more humid conditions, probably during interglacial periods. Overall, however, those features of modern soils that have been inherited from a previous soil-forming phase are difficult to quantify, and relict soils cannot always be identified with certainty. In general, they are of less value to the student of Quaternary environments than buried palaeosols.

Exhumed palaeosols.   Exhumed palaeosols are those that were formerly buried but have since been exposed by erosion of the covering mantle. They are, therefore, undergoing modification by current pedogenic processes and grade into both relict and buried palaeosols. They have not been widely recognised for in only a few areas, such as the loess country of the American Midwest (Ruhe 1965), can they be shown to have emerged from beneath an eroding overburden and the exhumed nature of the palaeosol can be clearly demonstrated. In the absence of such stratigraphic evidence, it is often impossible to differentiate between relict and exhumed palaeosols. The value of exhumed palaeosols in Quaternary research may therefore be limited.

Palaeosols and Quaternary environments

The interpretation of former environments from palaeosols rests, as in other fields of Quaternary investigation, on the uniformitarian principle of inferring past conditions from the perceived relationships between present-day soils and environments. Although this approach seems to work reasonably well with most types of biological evidence (Ch. 4) it is perhaps less satisfactory in geomorphological and pedological contexts where the dangers of **equifinality** (different processes leading to the production of similar forms) are always present. Soils with very similar physical and chemical characteristics can develop through a variety of genetic pathways (Ruhe 1965), and these may be impossible to detect in the fossil soil profile. Thus although buried palaeosols may be similar in morphology and in other characteristics to soils forming at the present day, they may not necessarily be analogous in terms of palaeoenvironment and regional conditions (Pawluk 1978). A further difficulty arising from the use of modern soils as analogues for Quaternary pedogenesis is that it is seldom possible to quantify the interaction between soil-forming factors, site and environment. Hence, extrapolations back in time are based at best on partial data, and at worst on often very broad generalisations.

In spite of these problems, a number of interpretations have been made of aspects of Quaternary environments based on the evidence of palaeosols. The most successful have undoubtedly been where zonal soils have been employed, where inferences have been based on a number of localities within a particular area, and where pedological data have been supported by evidence from palaeoecological studies of, for example, pollen and mollusca. Ruhe (1970) developed a chronology of landscape changes on the Great Plains of North America using relict and buried palaeosols. By using present-day soils (whose environmental characteristics are reasonably well known) as analogues, and by supporting the pedological evidence with data from faunal and floral analyses, he was able to demonstrate that the present semi-arid climatic régime dates from *ca.* 11 400–9100 years ago in the central regions of the Great Plains, while further north in Iowa, the prairie environment began between 7000 and 8000 years ago. Prior to that time, cooler and moister conditions had prevailed throughout the Wisconsinan period, with coniferous forest extending over much of Iowa. In northern Canada radiocarbon dates on buried podzols and Arctic brown soils have been used as the basis for a reconstruction of the Holocene migration of the forest/tundra boundary in the Keewatin area of the North West Territories (Sorensen *et al.* 1971; Sorensen 1977). Podzols are associated today with a spruce vegetation while the Arctic brown soils are found principally in tundra regions. Arctic brown palaeosols are often altered by podzolisation, but can be recognised by polygonal patterns resulting from the action of ground ice. The evidence suggests that during the Holocene, the position of the forest border ranged from 280 km north to a minimum of 50 km south of the present tree-line in south-west Keewatin (Fig. 3.20). These changes were often rapid and appear to have been in response to changes in the incidence of Arctic and Pacific-derived air masses. The presence of Arctic air during the summer months seems to be the most important vegetational and pedological control and this implies that the position of the forest border is thermally determined.

A characteristic of many buried soils is the presence of **Opal Phytoliths**. Most plants secrete opaline silica bodies and these assume the shape of the cell in which they are deposited. These forms are known as phytoliths and many are characteristic of the plants in which they are found (Lutwick 1969). As phytoliths can persist long after the soil organic matter has been removed by processes following burial, they may provide important clues to past vegetation. In Illinois phytolith remains have been used to demonstrate changes from forest to grassland during the Holocene

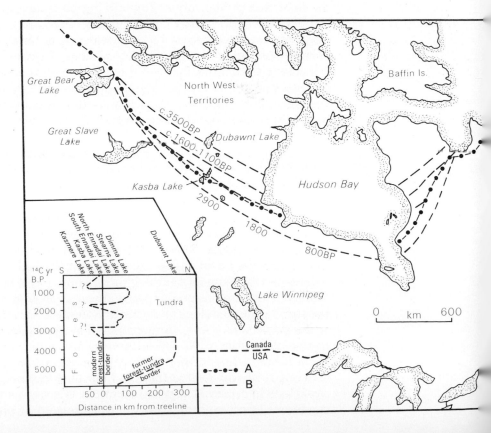

Fig. 3.20: Holocene migrations of the forest/tundra border in south-west Keewatin, NWT, Canada, based on radiocarbon dates and soil morphology.
A: present forest border;
B: estimated palaeoforest border location (after Sorensen *et al.* 1971; Sorensen 1977).

(Wilding and Dress 1969). They have also been employed in the identification of buried palaeosols (e.g. Dormaar and Lutwick 1969).

CAVE SEDIMENTS

Caves form natural sediment traps in which the deposits are largely protected from the effects of sub-aerial weathering agencies. As a consequence, cave sediments often contain a considerable part of the late Quaternary stratigraphic record. Moreover, as caves acted as refuges for animals, particularly the large carnivores, and as occupation sites for prehistoric man, many cave deposits are rich in animal bones and cultural remains. From these fossil and artefact assemblages it is sometimes possible to make inferences about former environmental conditions in the vicinity of the cave site (Ch. 4), but analysis of the cave sediments themselves can also provide valuable palaeoenvironmental information. Although caves are found in many different rock types, they are best developed in hard limestones, and the majority of work carried out to date on cave deposits has been in limestone regions.

A major factor governing the processes of sedimentation in caves is the shape of the cave itself. Schmid (1969) makes a distinction between **exogene** caves which are shallow niches in the hillside, often referred to by archaeologists and anthropologists as rock shelters, and **endogene** caves, which penetrate deep into the ground as chambers or passages. In exogene caverns, and at the entrance to endogene caves, sedimentation is directly influenced by prevailing weather conditions outside and the sedimentary sequence may record even minor climatic fluctuations. Moreover, as the deposits are derived both from within the cave and from the surrounding area, the stratigraphy in these situations is often complex. In the interior passages of endogene caves, however, sedimentation often operates under different laws. Daily and seasonal changes in weather and climate rarely penetrate, at least not directly, and therefore only the long-term climatic changes can affect the mode of sedimentation. Unless endogene passageways are near the ground surface, in which case exotic material may be introduced into the system through clefts in the cave roof, the majority of sediments are derived from within the cave itself, the principal exception being water-lain materials that may be transported from further afield.

Sediments in caves can be divided into those originating outside the cave system (**allochthonous**) and those that are derived from processes acting within the cave itself (**autochthonous**). The former include wind-blown sand and silt, water-lain sediment, colluvial and soliflucted deposits and glacial sediments (Fig. 3.21). In addition, there are materials brought in by animals and man, including the decayed parts of plants carried in for bedding, litter or food, animal excreta and carcasses, charcoal deposits from former hearths, and various artefact remains. Autochthonous sediments consist principally of rock rubble, cave earth and speleothems. Angular fragments of rock are common deposits in many caves and have been weathered from the roofs and walls to form **thermoclastic scree**. In

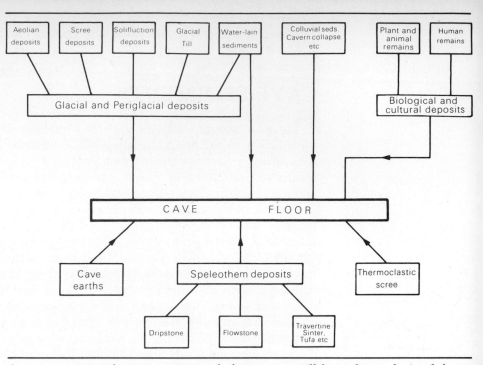

Fig. 3.21: Nature and origins of sediments on cave floors.

the outer parts of caves, some rock fragments will have been derived from insolation weathering (expansion and contraction at the rock surface due to temperature changes), while in limestone areas, solutional weakening by percolating groundwaters will result in pieces of rock breaking off from the walls and roofs of a cave. Where the roof of a cave lies near to the ground surface, talus cones of both angular and fine material may form beneath open fissures, as described by Sutcliffe (1960) in the Joint Mitnor Cave in south Devon, England.

Cave earth is composed of much finer materials (sand size and less) and may have a variety of origins. Near the cave entrance it is often largely allochthonous, being composed of wind-blown or water-lain sand or silt, or even inwashed colluvial sediment. In the deeper parts of limestone caves, however, cave earths are formed either from the acid-insoluble residues left by the solutional breakdown of the country rock, or from the secondary weathering of angular rock fragments that have accumulated on the cave floor. The latter involves both physical and chemical processes and includes secondary thermoclastism, freeze-thaw activity, chemical alteration and the action of man (Laville 1976). Cave earths are frequently red or brown in colour, due partly to the presence of oxides of iron and alumina, but in some cases reflecting the influence of phosphate derived from fossilised faecal matter, particularly the red or red-brown bat guano (Schmid 1969).

Speleothems are secondary mineral deposits formed in caves in karst regions and can be considered either as **dripstones** or **flowstones**. Dripstones are deposits of calcium carbonate (although some may be composed of varying quantities of aragonite or gypsum) formed by water dripping from the ceiling or walls of a cave, or from the overhanging edge of a rock shelter. The most common features developed in this way are the

Fig. 3.22: Giant speleothem (see figure on the left for scale) from the Gua Batu caves, Malaya.

columnar **stalactites** that hang from cave roofs and **stalagmites** that rise from the cave floor (Fig. 3.22). Flowstones are deposits of calcium carbonate, gypsum or other mineral matter that have accumulated on the walls or floors of caves in places where water trickles or flows over the rock (Monroe 1970). Upon reaching the cave floor, water may percolate into the interstices of clastic sediments, cementing them into a coherent, often very hard porous rock known as **cave breccia** (Cornwall 1958). In some cases a complete cover of precipitated calcium carbonate blankets the floor of the cave where it may become interbedded with the screes, breccias and cave earths. Such a flowstone cover has been referred to by some workers as a **stalagmite floor**. Other precipitated calcium carbonate deposits occasionally found in caves in karst regions include **travertine**, a light, compact and generally concretionary substance, extremely porous or cellular varieties of which are known as calcareous **tufa**, calcareous **sinter** or **spring deposit**. Compact banded varieties are termed **cave marble** or **cave onyx** (Monroe 1970).

Palaeoenvironmental significance of cave deposits

Both allochthonous and autochthonous cave deposits provide evidence of regional environmental changes. Full glacial conditions may be reflected by the occurrence of deposits of till which, in certain cases, may block the cave entrance almost completely (e.g. Rowlands 1971), but which are seldom found in the deeper cave passages. Former periglacial environments are indicated by frost-shattered scree and solifluuted and frost-riven material in and around the cave mouth, and by the occurrence of wind-blown sediment. Aeolian material often forms a major component of cave earths in the outer parts of caves (Ford 1975), and while its provenance is often less easy to establish than other periglacial sediments, the particle size distribution and characteristic surface weathering forms of the sand grains will often reveal the wind-blown origin of the deposits. Water-lain sediments may also be indicative of cold-climate conditions for many fluvial sediments found in caves are believed to be glaciofluvial in origin, having been transported and subsequently deposited by melt water streams (Ford 1975), while others may have resulted from increased runoff from melting snowbanks on the ground surface.

In some cases, it has proved possible to use fluvial sediments on the floors of caves to reconstruct regional environmental changes. Detailed study of the Agen Allwedd cave system of South Wales (Bull 1978a, 1980) has revealed a complex series of water-lain sediments which appear to date back to the last full glacial (Late Devensian) period. The typical sequence consists of a basal, glacially-derived unit representing either the onset or waning of the Late Devensian glacial phase. A period of stream abandonment led to the drying and cracking of the surface muds of this member prior to the next period of sedimentation. This is represented by a laminated deposit (known as the 'cap mud') which has been found in a comparable stratigraphic position in a number of cave systems in South Wales. Particle size analysis and examination of quartz grains from the deposit under the electron microscope (Bull 1976, 1978b) showed the sediments to have been derived from a surface environment that was

Approximate years BP	Event in cave	Palaeoenvironmental inference
ca. 5000 to present	Quiescent period and sediment winnowing. Downcutting of stream in adjustment to base level changes.	Dry conditions. Climate continually ameliorating.
7000–5000	Peat deposition in cave. Calcite speleothem development.	Peat moor development. Generally wetter conditions.
10 300–700	HIATUS	Climatic amelioration. Drier.
ca. 10 300	Water drains from cave Collapse of cave entrances Scree development	Climatic amelioration Relatively cold
ca. 10 800	Cap mud laid down. Cave flood	General climatic deterioration.
	HIATUS	
ca. 17 000(?)	Wedges glaciofluvial sediments Angular sediment phase. Water flushes the rounded sediments through the cave	Corrie and gully glaciers and snow Periglacial and Glacial(?) conditions
	HIATUS	
35 000–14 000	Rounded sediment phase	?
ca. 40 000	Doline 'controlled' cave boulder collapse	?
?	Various unknown sedimentary events	?
?	Cave passage completion	
?	Cave passage development	

Table 3.3: Clastic sedimentation events in Agen Allwedd cave, South Wales, and surface palaeoenvironmental inferences (after Bull 1980).

markedly cooler than that in the period preceding deposition. The laminated unit was therefore interpreted as representing sedimentation by percolating groundwaters from surface run-off during the cold phase of the Loch Lomond Stadial (Younger Dryas). The overlying peat lenses forming the uppermost unit in the sequence are therefore believed to be of Flandrian age. A suggested environmental reconstruction based on this sedimentary evidence is shown in Table 3.3.

In coastal areas, marine deposits reflecting former high stands of interglacial sea-level may be found interbedded with terrestrial sediments in the relatively-protected environment of cave entrances, and such sites are of considerable importance in the study of Quaternary sea-level changes. On the Gower coast of South Wales, for example, the cave known as Minchin Hole contains a sequence of deposits (Fig. 3.23) that includes two raised beaches separated by a cave earth, lying beneath a thick cover of talus and cave breccias (Bowen 1973a; Sutcliffe and Bowen 1973; Sutcliffe 1981). The stratigraphic position of the beaches supported by relative dating using amino-acid racemisation (Ch. 5) indicates former high sea-levels of two different ages (Andrews *et al*. 1979). The upper (*Patella*) beach is considered to be of last (Ipswichian) interglacial age and the lower beach must therefore belong either to an earlier stage within the Ipswichian or to a previous interglacial. This is the only published locality in the British Isles where such a clear stratigraphic relationship can be demonstrated.

Of the sediments that form within caves, there is a tendency to associate thermoclastic scree with cold climatic conditions and dripstone and flowstone with milder and wetter phases. With reference to the process of thermoclastism, Laville (1976) differentiates between **macrothermoclastism** which he associates with an annual freeze-thaw cycle in which rocks are subjected to prolonged freezing resulting in large, angular scree fragments, and **microthermoclastism**, which he believes is

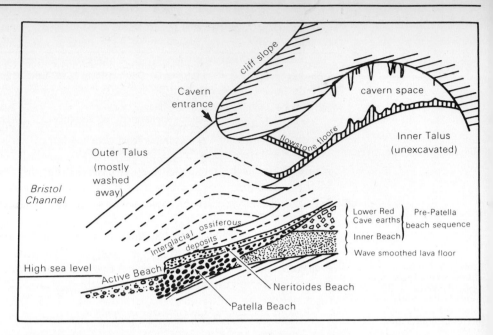

Fig. 3.23: Schematic section of Minchin Hole, South Wales. The *Patella* and *Neritoides* beaches and the overlying ossiferous deposits are considered to be of last (Ipswichian) interglacial age, but the ages of the Lower Red Cave Earths and the Inner Beach are not known (after Sutcliffe 1981).

related to a daily cycle with frequent freezing and thawing. Under the latter régime, finer debris is produced for conditions are less extreme and therefore ice does not penetrate so deeply into the rock. The size of the material in thermoclastic scree deposits, therefore, reflects the efficiency of operation of freeze-thaw processes and provides an indication of the severity of the former glacial or periglacial climatic régime.

Speleothem development in caves in periglacial environments is restricted by the often low levels of precipitation in such areas, by the perennially frozen subsoil that impedes downward penetration of groundwater and by generally low temperatures that restrict evaporation and subsequent deposition of calcium carbonate. In warmer and wetter environments, the more widespread percolation of groundwater and higher rates of evaporation within the caves themselves will lead to a more rapid build-up of dripstone and flowstone formations. Not surprisingly, therefore, the most spectacular speleothems are to be found in the tropical and subtropical regions of the world (Sweeting 1972; Fig. 3.22). Moreover, radiometric dates on late Quaternary speleothems in Britain and North America suggest that the widespread deposition of calcite takes place under non-glacial climatic conditions (Atkinson *et al.* 1978; Harmon 1977). Nevertheless, a straightforward equation between cold climate and lack of speleothem formation may not always be correct, for where snowfall is high, dripstone and flowstone may form. In the Causse Méjeau of southern France, for example, Enjalbert (1968) found very large stalagmites which he believed to have formed during glacial times from groundwaters highly charged with calcium carbonate emanating from snowbanks. Overall, however, the evidence from many parts of the world suggests that low levels of speleothem development are associated with *cold* and *dry* conditions, and that the major phases of dripstone and flowstone formation are associated with *wetter* and probably *warmer* environments. Other indications of mild and moist conditions, often supporting the evidence from speleothems, can be obtained from detailed sedimentary analyses of cave sediments. Greater rounding of rock fragments, increased porosity of sediments, presence of soil carbonate concretions, and variations in clay mineral abundance, particularly decreases in montmorillonite, all reflect weathering during warmer and wetter phases (Farrand 1975).

More precise climatic data can be obtained by using the oxygen isotope[6] composition of calcite deposited on cave speleothems. In deep caves where the chamber is removed from direct contact with the external atmosphere and there is little or no air circulation, speleothems will tend to form in isotopic equilibrium with their parent seepage waters. As the isotopic characteristics of these waters are determined largely by air temperature (*i.e.* variations in air temperature lead to alterations in the $^{18}O:^{16}O$ ratio in ground and surface waters), and because speleothems form through successive increments of calcium carbonate, the ^{18}O trace through a cave speleothem will contain a record of temperature fluctuations over time. The temperature in the inner parts of caves is often equal to the regional mean annual air temperature and thus stable isotope studies of cave speleothems represent a powerful new tool in palaeoclimatology. The principles of the technique are similar to those by which palaeoclimatic records have been obtained from ice cores and from carbonate remains in deep-sea sediments (p. 145). However, because the accumulation of calcium carbonate per unit area in speleothems is considerably faster than in sediments on the ocean floor, larger samples of material can be obtained from cave sites. In theory, therefore palaeotemperature data from cave speleothems should provide considerably better resolution in time than those obtained from ocean sediments, and perhaps also from ice cores. Details of the method are outlined in Hendy and Wilson (1968), Schwarcz *et al.* (1976) and Gascoyne *et al.* (1978).

The applications of the technique have been demonstrated by Harmon *et al.* (1978) who obtained synchronous warming and cooling trends over the past 200 000 years at six widely-separated localities in North America (Fig. 3.24). The sites investigated ranged from the Rocky Mountains in Alberta to eastern Mexico, and the time scales were established by $^{230}Th/^{234}U$ dating (Ch. 5) of the cave speleothems (Gascoyne *et al.* 1978). The evidence showed warm periods from 195 000 to 165 000 BP; from 120 000 to 100 000 BP and at 60 000 and 10 000 BP; cold intervals were dated to 95 000 to 65 000 BP and 55 000 to 20 000 BP. Although the correlations are not perfect, these periods of thermal maxima and minima can be related to the independent palaeoclimatic records obtained from deep-sea cores and from dated high-stands of sea-level. Of particular significance are the conclusions that both interglacial and interstadial phases appear to have begun very rapidly, with the transition from full glacial to full interglacial conditions occurring in as short a time-span as 2000 years, and that warmings take place at maximum rates of about 15 °C per 1000 years – whereas maximum cooling rates are a little less, about 10 °C per 1000 years.

Cave sediments, therefore, are of considerable interest to the student of Quaternary environments. The protective nature of the cave ensures that

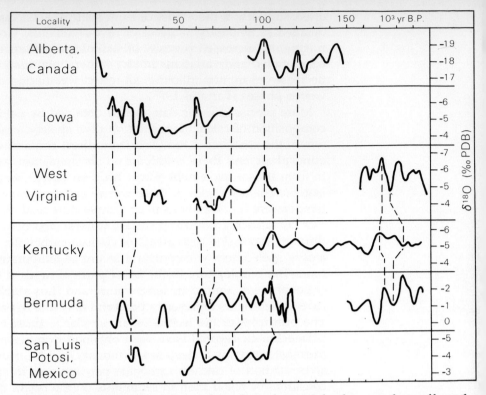

Fig. 3.24: Generalised record of δ¹⁸O variations in speleothems from six sites in North America. Peaks in the curves reflect warm episodes and troughs colder intervals. Tentative correlations between the curves are shown by dashed lines (after Harmon *et al.* 1978).

sediments remain relatively unaltered, and certainly they are less affected by sub-aerial processes than almost any other terrestrial deposit. The full potential of cave sediments themselves, as opposed to their fossil content, in the analysis of palaeoenvironments has perhaps yet to be realised.

LAKE, MIRE AND BOG SEDIMENTS

Preserved within sediments that have accumulated in lakes, mires and peat bogs is a diverse and often detailed record of environmental change. Because, given sufficient time, all lakes become infilled to form mires and bogs, lake and bog sediments are genetically related and frequently grade into one another. For this reason, the sediments are considered under a single heading. Lake, mire and bog deposits are important in a number of respects. First, the contained fossil flora and fauna provide an insight into both local and regional ecological changes. Secondly, the character of the sediments offers clues about former environmental conditions. This is particularly true in the case of lake sediments where variations in the physical and chemical properties reflect developments in the lake ecosystem, and also changes in the rates at which processes operated around the lake catchment. In both cases, the observed variations may be interpreted in terms of environmental change. Thirdly, fossil lacustrine sediments and shoreline features in many of the semi-arid regions of the world reveal a record of fluctuations in the level of these so-called 'pluvial lakes' in response to climatic changes during the later part of the Quaternary period.

In many lake basins in the mid- and low-latitude regions of the world, sedimentation has been intermittent or continuous throughout much of the Quaternary, and in some instances extends back into the Tertiary Period. Present-day lakes in which long sedimentary sequences are known to exist include Lake Biwa in Japan, Lake Titicaca in Bolivia, and a number of the lakes in and around the East African Rift Valley. Sedimentary records spanning long periods of the Quaternary are also found in many 'pluvial' lake basins. Until recently, however, few of these old lake sediment sequences had been investigated, due partly to the considerable logistical difficulties involved in raising long sediment cores from very deep water, and partly to the high cost of such operations. In the higher latitudes of western Europe and in parts of North America, on the other hand, many existing lakes are no older than *ca.* 15 000 years, as they formed following the retreat of the last ice sheets, and the sediments in mires and bogs are usually considerably younger. A remarkable exception to this general rule is the site at Grande Pile in the French Jura which contains a continuous sequence of peat deposits stretching back to the Eemian (last) Interglacial (Woillard 1978). Fragmentary remains of lake and peat deposits from the earlier stages of the Quaternary are, however, found in many localities. These are often intercalated between tills and other terrestrial sediments and usually represent the remains of materials that accumulated during interglacial and interstadial periods (Fig. 3.25).

Fig. 3.25: Clays and compressed peats (dark layer) of Reuverian (late Pliocene) age, overlain by sands and gravels of Tiglian (early Pleistocene) age at Oebel, Netherlands-German border. The organic layers are often very rich in plant macrofossils and provide a valuable data source for environmental reconstruction during the Pliocene-Pleistocene transition.

Although they are useful in stratigraphic subdivision, the palaeoenvironmental significance of these deposits usually lies more in their contained fossils than in the nature of the sediments themselves. The faunal and floral remains that are commonly found in lake and bog sediments are discussed in Chapter 4.

In this section we are concerned with those lithological characteristics of lake, mire and bog deposits that can be used as a basis for palaeoenvironmental reconstruction. The emphasis is placed on sediments that have accumulated during the late Quaternary (particularly during the Lateglacial and Flandrian periods), for these are often relatively accessible and can be sampled either in sections or with hand-operated corers, and are therefore frequently known in greater detail than the older deposits. In view of their distinctive nature and significance as palaeoclimatic indicators, however, 'pluvial' lake sediments are considered in a separate section.

The nature of lake and bog sediments

Lake sediments are both allochthonous and autochthonous in origin, being derived partly from organic production within the lake ecosystem, and partly from the inwash of both organic and inorganic material from around the lake catchment. If the lake is rich in mineral nutrients organic productivity will be high and the conditions are described as **eutrophic**. The typical deposit will be a green-brown organic-rich sediment known as **nekron mud** or by the Swedish name **gyttja**. In deeper waters, this will be extremely fine in texture and will consist of comminuted and largely-unrecognisable plant material, but will grade into **detritus gyttja** with recognisable plant macrofossils (fruits, seeds, leaves etc) in shallow waters. Where the lake substrate is calcareous, lime may be precipitated from the water by aquatic plants (e.g. by some of the pondweeds – *Potamogeton*) and other organisms, and a fine cream-white clay-rich sediment known as **marl** will accumulate. In general, sediments deposited under eutrophic conditions are predominantly autochthonous. Where the lake is poor in mineral nutrients and organic productivity is low (**oligotrophic**), allochthonous sediments will often predominate. If the inwashed materials are low in organic content, sands, silts and clays will accumulate, the finer grades of sediment being encountered in deeper waters. Where organic productivity is low, but the inwashed materials are dominated by humic substances from, for example, peats around the lake catchment, the lake waters are typically brown in colour due to the dissolved humic acids. In some lakes a dark-brown **gel-mud** composed largely of colloidal precipitates will accumulate. Such conditions are often described as **dystrophic** and the deposit is known by the Swedish term **dy**. Finally, where the lake waters support a rich diatom flora the sediments are sometimes characterised by a white silicious mud composed almost entirely of diatom frustules (Ch. 4) and which is termed **diatomite**. These deposits may form under either eutrophic or oligotrophic conditions depending on the ecological affinities of the diatom species.

Over time lakes silt up, plants encroach from the marginal zones, and areas of open water are progressively eliminated. The succession from open water to mire and bog is known as a **hydrosere** and the sediments gradually change in character from muds to peats. Three broad categories of peat can be identified and each is characteristic of a particular stage in the hydroseral succession. These are:

1. **Limnic peats** which form beneath the regional water table and which are composed partly of transported plant debris and partly of decayed vegetation formerly growing *in situ*.
2. **Telmatic peats** which form in the swamp zone between high and low water levels and which are largely autochthonous in origin.
3. **Terrestrial peats** which accumulate at, or above, the high water mark and which are entirely autochthonous in derivation.

Each of these peat types will be composed of the remains of particular peat-forming plants, depending on the stage in the hydrosere and the trophic status of the lake water (Fig. 3.26).

The terminology applied to peat-forming environments can be confusing. All waterlogged areas where peat develops as a result of

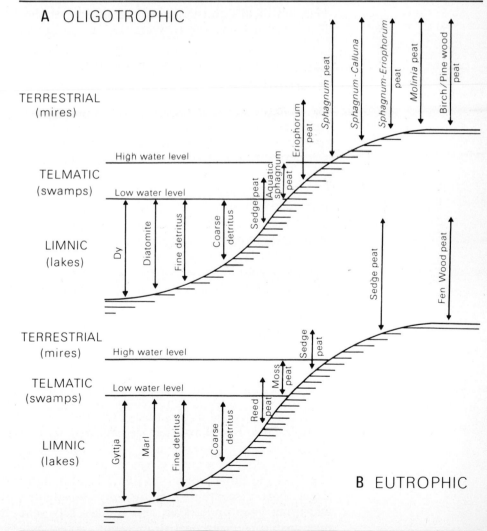

Fig. 3.26: Some sediment types deposited with increasing depth of water under oligotrophic (A) and eutrophic (B) conditions (after Birks and Birks 1980)

reduced vegetal decay under anaerobic conditions are termed **mires**. Mires can be divided into those in which the high water table that induces peat formation is a consequence of groundwater conditions, either where drainage is impeded (**soligenous mires**) or where water accumulates in enclosed basins (**topogenous mires**), and those in which the water table is maintained by high atmospheric moisture levels (**ombrogenous mires**). Topogenous mires are, of course, part of the hydroseral sequence and are usually referred to as **fens** if eutrophic and **valley bogs** if oligotrophic. Soligenous mires are almost always oligotrophic. Ombrogenous mires (usually termed **bogs**) can be subdivided into **raised bogs** and **blanket bogs**. Raised bogs develop mainly in lowland areas where the peat-forming plants, principally *Sphagnum* mosses, produce a domed surface above the level of the surrounding ground. In most areas of the British Isles, raised bogs form the final stage of topogenous hydroseral successions (Walker 1970). Blanket bogs, on the other hand, are typical of upland areas and develop as a continuous cover over the landscape where rainfall is high.

Further details on sediments in lakes, mires and bogs can be found in Moore and Bellamy (1974), West (1977a), and Birks and Birks (1980).

Palaeoenvironmental evidence from lake and bog sediments

The ecological history of lakes, mires and bogs can often be reconstructed from a careful examination of the variations in sediment type. In topogenous mires and bogs, the different types of peat provide indications of former nutrient status and hence of levels of productivity within the basin, while the various hydroseral stages can be identified by changes in the peat stratigraphy. The rate at which hydroseral succession has proceeded cannot usually be interpreted in terms of climatic change as local site factors will often be the principal determinants, but in certain circumstances it may be possible to make inferences about the height of the water tables and hence about former precipitation levels. A good example would be where there is evidence for disturbance of the hydrosere (e.g. a transition from bog peat to reedswamp peat) reflecting some fairly catastrophic environmental change, such as a rise in silty water, possibly resulting from a change to wetter climatic conditions (Walker 1970).

Bogs and mires

Evidence for climatic change can often be found in ombrogenous mires. In the early years of the present century, detailed analyses of peat bogs in Scandinavia enabled the botanists Blytt and Sernander to divide the Holocene into five climatically distinct periods on the basis of marked changes in peat bog stratigraphy (Table 3.4). Subsequently, the Blytt-Sernander climatic sequence was related to a scheme of pollen zones by von Post and other workers (Table 3.5), although this practice has now largely been discontinued, as the very considerable amount of pollen work which has been undertaken over the last fifty years has shown the relationship between peat stratigraphy, pollen assemblage zones and

Period	Climate	Evidence
Sub-Atlantic	cold and wet	poorly-humified *Sphagnum* peat
Sub-Boreal	warm and dry	pine stumps in humified peat
Atlantic	warm and wet	poorly-humified *Sphagnum* peat
Boreal	warm and dry	pine stumps in humified peat
Pre-Boreal	subarctic	macrofossils of subarctic plants in peat

Table 3.4: The Blytt–Sernander scheme of peat bog stratigraphy.

Years before present	Pollen zone	Blytt-Sernander period	Climate
— 1000	VIII	Sub-Atlantic	Deterioration
— 2000			
— 3000			
— 4000	VIIb	Sub-Boreal	Climatic optimum
— 5000			
— 6000	VIIa	Atlantic	
— 7000			
— 8000	VI	Boreal	Rapid amelioration
— 9000	V		
	IV	Pre-Boreal	
— 10 000	III	Younger Dryas	Cold
— 11 000	II	Alleröd	Cool

Table 3.5: Holocene/ Flandrian pollen zones and Blytt-Sernander climatic episodes.

climatic change to be rather less straightforward than was initially envisaged in the 1920s (Ch. 4). Nevertheless, in many ombrogenous mires in north-west Europe, particularly in raised bogs, distinctive horizons are found separating dark, well-humified peats from overlying light, less humified *Sphagnum* peats and these are believed to reflect major shifts in climatic conditions. Five such boundaries were described by Granlund (1932) in southern Sweden to which he gave the name **recurrence surfaces**. Each was believed to represent a change from drier to wetter conditions and these were dated to *ca*. AD 1200, AD 400, 600 BC, 1200 BC and 2300 BC. Nine recurrence surfaces were later identified by Nilsson (1964) in the Scania region of southern Sweden, while a recent investigation of five raised bogs in Denmark (Aaby 1976a) shows perhaps more than twenty stratigraphic changes from dark to light peat formation within the past 5500 years (Fig. 3.27), and these are believed to represent shifts not only to wetter, but also to cooler climatic conditions. Aaby's work involved colorometric determinations and fluctuations in rhizopod species as means of obtaining independent measures of humification, and the very large numbers of radiocarbon dates enabled a time-scale to be established which showed that the inferred climatic changes fitted statistically a 260-year cycle. Although the periodicity of 260 years was found to be the norm, occasionally a 'double period' of 520 years was

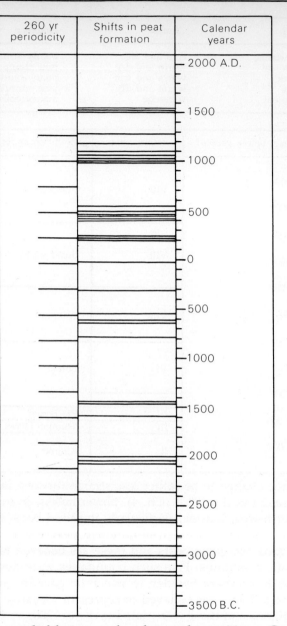

260 yr periodicity	Shifts in peat formation	Calendar years

Fig. 3.27: Record of climatically-conditioned shifts from dark to light peat formation, derived from sections in five Danish raised bogs (after Aaby 1976a).

recorded between the observed transitions. Curiously, a 520-year periodicity has also been noted in the curve for eustatic sea-level variations in south-east Sweden (Aaby 1976b). These results suggest that there may be considerable scope using this type of evidence for predicting future long-term climatic trends.

Probably the most widely-discussed recurrence surface is that which was first described by the German peat-stratigrapher Weber over 100 years ago. In many ombrogenous mires in Germany, Weber found, usually within a short distance of the bog surface, a clear and abrupt change from a dark, lower peat which contained stumps of pine, pieces of birch wood and fragments of *Calluna* and cotton grass, to an upper, less well-humified, lighter coloured peat. This level Weber termed the

	Below	(Trackway)	Above
Somerset			
Meare Heath track	3061	2840	
		2852	
Shapwick Heath track	3310	2470	2220
			2197
Westhay track		2800	
Blakeway track	2790	2600	
Viper's track		2520	
		2630	
Viper's platform		2410	
		2410	
		2460	
Nidons track	2642	2585	2628
	2482	2590	
Outside Somerset			
Chat Moss	3061		2661
Tregaron	3029		2669
	2879		2624
Whixall	3238	2307	
Llan Llywth	3178		3230
Flanders Moss	2712		
Average	2959	2604	2575

Table 3.6: Radiocarbon dates (yr BP) from a prominent recurrence surface in England and Wales (after Godwin 1975).

grenzhorizont (boundary horizon) and was interpreted by him and by generations of later workers as marking a change to more oceanic conditions. Most peat stratigraphers have correlated this stratigraphic boundary with the Sub-Boreal/Sub-Atlantic transition of the Blytt-Sernander scheme (Table 3.5). Radiocarbon dates from a large number of sites in north-west Europe appear to place the grenzhorizont at about 500 BC, the transition from the Bronze to the Iron Age on the archaeological time-scale. In many British raised bogs, a prominent recurrence surface has been dated to about the same time (Table 3.6).

Although recurrence surfaces in peat profiles are potentially useful in palaeoclimatic work, fundamental questions remain to be answered about their mode of origin and particularly about the synchronous nature of their development. Radiocarbon dating has shown that not only may the ages of the most prominent recurrence surfaces from nearby sites be at variance by several hundreds of years, but that there may be considerable discrepancies between the age determinations on a particular recurrence surface within a single site (e.g. Lundqvist 1962; Schneekloth 1968). This evidence strongly suggests that periods of peat growth may well differ from one part of a bog to another, and thus, while renewed growth of peat may partly be a response to increased climatic wetness, it may also simply be a function of local hydrological variations in the mire complex. If this is so, then many so called 'recurrence surfaces' may be nothing more than an expression of local drainage features in the course of the history of the mire (Moore and Bellamy 1974; Barber 1981).

Lake sediments

Lake sediments offer rather more scope for the student of Quaternary environments than do peat bogs, for as lakes are natural reservoirs of eroded material, within their sediments is preserved a record of changing processes acting upon the slopes around the lake catchment. Careful analysis of the physical and chemical properties of the lake deposits permits the observed changes in sediment type to be related to climatic variations or to the activities of man. It has long been customary for those working on the pollen content of lake sediments to draw parallels between inferred vegetation changes and variations in sediment stratigraphy. In north-west Europe, for example, Lateglacial lake deposits typically consist of a threefold sequence of organic-rich lake muds (often gyttja or clay-gyttja) which overlie and are underlain by mineral sediments with a very low organic content (see Fig. 7.5). The whole sequence often rests upon glacial gravels and commonly lies beneath Flandrian peats or organic muds (Fig. 3.28). Palaeobotanical evidence suggests that the two minerogenic horizons accumulated during periods of reduced vegetation cover, the former during the pioneer phase immediately following local

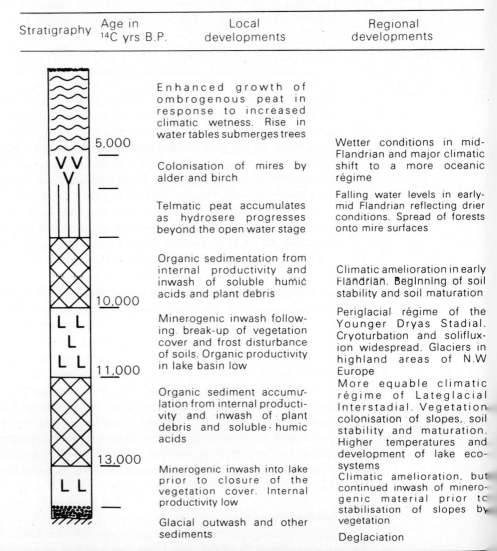

Stratigraphy	Age in ^{14}C yrs B.P.	Local developments	Regional developments
		Enhanced growth of ombrogenous peat in response to increased climatic wetness. Rise in water tables submerges trees	
	5,000		Wetter conditions in mid-Flandrian and major climatic shift to a more oceanic régime
		Colonisation of mires by alder and birch	
			Falling water levels in early-mid Flandrian reflecting drier conditions. Spread of forests onto mire surfaces
		Telmatic peat accumulates as hydrosere progresses beyond the open water stage	
		Organic sedimentation from internal productivity and inwash of soluble humic acids and plant debris	Climatic amelioration in early Flandrian. Beginning of soil stability and soil maturation
	10,000		
		Minerogenic inwash following break-up of vegetation cover and frost disturbance of soils. Organic productivity in lake basin low	Periglacial régime of the Younger Dryas Stadial. Cryoturbation and soliflux-ion widespread. Glaciers in highland areas of N.W Europe
	11,000		
		Organic sediment accumulation from internal productivity and inwash of plant debris and soluble humic acids	More equable climatic régime of Lateglacial Interstadial. Vegetation colonisation of slopes, soil stability and maturation. Higher temperatures and development of lake ecosystems
	13,000		
		Minerogenic inwash into lake prior to closure of the vegetation cover. Internal productivity low	Climatic amelioration, but continued inwash of minerogenic material prior to stabilisation of slopes by vegetation
		Glacial outwash and other sediments	Deglaciation

Fig. 3.28: Lateglacial and early Flandrian environmental changes inferred from lithostratigraphy of a typical north-west European lake and mire sediment sequence.

deglaciation, while the latter represents the cold phase of the Loch
Lomond Stadial or Younger Dryas at which time a severe periglacial
régime prevailed (Ch. 7). Minerogenic material was therefore transferred
from the catchment to the lakes, especially during the Younger Dryas
phase when surrounding slopes were affected as much by freeze-thaw
activity and solifluction as by overland flow. The organic-rich sediments,
however, contain a fossil record derived from a vegetation dominated by
shrub or woodland which presumably developed under more stable
conditions, as indicated, in turn, by a substantial reduction in the inwash
of minerogenic material.

A more sophisticated approach to the investigation of lake deposits
using variations in the chemical composition of lake sediments has been
outlined by Mackereth (1965, 1966). On the basis of work on the lakes of
north-west England, he proposed that changes in the chemical
composition of lake sediments could most readily be explained if the
sediments were regarded as sequences of soils derived from the catchment
areas. Increased soil erosion would result in the transfer of large amounts
of relatively unweathered material into the lake basins, and the mineral
fraction of the deposited sediments would be characterised by high
proportions of certain elements, most notably sodium, potassium, and
magnesium and, in certain cases, calcium, iron and manganese. During
periods of reduced erosive activity and soil maturation under a vegetation
cover, the mineral material transported into the lakes would have been
leached of its content of potassium, magnesium and sodium in particular,
and such 'stable' phases would be represented in the lake sediment record
by lower concentrations of those bases. Periods of reduced erosion would
coincide with higher values for organic carbon which result partly from
increased aquatic productivity within the lake basins, and partly from the
increase in organic matter washed in from the catchment. Subsequent work
has expanded and developed these ideas. Pennington and Lishman (1971),
for example, found that low iodine to carbon ratios in lake sediments
were characteristic of neutral forest soils around the lake catchments while
high iodine: carbon ratios were more typical of acid organic soils and
peats. Similarly Pennington et al. (1972) were able to establish a
relationship between an increase in iron and manganese content of lake
sediments and an extension of waterlogged soils and peat formation
around the basin catchment. Thus although in some cases modified by
more recent work (e.g. Pennington and Lishman, 1971; Pennington et al.,
1976), Mackereth's major conclusions continue to form the basis for much
current palaeolimnological research (Oldfield 1977), and, when used in
conjunction with palaeobotanical evidence, can provide new insights into
landscape change.

Pennington (1970) showed how chemical analysis of lake sediments
could be used to augment pollen analysis in the reconstruction of the
Lateglacial environment around a number of upland tarns in the Lake
District of north-west England. Pollen analysis was accompanied by
chemical investigations in which data were obtained for the elements
carbon, iodine, iron, manganese, calcium and sodium (Fig. 3.29). The
evidence indicated that during the period 14 000–12 500 years ago (the
early part of the Lateglacial Interstadial) increasing soil stability and

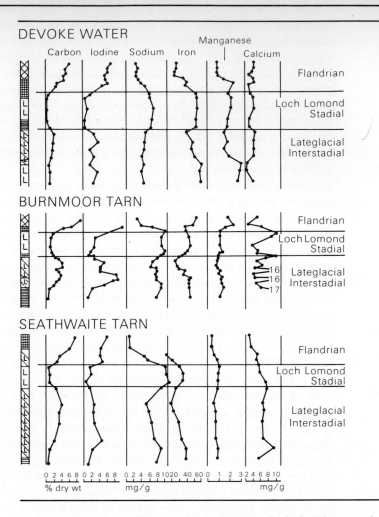

Fig. 3.29: Chemical analyses of Lateglacial and early Flandrian sediments from three upland tarns in the English Lake District (after Pennington 1970).

maturation occurred, as indicated particularly by the trends in the sodium and carbon curves, and no evidence could be found of severe soil erosion during that period. Falling manganese and iron concentrations during the middle and later Interstadial were interpreted as showing the presence of humic acids and reducing conditions in the Lateglacial soils, leading to podsolisation towards the end of the Interstadial. Declining iodine values may indicate decreasing precipitation levels. The significant increases in a number of the metallic elements, particularly sodium and iron, during the following cold period, the Loch Lomond Stadial, reflects the transport into the lakes of previously unweathered drift from deep within soils around the catchment, indicating more severe periglacial erosion than at any time since the wastage of the last ice sheet. A similar chemical history has been found in lake sediments of Lateglacial and early Flandrian age at sites in Scotland (Pennington 1977a).

The effect of man on the landscape during the Flandrian is also reflected in the chemical record of lake sediments. Mackereth (1966) discussed the increase in concentration of base elements in several lake cores following the decrease in woodland cover shown in the pollen records and dated to *ca.* 5000 BP. This, he suggested, was a reflection of increased soil erosion around the lake catchments following forest clearance by Neolithic man. A similar relationship was found in the sediment record from Loch Tarff near the Great Glen in Scotland, but there the increase in sodium, potassium, magnesium and iron above the clearance levels was accompanied by a very significant increase in the amount of iodine in the sediments. These chemical changes were interpreted as evidence for the development of a highly acid mor-humus soil following the clearance of the pine and birch woodland and its replacement by *Calluna* moorland. In the Lake District profiles, by contrast, the mid- and Late Flandrian vegetational change was from upland oak-elm-birch woods to hill grassland (Pennington *et al.* 1972).

Studies on more recent lake sediments have shown how man's activities have affected rates of erosion around the catchments and sediment yield in the lakes. At Frains Lake in south Michigan, Davis (1976) demonstrated the contrast in erosion before and after deforestation of the catchment at around AD 1830. Prior to that date, the average sediment yield was estimated to be about 9 tonnes/km^2/yr, but after the initial phase of forest clearance and ploughing, this increased by between 30 and 80 times. The present erosion rate was estimated to be around 90 tonnes/km^2/yr. At Braeroddach Loch in north-east Scotland, Edwards and Rowntree (1980) showed that the onset of pastoral activity after *ca.* 5390 BP resulted in a threefold increase in sediment accumulation while during the period of modern agricultural practices (dated to 370 ± 250 BP) about 25 per cent of the sediment deposited during the past 10 600 years appears to have accumulated. In view of the discussion above, the increase in the curve for sodium at these two horizons is particularly significant (Fig. 3.30).

Pluvial lakes

The concept of pluvial lakes has been introduced in Chapter 2 where the morphological evidence for variations in lake level was considered. Former lake levels can also be deduced, however, from careful analysis of the lake sediment sequences. High lake levels can be established by the mapping and measurement of exposures in the sediments and by the relationship of lacustrine sequences of both deep and shallow water deposits to strandlines, while phases of lake recession are shown by unconformities caused by sub-aerial erosion and by palaeosols, wedges of alluvium, colluvium and aeolian deposits between the lacustrine sediments (Morrison 1968a). The lake sediments themselves are highly variable and range from fine-grained silts, clays and marls characteristic of deeper waters, to increasingly more saline deposits indicative of lake later levels. The latter group includes algal limestones (often in the form of **stromatoliths**[7]) and evaporite deposits (salines such as borax, halite, gypsum etc) which reflect almost complete dessication of the lake floors. Where such deposits are interbedded with those of sub-aerial origin, a history of lacustral and interlacustral phases can be reconstructed. This type of work has been widely undertaken in the semi-arid and desert

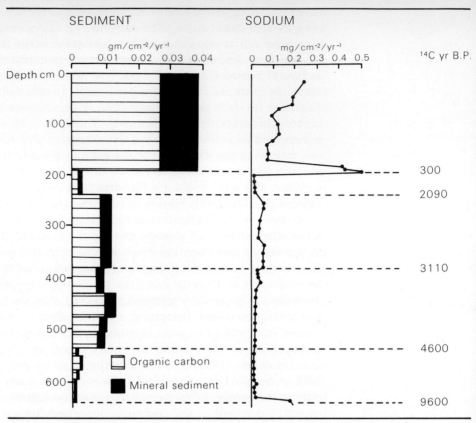

Fig. 3.30: Diagram showing sediment accumulation and sodium concentration in a Flandrian sequence from Braeroddach Loch, near Aberdeen, Scotland (after Edwards and Rowntree 1980).

regions of the American south-west, good examples being the reconstruction of the pluvial history of Lakes Lahontan and Bonneville (Morrison 1964, 1965).

In some areas, the stratigraphic history of pluvial lakes as revealed in surface exposures has been augmented by evidence from boreholes. A deep core from Searles Lake in California (Smith 1968, 1976) revealed alternating lacustrine, terrestrial and semi-terrestrial sediments extending over a vertical interval of 275 m. The upper 70 m or so are believed to be of Wisconsinan age (younger than *ca*. 130 000 years) and can be partially correlated with surface exposures, although the record of the lower part of the sequence has yet to be resolved. In the Great Basin of Utah, over 300 m of lacustrine sediments and interbedded palaeosols have been found in a core from the south shore of the Great Salt Lake. The sequence shows twenty-eight phases of expansion and contraction of pluvial Lake Bonneville over the past 800 000 years (Eardley *et al*. 1973). In Africa, cores have been raised from beneath the waters of a number of present-day lakes (Butzer *et al*. 1972), although the sequences are considerably younger than those so far obtained from North America. Nevertheless, many of the sediments contain palynological and chemical records providing evidence of lake-level fluctuations during the late Quaternary, and particularly during the Holocene period (Kendall 1969; Richardson and Richardson, 1972).

Pluvial lakes and Quaternary palaeoclimates

Before pluvial lakes can be used as barometers of climatic change, it is necessary to establish first that the lakes are, in fact, pluvial in origin. The expanded Caspian and Aral Sea systems (p. 78) were partly due to glacial meltwater influx, and it is equally possible for other non-climatic influences to cause fluctuations in lake water levels. These include tectonic activity and the creation of dams by avalanche debris, talus cones and lava flows. There are good reasons for believing that a number of lakes in the Rift Valley of East Africa, for example, have been affected by tectonic activity, and this has undoubtedly influenced the sedimentary and hydrologic history of the lake basins (Butzer *et al.* 1972). In general, however, the spatial consistency of evidence in many areas has been sufficient to exclude the non-climatic factors, at least for the later part of the Quaternary, and in the great majority of cases, lake-level fluctuations can be seen as a direct expression of surface water balance (Street and Grove 1979). Whether these changes in water level are a reflection of increased precipitation (in other words a pluvial phase) or reduced evaporation losses possibly influenced by groundwater flow, however, is still the subject of considerable debate (Morrison 1968a; Galloway 1970). In most regions of the world where there is evidence for fluctuating lake levels, it is still generally accepted that periods of maximum lake expansion probably reflect *both* reduced evaporation and increased precipitation. Nevertheless, it has been argued that the last phase of lake development in the American south-west could have been achieved with a precipitation little different from that of today, for evaporation would have been significantly reduced by markedly lower summer and winter temperatures during the Late Wisconsinan glacial period. Vegetational and geomorphological evidence have been provided to substantiate this view (Brakenridge 1978).

Reconstructing climatic changes from lake-level evidence relies heavily on closely-dated lacustral sequences as the basis for regional correlation. Chronologies have usually been based on radiocarbon dates from carbonate materials (algal limestones, calcite-cemented sands, gastropods, etc) and few dates are available on the more 'reliable' materials such as wood and charcoal. Radiocarbon age determinations on carbonates that have accumulated in lake waters are subject to a number of possible errors (outlined in Ch. 5) and these may perhaps account for some of the discrepancies that have arisen between individual radiocarbon chronologies for lake sequences in areas such as the American south-west (Broecker and Kaufmann 1965; Benson, 1978). In general it seems that more consistent and reliable radiocarbon dates have been obtained from Africa and Australia, by contrast with North America, perhaps reflecting the relative mobility of $CaCO_3$ in these different environments (Street and Grove 1979). Other dating methods (described in Ch. 5) which have been applied to pluvial lake deposits include the $^{230}Th/^{234}U$ technique which has not yet been widely used, but which has recently yielded a series of dates from the Searles Lake deposits in California that are consistent with those obtained by the radiocarbon method (Peng *et al.* 1978), and palaeomagnetism and tephrochronology. The last two methods have been used to establish a broad time-scale of events for the deep core obtained from Lake Bonneville discussed above.

Lake-level fluctuations in Africa and North America over the past 30 000 years are summarised in Fig. 3.31. The African data show intermittently high lake waters prior to *ca.* 21 000 BP, and falling lake levels to *ca.*

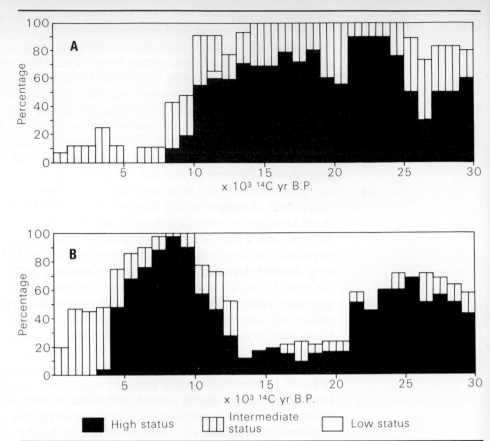

Fig. 3.31: Histograms of lake-level status for 1000-year time periods from 30 000 BP to the present day for the south-western United States (A) and intertropical Africa (B) (after Street and Grove 1979).

12 500 BP with the phase of maximum aridity being experienced around 13 000 BP. The highest water levels in almost all intertropical lakes are recorded at around 9000 BP and, following that maximum, lake levels fell intermittently to those of the present day. These changes are interpreted as reflecting major shifts in precipitation régimes and, when combined with palaeotemperature estimates derived from botanical and other data (Butzer *et al.* 1972; Richardson and Richardson 1972), allow inferences to be made about past rainfall. During the interpluvial from *ca.* 21 000 to 12 500 BP, for example, maximum precipitation estimates for East Africa range from 54–90 per cent of present-day levels, while for the Sahel, the figures may have been as low as 15–20 per cent. During the Holocene pluvial, rainfall estimates range from 165 per cent of modern rainfall in East Africa to 200–400 per cent in the Sudan and Mauritania (Street and Grove 1976). Using a hydrological and energy balance model for palaeolake Chad, Kutzbach (1980) derived precipitation estimates of at least 650 mm/yr for the early Holocene (10 000–5000 BP), at least 300 mm/yr more than the current rainfall in the Lake Chad basin.

In the American south-west, maximum lake levels appear to have been achieved prior to 10 000 BP, and most lakes either contracted markedly or disappeared completely during the Holocene. This pattern is exemplified by the closely-dated sediment record from the Lake Lahontan basin (Table 3.7) and from nearby Searles Lake in California. The water budgets of

Table 3.7: Chronology of lake-level fluctuations over the past 40 000 years in the Lake Lahontan basin, south-west United States (after Benson 1978).

Time period	Lake-level status
pre 40 000 BP	one high stand of unknown age
40 000–25 000 BP	low lake levels
25 000–21 500 BP	extremely high lake levels
21 500–13 600 BP	high lake levels
13 600–11 100 BP	extremely high lake levels
11 100– 9 000 BP	falling lake levels
9 000– 5 000 BP	very low lake levels – many lakes desiccated
5 000–present day	rising lake levels

these and other lakes have been used by numerous authors (Morrison 1965 and refs therein) to derive precipitation and temperature values for the periods of lake expansion. Estimates of increased precipitation at the lake maxima range from 18–23 cm above those of the present day, with a decrease in mean annual temperature of 2.7–5 °C (Morrison 1968a). Palaeotemperature estimates have also been obtained for the interpluvial intervals based on mineral assemblages of the saline layers in the calcic soils that formed on the desiccated lake floors (Smith 1968). The composition of these mineral assemblages reflects the temperature at which crystallation took place. Some saline units would crystallise during the summer, while others would form during the winter months. Palaeotemperature data from this type of analysis on the Searles Lake deposits are shown in Fig. 3.32.

The marked contrasts between lake-level fluctuations that are manifest in the data from North America and Africa clearly demonstrate that, while the traditional equation between high-latitude glacial periods and mid- and low-latitude pluvial phases may hold true for the Basin and Range region of the south west United States, it is scarcely applicable to the African situation. Nor is such a straightforward relationship observed in other parts of the world. Indeed, in many tropical and subtropical regions, the cold stages of the Quaternary appear to be related to periods of

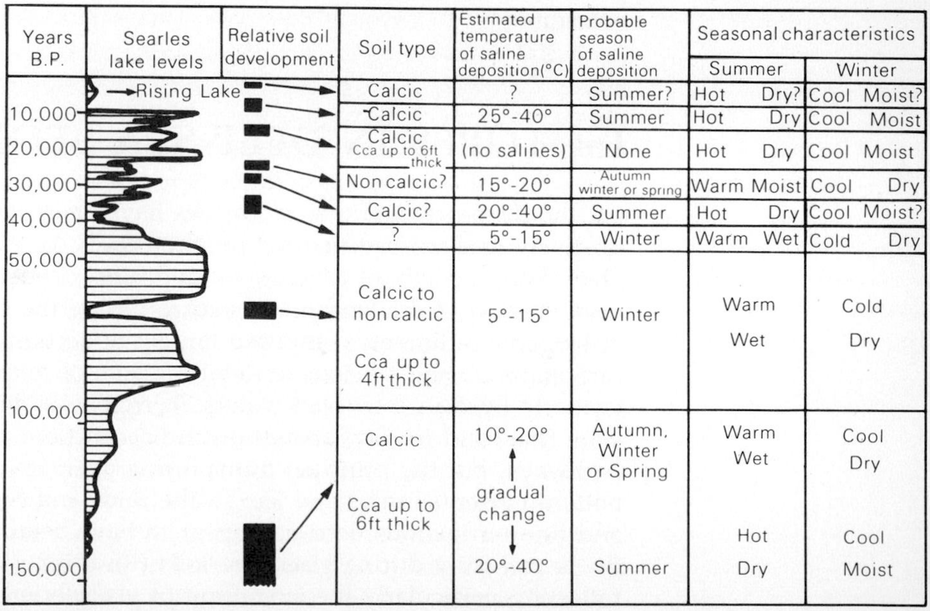

Fig. 3.32: Diagram from Searles Lake, California relating lake history, degree of development of fossil soils, crystallisation temperatures and season of correlative salines, and inferred climatic characteristics of selected interpluvial intervals (after Smith 1968).

increased aridity. A correspondence between full glacial episodes in higher latitudes and drier conditions elsewhere is, however, scarcely surprising in view of the greater continentality and reduced evaporation that would result from a eustatic fall in sea-level of well over 100 m, from the expansion of sea-ice cover in the Arctic and Antarctic oceans, and from a cooling of ocean waters by an average of 5 °C (Goudie 1977). By the same token, higher lake levels between 8000 and 9000 BP in many tropical and subtropical regions can be attributed directly to higher rainfall resulting from increased oceanic temperatures and the world-wide rise in sea level during the early Holocene (Street and Grove 1976). The desiccation of many North American saline lakes during the same time period is also indicative of warmer conditions, but by contrast with other areas, the American south-west in the period around 9000 BP does not appear to have been characterised by pluvial activity. Such variations reflect fundamental alterations in the patterns of global moisture brought about by shifting wind belts and pressure systems. A proper understanding of these spatial and temporal fluctuations awaits the development of palaeotemperature models in which the ocean/atmosphere system can be simulated for particular time periods of the Quaternary.

Overall, however, fluctuations in water level in closed basin lakes offer a useful means of investigating late Quaternary climatic changes. Pluvial lakes appear to respond rapidly to climatic variations, and are more direct indicators of water balance than vegetation composition and soil development. Some contain a long and detailed record of lacustral and interlacustral phases, and many of the more recent sequences can be dated by radiometric methods. Moreover, pluvial lake deposits are often found in areas where biological and other indicators of former environmental conditions are scarce. Although precise climatic data may be difficult to obtain as palaeolake budgets cannot easily be quantified, the value of pluvial lake records in the analysis of Quaternary environments, particularly in the area of macroscale climatic change, is clearly considerable.

ISOTOPES IN DEEP-SEA SEDIMENTS

On the deep-ocean floors, sediments have been accumulating in a relatively undisturbed manner for thousands, or even millions, of years. They consist partly of terrigenous deposits, *i.e.* detrital material derived from erosion of the land masses surrounding the ocean basins, and partly of biogenic sediments composed largely of accumulations of the carbonaceous and silicious skeletal remains of micro-organisms that formerly lived in the ocean waters. Terrigenous detritus (ranging in size from fine sand to clay) arrives on the ocean floor by a number of different pathways, but the principal transporting agencies are turbidity currents, bottom currents, wind and ice. In the mid- and high-latitudes, both coarse and fine terrigenous detritus appear to have been delivered to the ocean floors primarily during glacial periods (Ruddiman and McIntyre 1976), reflecting particularly the ice-rafting of glacially-eroded debris and, to a lesser extent, the transport of aeolian sediments from the greatly expanded periglacial regions. Wind-blown sediment may also have constituted a major proportion of the fine detritus input in low latitudes during glacial time (Folger 1970), but sea-level lowering of over 100 m would also have resulted in the discharge of large quantities of terrestrial debris from the major rivers as they flowed across the continental shelves, and this material would subsequently have been spread down the continental slopes and across the abyssal plains in gravity-controlled sediment flow (Bé et al. 1976). In many ocean sediment sequences, therefore, a broad correlation can be detected between increased deposition of terrigenous debris and former glacial episodes.

Biogenic sediments, on the other hand, are frequently characteristic of interglacial or warmer episodes. Such sediments, known as oozes, contain recognisable fossil remains and these provide a record of changes in ocean water temperature and, by implication, in atmospheric temperatures throughout the Quaternary. These are considered in more detail in Chapter 4. Many fossils, however, contain other indices of environmental change, namely variations in isotopes (see Ch. 5) of common elements bound within their skeletal remains. In calcium carbonate structures, isotopes of both carbon and oxygen vary with changing climatic and other environmental conditions (see Buchardt and Fritz 1980). Of paramount importance in Quaternary research, however, are variations in the ratios of oxygen isotopes. Oxygen isotope analysis of cave speleothems has already been discussed, but the method is also applicable to ice cores (next section), molluscan remains (Abell 1982) and particularly deep-ocean sediments (Berger et al. 1981). Since variations in oxygen isotope ratios in sediments on the ocean floors are considered to be one of the principal indices of global environmental change during the Quaternary, and also serve as the basis for schemes of stratigraphic subdivision and correlation (Ch. 6), the technique of oxygen isotope analysis will now be considered in some detail.

Oxygen isotope ratios and the ocean sediment record

Oxygen can exist in several isotopic forms, but only two, ^{16}O and ^{18}O, are of importance in oxygen isotope analysis, for the others either have very short half-lives or are found in nature in extremely low concentrations. The $^{18}O/^{16}O$ ratio in the natural environment varies between 1: 495 and 1: 515 with an average of approximately 1: 500 (Dansgaard 1954). This means that only about 0.2 per cent of oxygen in natural circulation is ^{18}O. Ratios of oxygen isotopes are measured not in absolute terms but as relative deviation ($\delta^{18}O$ per ml) from the mean ratios of a standard:

$$\delta^{18}O \ = \ 1000 \ \left(\frac{(^{18}O/^{16}O) \text{ sample}}{(^{18}O/^{16}O) \text{ standard}} \right) - 1$$

(Shackleton and Opdyke 1973).

The usual standards are PDB (belemnite shell – see Radiocarbon Dating section) for the analysis of carbonates and SMOW (Standard Mean Ocean Water) for the analysis of water, ice and snow (Craig 1961). The latter is used in the isotopic analysis of glacier ice cores (see below). PDB is $\pm$ 0.2 per cent in relation to SMOW. Analyses are carried out using a mass spectrometer on CO_2 gas prepared from the sample. Oxygen isotope ratios are expressed as positive or negative values relative to the standard (δ = zero). Thus, $\delta^{18}O$ value of -3% indicates that the sample is 0.3 per cent or 3.0 parts per millilitre deficient in ^{18}O relative to the standard.

The variation in isotopic composition of ocean waters over time is revealed by the changing ratios of $^{18}O/^{16}O$ in carbonate shells and skeletons found in deep sea sediments. Many marine organisms secrete carbonate structures and oxygen is abstracted from sea waters for this purpose. Thus, the oxygen isotope ratios in fossil carbonates buried in sediments on the ocean floors should reflect the ratios prevailing in the oceans at the time of their secretion. Analyses have been carried out on the remains of a range of marine micro-organisms, but by far the most widely used fossils are the tests of the planktonic and benthonic Foraminifera (Ch. 4). There is now a considerable body of evidence to suggest that the ratios of ^{18}O to ^{16}O in the ocean waters varied in a quasi-cyclic fashion with succeeding glacial and interglacial periods.

During evaporation of sea water, a natural **fractionation** (see p. 276) of oxygen isotopes occurs. In other words, there is a preferential evaporation of the lighter $H_2 {}^{16}O$ molecules so that atmospheric water vapour becomes relatively enriched in ^{16}O. Fractionation is particularly marked at higher latitudes where colder air masses are increasingly less able to support the heavier isotope ^{18}O. Thus the moisture-bearing winds which nourish the polar glaciers contain relatively higher quantities of the lighter ^{16}O, and this, in turn, is reflected in the isotopic composition of glacier ice. During the cold phases of the Quaternary, with markedly expanded ice masses in both northern and southern hemispheres, large quantities of ^{16}O were trapped in the ice sheets leaving the oceans relatively enriched in ^{18}O and thus isotopically more positive. Conversely, the melting of the ice masses in the interglacials liberated large volumes of water enriched in ^{16}O back into the oceans. Analysis of $^{18}O/^{16}O$ ratios in Foraminifera has revealed that the glacial/interglacial variation in isotopic composition of ocean waters was, in fact, extremely small. For example, in the most widely-discussed ocean core so far investigated, V28–238 (Ch. 6), the isotopic differences between glacial and interglacial stages range from 0.47 to 1.37 parts per millilitre relative to standard. Some parts of the oceans were, it seems, only a little over 1% more positive than at the present day (Shackleton and Opdyke 1973).

An oxygen isotope trace through a core of deep ocean sediment therefore reveals a record of glacial/interglacial changes spanning, in many instances, the whole of the Quaternary. In this section we are concerned principally with the palaeoenvironmental significance of these records, but the division of the ^{18}O trace into isotopic stages also provides a unique basis for global correlation. Oxygen isotope *stratigraphy* is discussed fully in Chapter 6.

The proportion of ^{18}O in carbonate shells is dependent upon two main factors, temperature and isotopic composition of sea water during secretion. Considerable controversy has arisen, however, over which of these variables is primarily responsible for the observed variations in isotopic ratios in marine carbonate fossils. The relationships between carbonate oxygen isotopic ratios and temperature were first established by Urey (1947) who reasoned that, as a small fractionation that is temperature-dependent occurs when carbonate is precipitated slowly in sea water, it should, at least in theory, be possible to estimate former temperatures by measuring the extent of isotopic fractionation in marine carbonate fossils. The measurement of such a process requires extremely sensitive instrumentation for the ^{18}O enrichment in respect to ^{16}O increases by only about 0.02‰ per 1 °C temperature fall. In order to derive palaeotemperatures by this method, however, it is necessary to know the former isotopic composition of sea water, and as this cannot be established directly, it must be estimated. Emiliani, who pioneered the stratigraphic application of oxygen isotope analysis, suggested that the isotopic content of ocean waters had varied by *ca.* 0.4‰[8] between glacial and interglacial stages on the basis of (a) an estimate of the average isotopic composition of Pleistocene glacier ice (–15‰), (b) the volume of ocean waters required to generate the Pleistocene ice sheets (*ca.* 2.6 per cent of present ocean volume), and (c) the likely effects that the removal of that volume of water with an isotopic content of – 15‰ would have on the $^{18}O/^{16}O$ ratios of the remaining ocean waters (Emiliani 1955). Using the value of – 0.4‰, Emiliani interpreted the fluctuations in isotopic content of planktonic Foraminifera in cores from the Caribbean and Equatorial Atlantic as reflecting palaeotemperature changes of 6 °C between the last glacial and the postglacial periods.

Subsequently, however, controversy has arisen over both the value adopted by Emiliani for the former isotopic composition of ocean waters and, more fundamentally, over his assertion that the observed isotopic changes are primarily a reflection of palaeotemperature changes. A number of authors (e.g. Olausson 1965; Shackleton 1967; Dansgaard and Tauber 1969; Shackleton and Opdyke 1973) have argued that fluctuations in isotopic composition of ocean waters were much greater than envisaged by Emiliani. Revised estimates ranging between 1.2 and 1.8‰[9] called into question palaeotemperature values derived originally for the Caribbean and Equatorial Atlantic. More important, however, was the appreciation that observed fluctuations in isotopic profiles from foraminiferal remains reflected a factor other than the change in sea-water temperature, namely variations in the volume of land ice. As temperature and other environmental changes in the deep oceans of the world are likely to be minimal, observed variations in $\delta^{18}O$ values in benthonic (bottom-dwelling) foraminiferal fossils are only explicable in terms of changes in land-ice volumes. Even in planktonic (surface-dwelling) Foraminifera, isotopic profiles appear only to be a partial reflection of surface water temperatures, for they too are influenced by the expansion and contraction of the continental ice sheets. For example, independent faunal evidence in Atlantic and Caribbean cores showed that the shift from last full glacial to postglacial conditions was marked by sea surface temperature changes of only 2.2 °C and that of the calculated isotopic change of 1.8‰, as much as 1.4‰ could be attributed to ice volume changes (Imbrie *et al.* 1973). Dansgaard and Tauber (1969) suggested that on average 70 per cent of the observed isotopic changes in ocean waters are due to glacial control, while van Donk (1976) has argued that in the Atlantic Ocean, at least 90 per cent of the recorded changes in isotopic composition in a large number of sediment cores are attributable to variations in the isotopic composition of ocean waters consequent on the waxing and waning of the large continental ice sheets.

There is, therefore, a growing body of opinion that the changing ratios of $^{18}O:^{16}O$ in biogenic carbonates in deep-sea sediments do not provide a straightforward history of palaeotemperature variations, but are more readily interpreted as a record of **palaeoglaciation**. However, while changes in ice volume now appear to be the dominant control affecting isotopic profiles in planktonic as well as benthonic Foraminifera, the effect of other factors, including surface water temperatures, evaporation/precipitation ratios and infusion of isotopically 'light' water means that the planktonic foraminiferal records from various parts of the oceans agree only so far as the main features are concerned (Dansgaard and Duplessy 1981). In order to derive an estimate of palaeotemperatures of surface waters and the extent of glacially-induced changes based on planktonic Foraminifera, it is necessary to make corrections for the influence of regional water masses in each particular area (e.g. Kahn *et al.* 1981). Benthonic foraminiferal records on the other hand are less ambiguous, and although these fossils are less numerous, they provide a more reliable method for establishing the extent of former global ice cover.

If changes in the isotopic composition of benthonic Foraminifera can be taken as an index of land-ice volumes, then the oxygen isotope curve can also be regarded as an indicator of eustatic sea-level changes. It has been suggested that a sea-level change of *ca.* 10 m will be represented by a 0.1‰ change in the oxygen isotope ratio (Shackleton and Opdyke 1973). At the height of the last glacial, deep-ocean waters in many parts of the world appear to have been enriched in ^{18}O by about 1.6‰, which would be equivalent to a sea-level lowering of *ca.* 165 m at the height of the Devensian (last glaciation) by comparison with the present day (Shackleton 1977). In view of the problems that have been encountered in deriving a meaningful eustatic sea-level curve for the Quaternary period from terrestrial evidence (Ch. 2), the oxygen isotope record from the deep-ocean floors may provide an independent monitor of glacio-eustatic sea-level change.

Limitations in oxygen isotope analysis

There are a number of limitations in oxygen isotope analysis. These include:

Resolution.   Sedimentation rates vary substantially throughout the oceans. In areas of low sedimentation rates a long record is preserved, but resolution of individual isotope stages is often poor. Even in areas of relatively high sedimentation rates, resolution, at best, is such that an average sample used for oxygen isotope analysis represents a time-span of some 1000 years.

Sediment mixing.   Where mixing of sediments has occurred as a result of bottom-dwelling burrowing organisms or the action of turbidity currents, then the clarity of the oxygen isotope record will be blurred. Shackleton and Opdyke (1973, 1976), for example, have suggested that in a number of Pacific cores the peak-to-peak amplitude in oxygen isotope values has been reduced as a result of sediment mixing.

Isotopic equilibrium between test carbonate and ocean water.   Many species of benthonic Foraminifera do not secrete carbonate that is in isotopic equilibrium with the ocean water that they inhabit, and some knowledge is required of the extent to which different species undergo isotopic fractionation (Hecht 1973). When benthonic tests are being measured, therefore, more importance is attached to species such as *Uvigerina senticosa* and *Globocassidulina subglobosa* which are known to deposit carbonate in isotopic equilibrium with ocean waters (Shackleton and Opdyke 1977).

Synchroneity of ^{18}O changes.   It is unlikely that changes in ocean surface temperature were synchronous on a global scale, and this may therefore limit the applications of isotopic records from planktonic Foraminifera in global correlation. However, deep-sea temperature changes that occurred during the Quaternary were probably small, and the stability requirements of the oceans suggest that they occurred essentially synchronously (Shackleton and Opdyke 1976). For this reason, world-wide comparisons between individual isotopic records should be based on benthonic rather than planktonic Foraminifera.

Species dependence.   The interpretation of oxygen isotope ratios is dependent on the specific foraminiferal tests employed in analysis. Where mixed populations have been used, the isotopic ratios are average values from tests that occupied different oceanic habitats, each characterised by slightly different temperature and salinity conditions. Clearly, this can pose problems in the assessment of different isotopic profiles. Even where records have been based on the tests of a single species problems arise, for many Foraminifera exhibit a marked seasonality in their life cycles. Some, for example, have growth periods of only a month or so, some species secrete calcite preferentially in warmer seasons, and some migrate through the water column during their life cycle.

Carbonate dissolution.   After death, carbonate microfossils sink within the water column, and many will become dissolved or disaggregated before reaching the ocean floor. Some foraminiferal species are more susceptible to dissolution than others, so that in settling there is a selective removal usually of the species that lived closer to the surface. Between 3 and 5 km depth, $CaCO_3$ solution equals $CaCO_3$ supply, the level defined as the **carbonate compensation depth** (CCD). Below that depth only the most robust microfossils arrive on the sea bed. The stratigraphic studies of foraminiferal assemblages are therefore usually restricted to sediments that have accumulated in waters that are shallower than the CCD.

In spite of these limitations, however, it is apparent that oxygen isotope analysis of deep-sea sediments is a technique of considerable importance in Quaternary research. Although perhaps not as secure an indicator of palaeotemperatures as was at one time believed, oxygen isotope ratios do, nevertheless, provide a unique record of glacial/interglacial cycles, of changing global ice volumes, and of glacio-eustatic oscillations of sea-level. Moreover, the recognition of comparable isotopic stages in cores from different areas of the world's oceans provides a means of correlating environmental changes on a global scale (Ch. 6).

ICE CORE STRATIGRAPHY

Since the Second World War, work by groups such as CRREL (United States Army Cold Regions Research Engineering Laboratory) and the British Antarctic Survey has led to the development of rotary drilling rigs capable of raising undisturbed cores from deep within the world's ice sheets. Analysis of cores from Greenland, Antarctica and some of the smaller polar ice caps have revealed a record of annual increments of snow and ice accumulation extending back, in several cases, to the last (Ipswichian) interglacial. A number of properties of these annual ice layers have attracted the interest of Quaternary scientists. These include the melt features that can be detected in pits and in the upper layers of ice cores and which can be related to periods of summer melting on the former glacier surfaces, the presence of extra-glacial dust particles which may provide evidence of former wind directions, and the variations that can sometimes be detected in certain chemical inclusions, such as chloride concentrations which show a contrast between winter and summer ice layers (see e.g. Langway Jr 1970; Mosley-Thompson and Thompson 1982). Of particular significance to Quaternary research, however, has been the analysis of oxygen isotope variations. During cold phases with significantly expanded glaciers in high latitudes, relative increases are recorded for ^{16}O in the ice cores, while reduced quantities of the lighter oxygen isotope are indicative of warmer climatic intervals. Oxygen isotope traces through present-day ice caps can therefore provide a continuous record of global climatic changes.

Climatic variations over the past 100 000 years are revealed by oxygen isotope analyses from Camp Century in Greenland and from Devon Island in the Canadian Arctic (Fig. 3.33). The curves represent changes in the $^{18}O/^{16}O$ ratios (δ) and are expressed in parts per millilitre relative to SMOW. The records are similar in a number of respects: lower δ values prior to *ca.* 60 000 BP during what are interpreted as 'interglacial' conditions, generally higher δ values during the last glacial stage with maxima between *ca.* 20 000 and 10 000 BP, and lower δ values once more in the Flandrian (Holocene) post 10 000 BP. This general pattern is also reflected in oxygen isotope profiles from the Antarctic ice sheet (Epstein *et al.* 1970; Johnsen *et al.* 1972; Robin 1977). Two principal problems are encountered in the interpretation of these records, however. The first relates to the establishment of a reliable time scale for oxygen isotope profiles in glacier ice, while the second concerns the relationships that can be inferred between the observed δ values and former temperatures.

Satisfactory dating of ice cores is difficult, for no radiometric method currently exists for the dating of ice older than *ca.* 1000 years (Johnsen *et*

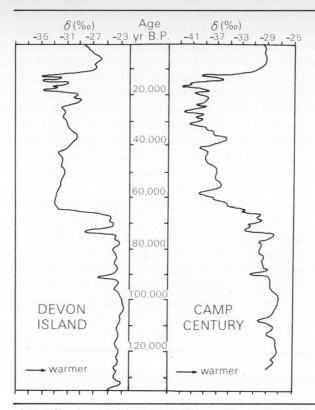

Fig. 3.33: Oxygen isotope profiles through the ice cap on Devon Island, NWT, Canada, and at Camp Century, Greenland (after Paterson et al. 1977).

al. 1972). Annual layers in ice cores can be detected by using melt features, density measurements or light transmissivity of the ice (Langway Jr 1970), but the distinctive features of the original macrostratigraphy are usually obscure and almost impossible to detect at depth. An alternative approach is to use the seasonal variations of the stable isotope concentrations in the accumulating snow and ice, by counting summer layers characterised by relatively high δ values. This method has been employed with some success in the upper levels of the Camp Century ice core (Johnsen et al. 1972), and in that particular case enabled a time-scale to be established back to ca. 8300 BP. The technique is difficult to apply in deep ice, however, and even in the surface layers, years with very low snow accumulation may not be detectable in the δ record. By far the most widely used method of dating has been by glaciological means. Age/depth relationships in ice cores have been calculated on the basis of generally accepted models of glacier flow, and on estimates of various parameters that affect the mode of ice movement (Dansgaard et al. 1969, 1971). However, because the various components of glacier movement at any one point in a glacier system are difficult to quantify, there are doubts about the applicability of this method also (Robin 1977). There is, therefore, no universally acceptable technique for establishing a reliable time-scale for ice cores, particularly in deep ice older than ca. 10 000 years, and this clearly imposes major constraints on the interpretation of the oxygen isotope records from glaciers and ice caps. Further details of the methods used in the dating of ice cores can be found in Hammer et al. (1978).

The second problem involves palaeotemperatures. Unfortunately, over much of the ice core record, the observed δ values cannot be translated directly into temperature values. This is due principally to the fact that many of the deeper layers may have originated up-glacier from the sampling point where slightly different climatic, and therefore isotopic conditions prevailed (Johnsen *et al.* 1972). Moreover, variations in the observed δ values may reflect environmental changes other than temperature. These include changes in (a) the isotopic composition of sea water that provides moisture for precipitation; (b) the ratio of summer to winter precipitation; (c) the meteorological wind pattern; (d) the flow pattern of ice in the accumulation areas; and (e) the variations in thickness of the ice sheet itself (Dansgaard *et al.* 1969). In spite of these problems, however, some general palaeotemperature estimates have been obtained from ice core evidence. For example, Paterson *et al.* (1977) have interpreted the Devon Island and Camp Century records as indicating a cooling of more than 1 °C since 5000 BP, while Robin (1977) has inferred a temperature in Antarctica around 20 000 BP of around 6 °C to 8 °C lower than at present. Robin also suggested that, prior to 100 000 BP, the oxygen isotope record in ice cores from both Greenland and Antarctica indicates a climate 2–3 °C warmer than at the present day.

An alternative approach to the derivation of palaeotemperatures from oxygen isotope traces in ice cores spanning the historical period has been outlined by Dansgaard *et al.* (1975). Variations in the $^{18}O/^{16}O$ ratio in a core

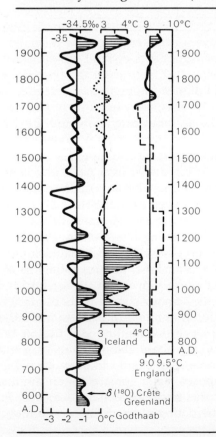

Fig. 3.34: Comparison between $\delta^{18}O$ concentration (left) in snow fallen at Crête, central Greenland, and temperatures for Iceland and England. The full curves are based on direct observations, the dotted part of the Icelandic record is estimated form sea-ice observations, while the dashed curves depend on indirect evidence (after Dansgaard *et al.* 1975).

from Crête on the crest of the Greenland ice sheet revealed a climatic record for the past 1420 years (Fig. 3.34). The core was dated by glaciological means, and also by using seasonal variations in δ. The temperature scale for Godthaab in west Greenland (shown below the curve in Fig. 3.34) was calculated by comparing the 0.9‰ shift from 1890 to 1935 with the corresponding 2.1 °C change in Godthaab temperatures over the same time period. On this basis a relationship was established between temperature and down-core variation in δ (correlation coefficient 0.94), and this enabled the δ curve to be read in terms of Godthaab temperatures. Observed and estimated temperatures are also shown on Figure 3.34 for Iceland and England. The significant correlation (0.88 for Iceland; 0.83 for England) between the upper parts of the curves suggests that the δ profile is representative of climatic change far beyond the Greenland area. However, this conclusion applies only to climatic changes of medium frequencies (60 to 200 years), for longer term climatic changes appear to have occurred at least 250 years later in western Europe by comparison with Greenland.

Although this approach can only be used for the relatively recent past, the results serve to demonstrate the close relationship that can exist between temperature variations and observed fluctuations in oxygen isotope ratios in ice cores and, moreover, that the ice core record reflects climatic changes of both a local and regional nature. Providing that a satisfactory means can be found for dating ice cores, this line of research clearly has considerable potential as a means of establishing a relatively uninterrupted record of late Quaternary climatic change.

CONCLUSIONS

In spite of the impressive nature of the evidence that has been obtained from the sediments on the deep-ocean floors, the terrestrial stratigraphic record, albeit diverse and highly fragmented, continues to form the principal source of data for the reconstruction of Quaternary environments. Almost all Quaternary sediments contain within their matrix important clues about their mode of deposition and often about the climatic régime under which the sediments accumulated. This information can frequently be extracted from the stratigraphic record by the judicious application of the various physical and chemical techniques outlined at the beginning of the chapter, and by analogy with geological and sedimentological processes that can be observed in operation at the present day. Moreover, because of the multifaceted nature of the lithostratigraphic record, a range of palaeoenvironmental indices can often be obtained from a single section or core. It must be emphasised, however, that the sedimentary evidence can seldom be properly evaluated in isolation. In the analysis of, for example, glacial deposits, periglacial sediments or pluvial lake sequences, the stratigraphic record should, wherever possible, be integrated with available morphological evidence to produce a synthesis of landscape or climatic change. The same applies equally to the fossil record, the third category of evidence used in the reconstruction of Quaternary environments, and which forms the subject matter of the following chapter.

NOTES

1. In particle shape analysis, the a axis is the longest axis of the particle, b is the intermediate axis, and c is the shortest axis.
2. An **electron microscope** consists of a cathode-ray tube through which a beam of electrons is passed. These are concentrated on, and pass through, the specimen to produce a magnified image on a photographic plate. After development, the electron photomicrograph shows the structure of the object in terms of its electron density.
3. The term **diamicton** refers to non-sorted terrigenous sediments and rocks containing a wide range of particle sizes, regardless of genesis (Flint 1971).
4. **Load structures** often form where sand is deposited over a hydroplastic mud layer. Under the weight of the sand, the mud layer becomes distorted and bent downwards. In some cases the sand layer sinks and forms lobes; in others the mud layer becomes pushed up to form tongues (see Reineck and Singh 1973).
5. Valentine and Dalrymple (1976 p. 210) use the term **diagenesis** in a broad sense to include 'the weathering of minerals by oxidation and reduction hydrolysis and solution, the biological action of anaerobic bacteria, compaction, cementation, recrystallisation, and the lattice alteration of clays by the expulsion of water and ion exchange'.
6. **Isotopes** are atoms of an element that are chemically similar but have different atomic weights (see Ch. 5). Oxygen consists principally of two isotopes, the heavier ^{18}O isotope and the lighter ^{16}O isotope. The ratios between ^{18}O and ^{16}O in the atmosphere, seas, groundwater and ice are often controlled by climatic conditions.
7. **Stromatoliths** are laminated sedimentary structures built by dense mats of blue-green algae which trap and bind sediment particles (Garrett 1970).
8. This value was subsequently modified to 0.5‰ (Emiliani 1966, 1971).
9. A major factor influencing these figures was that the average isotopic composition of land ice was subsequently found to be *ca.* -30‰ to -35‰ (see Dansgaard and Tauber 1969).

CHAPTER 4

Biological evidence

INTRODUCTION

Biological evidence in the form of fossil plant and animal remains has always been the cornerstone in the reconstruction of Quaternary environments. The approach has traditionally followed uniformitarian principles in that it has been assumed that the present can be used to unlock the history of the past, and thus by applying to fossil assemblages what is known about the environmental factors that influence the distribution of modern plant and animal populations, it is possible to make inferences about the types of landscapes and climates that existed in earlier times. The method has a sound, logical basis and can be defended philosophically (Rymer 1978). In practice, however, things are not quite so straightforward, for in order to derive meaningful information from fossil evidence, certain assumptions have to be made both about the fossil evidence itself and also about our present-day knowledge of plant and animal ecology. The most important of these assumptions are:

(a) that we fully understand and are able to isolate the environmental parameters governing present-day distributions of plants and animals;
(b) that present plant and animal distributions are in equilibrium with those controlling variables;
(c) that former plant and animal distributions were in equilibrium with their environmental controls;
(d) that former plant and animal distributions have analogues in the modern flora and fauna;
(e) that the ecological affinities of plants and animals have not changed through time;
(f) that a fossil assemblage is representative of the death assemblage and has not been biased by differential destruction of its original component parts or by contamination by older or younger material;
(g) that the origin of the fossil assemblage (**taphonomy**) can be established; and
(h) that the fossil remains can be identified to a sufficiently low taxonomic level to enable uniformitarian principles to be applied.

The extent to which these assumptions can be met varies both with the type of fossil evidence (pollen grains, animal bones etc), and the nature of the assemblage, but it is rare indeed for all of the above to be satisfied. In this chapter a number of different forms of biological evidence that have been used in the analysis of Quaternary environments are assessed in the light of the foregoing list. The purpose is to demonstrate the strengths and weaknesses of the different methods, the potential sources of error, and the types and levels of information that each method can be reasonably expected to provide. The list of techniques is by no means exhaustive, but constitutes a representative sample of some of the more widely-used methods in Quaternary research.

PLANT MICROFOSSIL ANALYSIS

Pollen analysis

Of all the methods currently employed in the reconstruction of Quaternary environments, undoubtedly the most widely-adopted and arguably the most successful is the technique of pollen analysis. The method is extremely versatile allowing, on the one hand, inferences to be made about changing vegetation patterns over broad spatial and temporal scales, while at the other extreme it is possible to detect the influence of man as an agent of landscape change over a very short time period within a limited geographical area. As a consequence, a great body of literature now exists on pollen analysis, and it is quite impossible to do full justice to the technique within the space available in this volume. The aim here, therefore, is to provide a general introduction to pollen analysis as a palaeoenvironmental technique and to outline the principal strengths and weaknesses of the method. Those who may wish to explore further into pollen analysis are directed towards the texts by Faegri and Iversen (1975), Moore and Webb (1978) and Birks and Birks (1980).

The nature of pollens and spores

Pollen grains are formed in the anthers of the seed-producing plants (**angiosperms** or **gymnosperms**). They contain the **male gamete** of the plant and aim to reach the stigma of the female part of the flower where fertilisation can take place. The lower plants (cryptogams) such as the ferns (Pteridophyta) and mosses (Bryophyta) produce spores during the sporophyte generation of the plant's life cycle whose function is to disperse the plant to suitable habitats where the gametophyte (the second stage in plant generation) can grow. During dispersal, large numbers of pollens and spores become incorporated and fossilised in sediments, and it is the extraction, identification and counting of these fossil grains which forms the basis of the technique of pollen analysis.

Most pollens and spores are extremely small, few exceeding 80 to 100 μm in diameter, with the majority falling in the size range 25 to 35 μm. A typical pollen grain consists of three concentric layers (Faegri and Iversen, 1975). The central portion is the living cell which is surrounded by a covering of cellulose know as the **intine**. Neither of these survive in the fossil form. The outer layer or **exine** consists of a highly resistant, waxy coat of material called **sporopollenin** whose prime function is to protect the young gametophyte from desiccation and microbial attack. However, it also has the effect of preserving pollen grains in sediments when almost all other organic constituents are reduced to structureless and unrecognisable substances. The wall of the grain is characterised by a variety of morphological features which, along with the number and distribution of germinal apertures, and the shape and size of the grain, form the basis for pollen and spore identification. Some examples of pollen grains with their diagnostic characters are shown in Fig. 4.1.

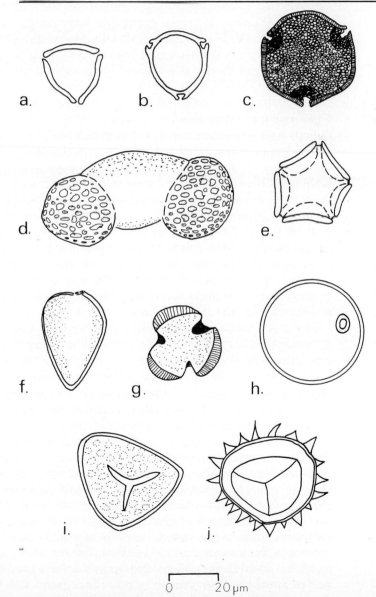

Fig. 4.1: Drawings from selected pollen grains (a–h) and spores (i and j). a. hazel (*Corylus*); b. birch (*Betula*); c. lime (*Tilia*); d. pine (*Pinus*); e. alder (*Alnus*); f. sedge (Cyperaceae); g. mugwort (*Artemisia*); h. grass (Gramineae); i. *Sphagnum*; j. lesser clubmoss (*Selaginella*). Scale in μm.

0   20 μm

Pollens and spores are disseminated by a variety of means. Spores are usually dispersed by wind, but pollen grains can also be spread by water, by insects, by animals, by birds and by man. Those plants that liberate wind-borne pollen are termed **anemophilous** and generally produce far greater numbers of grains than the **entomophilous** species, which rely on insects or other zoological vectors for pollination. Wind dispersal is facilitated by the small size, smooth surface features and low specific gravity of the grains, while in the Gymnosperms (e.g. pine – *Pinus*; spruce – *Picea*), air bladders or sacs have evolved enabling those particular pollens to stay airborne for very long periods of time, and also to travel considerable distances. The entomophilous grains possess a hardy, armoured surface, which often has prominent spines and a coat of sticky material that causes them to adhere both to each other and to the body of the animal. Generally, these grains are large (in excess of 60 μm) or very small (less than 15 μm) and are usually less well-represented in the fossil record than the wind-dispersed types. The whole field of pollen production and dispersal is extremely complex and is considered in more detail below.

Field and laboratory work

Pollens and spores are usually well-preserved in lake and pond sediments and in peats. They are also found in soils (Dimbleby 1957, 1961) and in cave earths (Campbell 1977), where they are less well-preserved, and they have even been discovered in ocean floor sediments (Heusser and Balsam 1977). The degree of preservation of pollen and spores depends on a variety of factors, the most important of which are the grain size of the sediment matrix and the anaerobic nature of the depositional environment. Pollen grains are, for example, less well-preserved where loosely-compacted peats allow aerobic or microbial attack. Similarly, grains can easily be damaged or destroyed by mechanical abrasion in coarse-grained sediments such as riverine deposits, soils and lake sediments near the point of stream inflow.

Samples containing fossil pollen can be taken from sections exposed in river banks, cliffs, road cuttings or building excavations, or by digging pits. Alternatively, cores can be obtained using the equipment described in Chapter 3. In the laboratory, following sieving or differential flotation procedures, the samples undergo various chemical treatments in order to remove as much of the sediment matrix as possible. Lignins and celluloses are removed from organic sediments by oxidation and acetolyis, while minerogenic sediments are treated with hydrofluoric acid to remove the silicious material. Carbonates and calcareous sediments are treated with hydrochloric acid. The residues containing the pollens and spores are stained with an organic dye such as safranin to bring out the surface characteristics of the grains, and then mounted on glass slides in a suitable medium such as glycerine jelly or silicon oil. These can then be examined under a microscope, usually at × 400 magnification, and by traversing backwards and forwards across the slide, a count can be made of all the identifiable pollens and spores until a predetermined total (the **pollen sum**) has been reached. Individual grains can be recognised by their exine characteristics (see above), identifications being based on poller keys, photographs and particularly reference slides made up from modern pollen. Good accounts of laboratory procedures and examples of pollen keys can be found in Faegri and Iversen (1975) and Moore and Webb (1978).

Pollen diagrams

Where samples have been taken from a stratified sequence of sediments such as a lake or peat deposit, analysis at different horizons will show changes in pollen content (and, by inference, in vegetation composition) over time, and such changes can be depicted graphically in the form of pollen diagrams. These can be based either on percentage values or on pollen concentration data, the latter frequently being termed 'absolute pollen diagrams'. **Percentage pollen diagrams** usually take two forms. In some cases a pollen sum of, perhaps, 300 or 500 grains is selected for each level, and individual pollen and spore types are then simply expressed as percentages of the total counted for that level (Fig. 4.2). It is usual practice, however, to exclude pollen of aquatic plants and spores from the initial pollen sum, the former because they are produced in a totally different environment from the terrestrial pollen with which the investigator is primarily concerned, and spores because they are formed in a different way from pollen grains. An alternative form of percentage diagram is one based on a pollen sum of arboreal pollen grains only. In this type of diagram, all pollen and spore types are counted until a total

Fig. 4.2: Percentage pollen diagram spanning the Early and Middle Flandrian from Craig-y-Fro, Brecon Beacons, South Wales (modified after Walker 1982b).

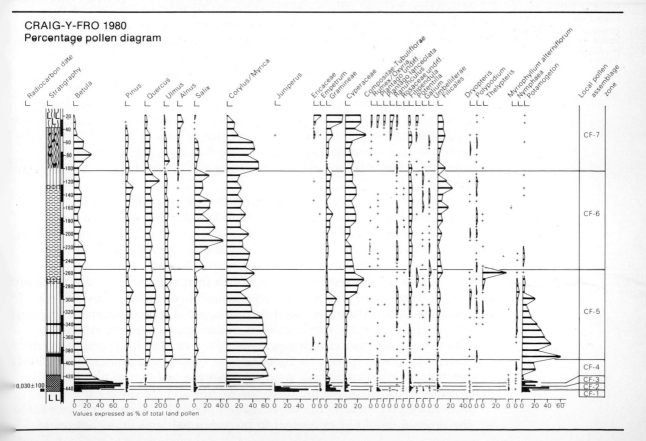

CRAIG-Y-FRO 1980
Percentage pollen diagram

Values expressed as % of total land pollen

CRAIG-Y-FRO 1980
Percentage pollen diagram

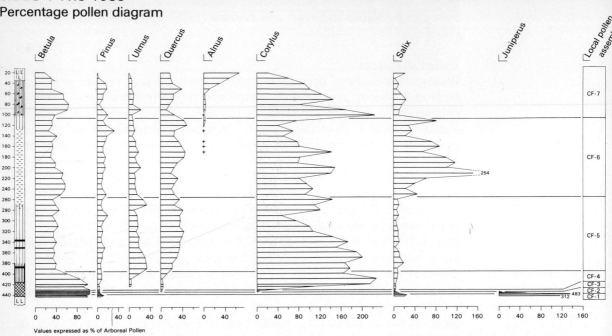

Values expressed as % of Arboreal Pollen

Fig. 4.3: Pollen diagram from Craig-y-Fro, Brecon Beacons, South Wales, based on arboreal pollen percentages (after Walker 1982b).

of perhaps 200 tree pollen grains have been identified, and then each individual pollen and spore type is expressed as a percentage of the tree pollen sum (Fig. 4.3). In the early stages of pollen analysis when workers were especially concerned with forest history, the latter type of diagram was invariably employed, and it is still widely used in archaeological investigations and in studies of Flandrian landscape change. More recently, however, with an increasing interest in the cold stages of the Quaternary where much lower frequencies of tree pollen are recorded, the emphasis has moved towards the percentage pollen diagram in which the pollen sum is based on frequencies of land pollen. A list of taxa commonly found in pollen diagrams from north-west Europe is shown in Table 4.1.

One problem arising from the representation of pollen data in percentage form is that the curves for individual taxa are, of necessity, interdependent. In other words, an increase in the influx of, for example, *Betula* pollen to a site will automatically lead to a suppression of relative percentages of other taxa represented in the diagram. Thus statistical fluctuations will be reflected in the pollen diagram which do not represent ecological changes. One way of overcoming this problem is by the use of **'absolute pollen diagrams'**, as these are based not on percentage changes, but on changes in the *total number* of pollen grains per unit volume of sediment. A number of techniques are now available for establishing the amount of pollen (pollen concentration) in a body of sediment (Peck 1974) but the one most commonly used involves the addition of a known quantity of exotic pollen grains to the fossil sediment during the laboratory preparation. The observed ratio of exotic to fossil pollen enable the original fossil pollen content to be computed, and thus changes in

Trees

Betula (birch)	*Alnus* (alder)	*Picea* (spruce)
Pinus (pine)	*Fagus* (beech)	*Populus* (poplar)
Ulmus (elm)	*Fraxinus* (ash)	*Carpinus* (hornbeam)
Quercus (oak)	*Tilia* (lime)	*Sorbus* (rowan)

Shrubs

Corylus (hazel)	*Hippophae* (buckthorn)	Ericaceae (heather family)
Salix (willow)	*Hedera* (ivy)	*Calluna* (ling)
Juniperus (juniper)	*Ilex* (holly)	*Empetrum* (crowberry)

Herbs

Gramineae (grass family)	*Rumex* (sorrel/dock)
Cyperaceae (sedge family)	*Ranunculus* (buttercup)
Caryophyllaceae (pink family)	*Thalictrum* (meadow rue)
Chenopodiaceae (goosefoot family)	Rosaceae (rose family)
Compositae (daisy family)	*Filipendula* (meadow sweet)
Artemisia (mugwort/wormwood)	*Potentilla* (tormentil)
Cruciferae (cabbage family)	*Dryas* (mountain avens)
Galium (bedstraw)	*Saxifraga* (saxifrage)
Epilobium (willow-herb)	Scrophulariaceae (figwort family)
Helianthemum (rock-rose)	*Succisa* (scabious)
Labiatae (labiate family)	Umbelliferae (carrot family)
Leguminosae (pea family)	*Urtica* (nettle)
Plantago (plantain)	*Valeriana* (valerian)

Aquatics   Spores

Aquatics	Spores
Littorella (shoreweed)	Filicales (ferns)
Myriophyllum (water milfoil)	*Dryopteris* (male fern)
Nuphar (yellow water-lily)	*Isoetes* (quillwort)
Nymphaea (white water-lily)	*Polypodium* (polypody fern)
Potamogeton (pondweed)	*Lycopodium* (clubmoss)
Typha (bulrush)	*Pteridium* (bracken)
	Selaginella (lesser clubmoss)
	Sphagnum (sphagnum)

Table 4.1: Some common taxa found in pollen diagrams from north-west Europe.

pollen concentration at different levels in a profile can be depicted diagrammatically (Fig. 4.4). If radiocarbon dates are then obtained on a series of samples from the same profile, it may be possible to calculate the rate of sediment accumulation. In this case, the **pollen concentration diagram** can be converted into a **pollen influx diagram** which expresses the data in the form of pollen grains/cm^2 of surface of sediment per year.

Although this technique will overcome the problem of statistical interdependence of taxon curves and is therefore potentially of considerable value in palaeoenvironmental reconstruction, it too has problems of application. Variations in rates of deposition within a body of sediment will tend to distort pollen concentration values, and great care must be taken in the comparative analysis of pollen concentration data. This is particularly true when comparisons are being made between different types of sediment as, for example, between lake muds and telmatic or terrestrial peats. Even within a seemingly uniform deposit, however, it cannot be shown unequivocally that the sedimentation rate has been uniform. To some extent this problem can be surmounted by radiocarbon control on the sequence, but here again difficulties can arise over the often thick lenses of sediment (up to 10 cm) which have to be

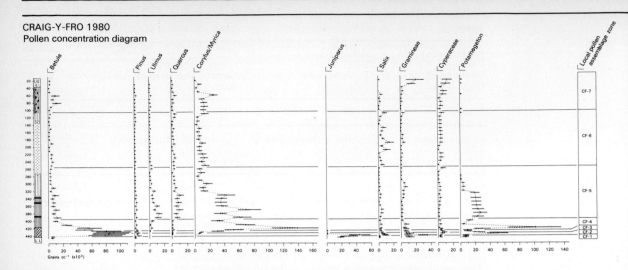

CRAIG-Y-FRO 1980
Pollen concentration diagram

Fig. 4.4: Pollen concentration diagram from Craig-y-Fro, Brecon Beacons, South Wales (after Walker 1982b).

extracted from cores in order to provide sufficient material for dating purposes, and the fairly large standard deviations on the radiocarbon dates (Ch. 5) which together may render the age determination meaningless in a stratigraphic context. Although potentially valuable, therefore, pollen concentration diagrams must be used with caution and are best employed in conjunction with, rather than as a substitute for, percentage pollen diagrams.

In order to make sense of the considerable amount of data shown in a typical pollen diagram, it is necessary to divide the diagram into pollen-stratigraphic units characterised by distinctive groups of pollen types. In this way, a series of **pollen zones** are created and these may have local or regional significance. Pollen diagrams are usually zoned subjectively, although recently multivariate classificatory techniques have been employed which largely overcome the subjective element in pollen boundary location (e.g. Gordon and Birks 1972, 1974).

In the early days of pollen analysis, it was customary to develop zonation schemes which were regionally based, those by Godwin (1940) for southern England, Mitchell (1942) for Ireland, and Firbas (1949) for Europe being fairly typical examples. This approach was essentially deductive in that pollen diagrams from individual sites were zoned on the basis of the zonation scheme that was operative for that particular area. Whilst the method had the advantage of simplifying the general picture of vegetation change that could be inferred from the pollen data, in practice, local pollen-stratigraphic variations tended to be masked in the search for regional uniformity. Moreover, an unfortunate and unforeseen outcome of the widespread application of the regional pollen zonation scheme was the links that were made between pollen zones, temporal changes, climatic phases and geological and geomorphological events. The equation by some authors of the Godwin-type zonation system with the Blytt/Sernander vegetation/climatic periods (Table 3.5) meant that many of the numbered pollen zones were given climatic connotations.

General dissatisfaction with the above zonation scheme led many workers to adopt a more inductive approach to pollen diagram zonation.

Following the work of Cushing (1967a) in Minnesota, it is now common practice to divide pollen diagrams into **local pollen assemblage zones** usually on the basis of the dominant terrestrial taxa. Comparisons can then be made between individual pollen diagrams and, on the basis of perceived similarities, **regional pollen assemblage zones** can be erected. An example of how this method operates is shown in Table 4.2 where local pollen assemblage zones from seven sites on Rannoch Moor in western Scotland were integrated to form a series of 'R-zones' that have regional applicability for the early and mid-Flandrian periods. Where such regional pollen assemblage zones can be identified, they can be used as a means of relative dating of events within a particular area. On the broader scale, the recognition of distinctive pollen assemblages has formed the basis for the relative dating of interglacial sequences in Britain and north-west Europe (see e.g. Godwin 1975; West 1977a).

The interpretation of pollen diagrams

The interpretation of a pollen diagram is the most difficult part of pollen analysis, and requires a knowledge of pollen production and dispersal, pollen source and pollen deposition, pollen preservation and the relationship between fossil pollen and former plant communities. Only when these factors have been carefully evaluated can sound inferences be made about former vegetation cover and, by implication, about former climates.

First, it is important to appreciate the fact that not all plants produce

Table 4.2: Correlation between local and regional pollen assemblage zones in the Rannoch Moor area of Western Scotland (after Walker and Lowe 1979).

Regional pollen assemblage zone		Rannoch Station 1		Rannoch Station 2		Corrour 1		Corrour 2	
R-8	Ericaceae zone							C-2f	Ericaceae
R-7	*Betula* *Pinus* *Alnus* zone	RS-1f	*Betula* *Pinus* *Alnus* Ericaceae			C-le	*Alnus* *Pinus* *Betula* *Corylus*	C-2e	*Betula* *Pinus* *Alnus* *Corylus*
R-6	*Betula* *Pinus* zone	RS-le	*Betula* *Pinus* *Corylus* Ericaceae			C-ld	*Betula* *Pinus* *Corylus* Ericaceae		
R-5	*Betula* *Corylus* Ericaceae	RS-ld	*Betula* *Corylus* Ericaceae			C-lc	*Betula* *Corylus* Ericaceae	C-2d	*Betula* *Corylus* Ericaceae
R-4	*Betula* *Corylus* zone	RS-lc	*Betula* *Corylus*	RS-2d	*Betula* *Corylus*	C-lb	*Betula* *Corylus*	C-2	*Betula* *Corylus*
R-3	*Betula* *Juniperus* zone	RS-1b	*Juniperus* *Betula*	RS-2c	*Betula*	C-la	*Betula* *Juniperus*	C-2b	*Betula* *Juniperus*
R-2	*Juniperus* *Betula* zone			RS-2b	*Juniperus* *Betula*			C-la	*Juniperus* *Empetrum* *Betula*
R-1	*Empetrum* zone	RS-la	*Empetrum* *Salix* Gramineae	Rs-2a	*Empetrum*				

the same amount of pollen. It has already been noted that entomophilous species usually produce less pollen than anemophilous plants and will therefore be under-represented by comparison in the fossil record. Even less well-represented are the **autogamous** plants such as wheat (*Triticum*) which are self-pollinating and which liberate very little pollen into the atmosphere. Even more extreme are **cleistogamous** plants (e.g. *Viola*), the flowers of which never open and thus pollen is very rarely released. Within each of these plant types, however, there is considerable variation. The lime (*Tilia cordata*) and ling (*Calluna vulgaris*), for example, are both insect-pollinated yet they usually liberate large quantities of pollen. On the other hand, beech (*Fagus sylvatica*) and oak (*Quercus robur*) are both wind-pollinated, yet are often low pollen producers. To date, insufficient work has been carried out in this field to allow corrections to be made to fossil pollen frequencies to account for the variability of pollen production, but Table 4.3 gives some indication of the relative pollen production of some typical trees and shrubs.

Secondly, it is necessary to know something about the source of fossil pollen in a body of sediment. It is important to establish whether plants were growing on the bog surface or within the lake basin, around the margins of the site, in the immediate vicinity, or some distance away. Moreover, it is necessary to know something of the mechanisms involved in the transport of pollen from its source to the eventual point of deposition. Initially, attention was focused on airborne transport of pollen. Tauber (1965), for example, suggested that in a forested region airborne pollen arriving at a bog or lake surface would have travelled by one of three pathways; either through the trunk space, or through the forest canopy, or by raindrop impact from the air above. Factors such as wind speed through the trunk and canopy, the density of woodland cover, thickness of foliage, time of pollination of the trees, and the size, shape and proximity to the source of the first major bog or lake will all play a part in determining the pollen composition of the final assemblage. More recently, however, research has been directed towards the transport of pollen grains by water, and it now appears that, in certain sites, a large proportion (in some cases up to 90 per cent) of the pollen arriving in a lake may be derived from inflowing streams and groundwater (Peck 1973; Bonny 1976). Moore and Webb (1978) add two

After Pohl (1937)		After Iversen (in Faegri and Iversen 1975)	
(*Fagus* = 1.0)		A. High:	*Pinus*
Alnus	17.7		*Betula*
Pinus sylvestris	15.8		*Alnus*
Corylus	13.7		*Corylus*
Betula	13.6	B. Moder-	
Picea		ate:	*Picea*
Carpinus			*Quercus*
Quercus	13.4		*Fagus*
Fagus	7.7		*Tilia*
	1.6	C.	
	1.0	Low:	*Ilex*
			Viscum
			Lonicera

Table 4.3: Relative pollen production for some European trees and shrubs.

further sources of pollen, namely a local input from aquatic plants growing in the lake or mire plants growing on the bog surface, and a secondary component from pollen which has been deposited around the site catchment and has subsequently been remobilised and incorporated into the lake sediments at a later date (see below).

In order to establish how far these various components contribute towards the development of a pollen assemblage in a lake or bog site, studies have been undertaken of modern pollen dispersal and these data have subsequently been compared to the species composition of the vegetation around the sampling site. Data on present-day pollen rain can be obtained from moss polsters on bog surfaces or from specially-designed pollen traps for collecting both atmospheric pollen and pollen settling in lake waters (e.g. Lichti-Federovich and Ritchie 1968; Birks 1973a; Peck 1973; Bonny 1976). This type of study can provide valuable quantitative data on the relationship between local, regional and long-distance components in the pollen rain and the modern vegetation cover.

In general, it would seem that most wind-borne pollen is deposited within a few kilometres of its source, and only a very small proportion of the pollen grains liberated into the atmosphere are likely to travel very far. Nevertheless, studies have shown that far-travelled pollen may, in certain circumstances, constitute an important element of the atmospheric pollen rain. Tyldesley (1973), for example, recorded tree pollen, largely of birch and pine, over the Shetland Islands off the north coast of Scotland in densities of up to 30 grains per cubic metre of air. The Shetland Islands today are treeless and the source of this exotic pollen is believed to be Scandinavia. Similarly, Fredskild (1973) reported significant concentrations of arboreal pollen (principally *Alnus, Betula, Pinus* and *Picea*) at sites in eastern Greenland whose source area is believed to have been north-east North America. The occurrence of pollen of long-distance origin in the fossil record is most likely to cause confusion where pollen assemblages are being examined from environments where the local pollen production was low, such as the close of the last glacial period, or in sediments of Loch Lomond Stadial age in the British Isles (Walker 1975a).

A third factor to be considered when pollen diagrams are being interpreted concerns the nature of pollen deposition. Differential settling velocities of pollen in lakes and ponds, coupled with the disturbance of sediment on the lake floor either by currents or by burrowing organisms can lead to complications in the fossil record. Equally misleading can be the occurrence of redeposited or secondary pollen that has been washed into the lake by stream flow, overland flow, solifluction, or by collapse of the basin edge sediments and the subsequent redistribution of material across the lake floor. These grains will clearly be of a different age from those arriving at the lake surface from the atmospheric pollen rain, and although they can often be distinguished from the primary pollen by signs of exine deterioration (see below), they are potential sources of confusion in the interpretation of the biostratigraphic record.

In general, fewer uncertainties are caused where terrestrial sites are used in preference to lakes, although here too complications may arise.

(a) Sequence of increasing corrosion susceptibility of selected pollens and spores

Lycopodium	(low)
conifers	
Tilia	
Corylus	
Alnus, Betula	
Quercus	
Fagus	(high)

(b) Sequence of increasing oxidation susceptibility of selected pollens and spores

Lycopodium clavatum	(low)
Polypodium vulgare	
Pinus sylvestris	
Tilia spp.	
Alnus glutinosa, Corylus avellana, Myrica gale	
Betula spp.	
Carpinus betulus	
Populus spp., *Quercus* spp., *Ulmus* spp.	
Fagus sylvatica, Fraxinus excelsior	
Acer pseudo-platanus	
Salix spp.	(high)

Table 4.4: Corrosion and oxidation susceptibility of selected pollens and spores (after Havinga 1964).

Studies have shown that when pollen arrives on the surface of a bog, there may be a tendency for both lateral and vertical mixing to occur, with the larger grains remaining on the surface while smaller grains may migrate downwards into the *Sphagnum* peats (Birks and Birks 1980). In general, however, these movements are believed to be relatively insignificant when set against the time-scales usually involved in peat bog stratigraphy. More problematical is the behaviour of pollens and spores in soils. The major difficulty with soil pollen analysis is that the processes of leaching and capillary action will have the effect of moving fossil grains up and down the profiles (Havinga 1974). Mixing by earthworms and other soil organisms further exacerbates the problem. In general, it would seem that only in very acid soils are the problems of mixing overcome to the extent that reliable interpretations can be made of the fossil pollen record.

Fourthly, many fossil pollen grains show signs of deterioration resulting from physical, chemical and biological attack on the exine. Experimental work by Sangster and Dale (1964) and Havinga (1964, 1967) has shown that pollen and spores vary in their susceptibility to such processes as oxidation and corrosion (Table 4.4) and that this variability may be partly attributable to the nature of the depositional environment. Some grains, for example, spores of the clubmosses (*Lycopodium*) and certain ferns (*Polypodium*) show remarkable resistance to deterioration, while others such as the more delicate grains of nettle (*Urtica*) and poplar (*Populus*) may be destroyed altogether. As a consequence, some pollen types may be under-represented in the fossil record while others may be over-represented. Cushing (1967b) has described four categories of deterioration in pollen grains:

(a) corrosion: where the exine is pitted or etched;
(b) broken: where the grains are ruptured or split, or pieces hav completely broken away;
(c) crumpled: where the grains are folded, twisted or collapsed;

(d) degraded: where the structural elements are fused together presenting a 'solid' or 'fossilised' appearance to the grain.

These different categories of deterioration reflect very closely the nature of the depositional environment. Corroded grains, usually indicate oxidation – often in poorly-compacted peats, while degraded grains are frequently indicative of secondary deposition, the exine surfaces having undergone structural modification through reworking. As such, deteriorated pollen diagrams may provide useful corroborative information on local environmental conditions (Lowe 1982). Soil pollens frequently show all of the above characteristics, and recognition is consequently made more difficult by the often poor state of preservation of these grains.

Finally, there is the vexed question of how far it is possible to relate pollen assemblages to plant communities, and how far we are justified in making inferences about former climatic and environmental conditions on the basis of pollen analytical data. It is now generally accepted that many former plant communities, especially those dominated by herbaceous taxa, which were characteristic of large areas of western Europe and North America during the cold phases of the Quaternary, have no analogues in the modern flora. One such example is the *Artemisia*-dominated association that appears in many pollen diagrams from north-west Europe during the Younger Dryas or Loch Lomond Stadial of the Lateglacial period. Although Pennington (1980) has recently described an *Artemisia borealis*- and *Silene*-rich community in Greenland which produces pollen rain superficially resembling that of the Lateglacial cold phase, latitudinal and altitudinal variations in association with seasonal and diurnal temperature fluctuations would probably have combined to produce significantly different plant communities during past cold phases in western Europe from those existing in present-day tundra regions.

These difficulties are further compounded by the limitations imposed on palaeoecological inference by the taxonomic imprecision of pollen identification. Under normal microscopy, it is occasionally possible to identify pollen grains to the species level; distinctions can usually be made, for example, between species of plantain (*Plantago*), saxifrage (*Saxifraga*) or clubmoss (*Lycopodium*). More frequently, however, it is only possible to make identifications to the generic level, for example, birch (*Betula*), willow (*Salix*) or mugwort (*Artemisia*), while in other cases it is not possible to subdivide beyond the family level. Grass (Gramineae) and sedge (Cyperaceae), for example, are rarely taken to the generic or specific level. A pollen diagram therefore consists of a data bank at a variety of taxonomic levels, and this obviously imposes major constraints upon the reconstruction of former plant communities, particularly as some plant families and genera contain elements with markedly contrasting ecological affinities.

Because of these difficulties, the reconstruction of environments and climate from pollen analytical data is often largely dependent on the behaviour of key species within the pollen diagram, whose autecology is reasonably well-known. This method is known as the **indicator species approach** and usually involves taking the distribution of a particular plant and then attempting to fit that distribution to selected climatic parameters. This technique is widely used in palaeoecological studies, but it is not without its problems (Birks 1981; Moore 1980), for it assumes an adequate knowledge of the various factors governing plant distribution and, moreover, in view of the obstacles imposed by the taxonomic imprecision of the majority of pollen data, it requires a certain intuition in the interpretation of pollen diagrams. This may often leave climatic values unquantified and somewhat vague, but it is, nevertheless, perhaps the least unsatisfactory method of analysis currently available.

The reader may be forgiven for thinking that, in the light of what has gone before, difficulties in the interpretation of pollen data render the technique of dubious value to the analysis of Quaternary environments. That this is clearly not the case is demonstrated by the very large number of pollen-based papers that have appeared in the Quaternary literature in recent years and by the remarkable degree of consistency that has been obtained in the results. There is a feeling, however, that in some quarters there has been a somewhat cavalier attitude to the use of pollen diagrams as palaeoenvironmental tools, and the above discussion is designed merely to instil a note of caution without necessarily overemphasising the difficulties. There is no doubt that the method is one of the most powerful at our disposal for reconstructing environments of the past, but as is so often the case, it is perhaps best used in conjuction with other sources of biological data, and it is to these that we now turn our attention.

Diatom analysis

The study of diatoms in Quaternary sediments predates pollen analysis. By the end of the nineteenth century, a considerable amount of work had been undertaken on diatom remains, and while most of these studies had little stratigraphic or palaeoecological value, they laid the foundations for later research on diatom taxonomy. Although not as widely employed in palaeoenvironmental work as pollen analysis, Quaternary diatom remains have proved extremely useful as indicators of local habitat changes, particularly in lake sediments, and also in both shallow and deep-sea marine deposits.

The nature and ecology of diatoms

Diatoms are microscopic unicellular algae that range in length from 5 μm to 2 mm, although the majority of species encountered are in the size range 20–200 μm (Brasier 1980). The living organism secretes a silicious shell or **frustule** which is often compared to a pill-box, as one half of the box (the **valve**) fits over the other and encloses the protoplasmic mass. The wall of the frustule may be a single layer of silica or it may be more complex, consisting of a double silica wall separated by vertical silica slats (Burckle 1978). The frustules are either circular (**centric**) or elliptical (**pennate**) and are perforated by tiny apertures (**punctae**). The arrangement of the punctae is one of the most important diagnostic characteristics of diatoms, although other structural details, including the reticulations, canals, and ribs (Fig. 4.5) are important for separation below the generic

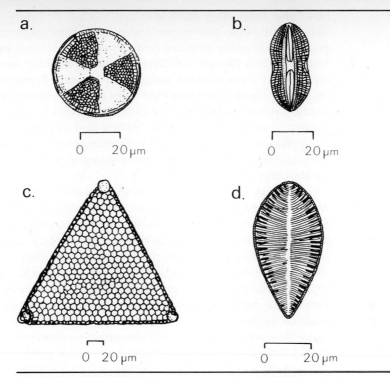

a.

b.

⌐0   20 µm⌐

⌐0   20 µm⌐

c.

d.

⌐0 20 µm⌐

⌐0   20 µm⌐

Fig. 4.5: Drawings of selected diatoms. a. *Actinoptychus undulatus* (Bailey) Ralfs.; b. *Diploneis didyma* Ehrenberg F.; c. *Triceratum favus* Ehrenberg D.; d. *Surirella ovalis* De Brebisson (after van der Werff and Huls 1958–74). Scale in µm.

level, and very careful scrutiny of the diatom valve is necessary for specific identification.

Diatoms are found in a wide range of aqueous environments. They exist in benthonic (bottom dwelling), epiphytic (attached), and planktonic (free-floating) forms, and while all species require light and are therefore limited to the **photic zone** (usually less than 200 m depth), they occupy a large number of ecological niches. In the sea they are found in lagoons, shelf seas and deep oceans; they are common in the inter-tidal zone in estuaries and salt marshes; and they are often abundant in ponds, lakes and rivers. Certain species even live on wetted rocks, in the soil or attached to trees. The freshwater, soil or epiphytic niches are dominated by pennate diatoms, while the centric forms are more common as plankton in marine waters, especially in the sub-polar and temperate latitudes. Marine benthonic habitats, however, are characterised by pennate diatoms. In certain aquatic environments diatoms may be so abundant that their frustules form a distinctive silicious sediment known as **diatomite**. Such deposits are currently accumulating in the sub-Arctic and sub-Antarctic oceans of the world (see below). Diatomite, however, is also found in freshwater lakes (Figs 3.1 and 3.26).

The distribution of each diatom species is controlled by a number of variables including acidity, oxygen availability, mineral concentration and especially water temperature and salinity. Changes in any of these parameters will have a major effect on the structure and composition of the diatom community and major environmental changes will therefore be reflected in fossil diatom assemblages. Further discussion on the morphology, ecology and distribution of diatoms can be found in Werner (1977), Burckle (1978), Brasier (1980), and Barber and Haworth (1981).

Field and laboratory methods

Diatom frustules, like pollen grains, are particularly well-preserved in fine-grained sediments, and are easily damaged or destroyed in coarse-grained materials. Samples for analysis can be taken from vertical sections exposed, in shallow water marine or estuarine deposits or, more frequently, from cores obtained from lakes, shelf seas or deep-ocean floors. In recent years there has been a trend towards more integrated studies in which, for example, pollen and diatoms are analysed from the same core (e.g. Evans 1970; Pennington *et al.* 1972; Walker 1978), an approach which clearly provides a much more secure basis for palaeoenvironmental inference.

In the laboratory, the diatoms are isolated from the sediment matrix. Diatom and pollen counts cannot be carried out on the same slide for during pollen preparation NaOH and HF treatments destroy the diatoms, while severe oxidation in the diatom preparation removes all traces of pollen. Sulphuric and nitric acids are used to clean the diatom frustules, a sodium carbonate solution is employed to disperse silicious deposits while organic sediments are usually oxidised by a mixture of potassium dichromate and sulphuric acid. The residues are mounted on slides and counted under a microscope using phase-contrast illumination and magnification of up to $\times$ 1000. Diatoms are often more abundant than pollen, and a count of 500 or 1000 valves may form the diatom sum. As with pollen grains, identifications are based on type collections, keys and photographs in diatom manuals and catalogues (e.g. van der Werff and Huls 1958–74; Barber and Haworth 1981).

Diatom counts through a stratified sediment sequence are most frequently presented in the form of a percentage diagram, although concentration diagrams (total number of valves per unit volume of sediment) and influx diagrams (based on radiocarbon dates from the profile) have also been constructed (e.g. Battarbee 1973). These diagrams can be divided into diatom assemblage zones either on the basis of visual inspection of the data, or on numerical analysis, and these zones then form the basis for subsequent palaeoecological interpretation. The strengths and weaknesses of these forms of data presentation have already been discussed with reference to pollen analysis. Other methods of data depiction that have been employed by diatom analysts include the tabular format where counts are listed in percentages (e.g. Haworth 1976), and the ratio diagram (e.g. du Saar 1978) in which all the species that belong to a particular chloride ion class – marine, brackish and fresh – are added together, and the sum is then expressed as a percentage of the total number of species in the sample (Fig. 4.6).

Diatoms and Quaternary environments

The problems attendant upon the interpretation of Quaternary diatom data are, in many ways, similar to those discussed above for fossil pollen. Diatom valves are light and easily transported, and thus in estuarine sediments, for example, a complex admixture of marine, brackish and freshwater forms are often encountered. Freshwater diatoms are not uncommon in deep-sea sediments, having been blown in by wind (Parmenter and Folger 1974), or brought in by rivers and subsequently

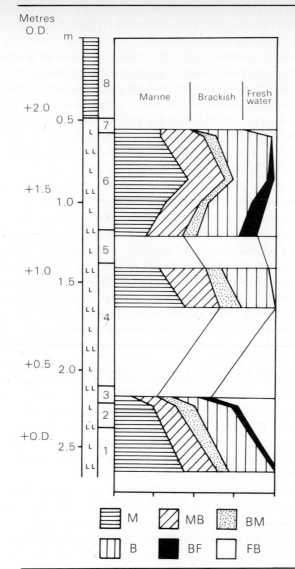

Fig. 4.6: Diatom diagram from Downholland Moss, Lancashire, showing changing proportions of marine (M), brackish (B) and freshwater (F) diatom groups (after du Saar 1978).

redistributed by turbidity currents off the edge of the continental shelf. In the deep oceans, diatoms are often found that have been transported many hundreds of miles from their source. Burckle and Biscaye (1971) for example, discovered that High Antarctic diatoms were being transported as far north as the equatorial waters off the coast of Madagascar.

Selective destruction of diatom remains is another potential error source in the interpretation of diatom assemblages. In the deep oceans, diatoms, along with other marine micro-organisms, are prone to dissolution under pressure at great depths (see below), but in brackish and freshwater situations also, the less robust and weakly-silicified diatom forms will tend to dissolve where conditions are highly alkaline or very acid (Round 1964). In these environments, the diatom death assemblages will be biased towards the stronger and more heavily-ornamented forms. Evidence of reworking of diatom frustules can sometimes be detected where valves are broken, partially dissolved, or demonstrate signs of mechanical abrasion, although 'secondary' diatoms within a body of sediment may not always be recognisable.

A further problem that arises from the use of diatoms in palaeoenvironmental work concerns the uncertainty surrounding the isolation of specific environmental parameters that govern the distribution of individual diatoms. It has often been suggested that certain freshwater diatoms are characteristic of particular thermal environments (cold, boreal, temperate etc), and thus the impression has been given that fossil diatoms can be used as indicators of former climatic conditions. Work by Haworth (1976) and Brugam (1980), however, has shown that most diatoms are, in fact, very tolerant of temperature changes and are much more sensitive to limnological changes, such as fluctuations in lake trophic status, changes in water chemistry and oscillations in lake water levels, than to the effects of climatic changes (such as temperature and rainfall). However, as biological variations are most frequently a consequence of changes occurring in the soils, vegetation and water budget of the lake catchment, and as these are often a reflection of long-term changes in the climatic régimes, diatom assemblages may be useful sources of palaeoclimatic information, particularly where the diatom record is related to other palaeoenvironmental indicators such as pollen and plant macrofossils. This type of approach is exemplified by the work of Brugam (1980) at Kirchner Marsh in Minnesota. Modern diatom assemblages from lakes in Labrador and Minnesota were compared with fossil diatoms from the former lake now occupied by Kirchner Marsh, and these data were assessed in the light of pollen and plant macrofossil evidence. The oldest sediments in the sequence dating from *ca.* 13 000 to 10 200 BP (spruce pollen zone) were found to contain diatoms resembling those from the deep, nutrient-poor lakes in northern Minnesota. The pine pollen zone (*ca.* 10 200 to 9500 BP) had a diatom flora with few modern analogues, but the oak zone (9500 BP to present) showed an initial brief pulse of diatom species indicative of eutrophication, followed by assemblages characteristic of shallow lakes, suggesting a fall in water levels during the relatively short prairie phase from *ca.* 7500 to 5300 BP. Significantly, the plant macrofossil data showed that the shift to shallow water diatoms occurred when the aquatic macrophytes appeared at the site in abundance. The study demonstrates that, although diatom fossils cannot be expected to yield the precise climatic information that can often be obtained from the remains of terrestrial vegetation (pollen, plant macrofossils etc), they do add significantly to the record obtained from these other fossil types by providing evidence on the limnological aspects of climatic change.

Not all of the changes that take place within a lake catchment are climatically determined, however, for different land-uses and land management practices will be reflected in the chemistry of lake waters (Ch. 3) and this, in turn, will affect the nature of the diatom assemblage. Pollen-stratigraphic work on Lough Neagh (the largest lake in the British Isles) in Northern Ireland, for example, showed that during the historic period, the catchment of the lake changed from a predominantly wooded landscape to one of arable and pastoral cultivation (O'Sullivan et al. 1973).

This trend was accompanied by changes in the diatom record from an acidophilous to a predominantly eutrophic flora, with the majority of changes taking place at the beginning of the seventeenth century following the wholesale destruction of forested areas by English and Scottish settlers (Battarbee 1973). Similarly, pollen, diatom and cladoceran analyses from recent sediments in Shagawa Lake, Minnesota, all indicate that significant changes occurred in the biostratigraphy following the arrival of European settlers and the economic development of the basin catchment (Bradbury and Waddington 1973). Clearly, therefore, great care is needed in the interpretation of the diatom record in lake sediments, particularly during the later part of the Flandrian when man's influence on the landscape was becoming increasingly marked.

An area of Quaternary investigations in which diatoms have been used with considerable success is that of sea-level changes. Diatoms are highly sensitive to salinity variations and thus in a transgressive marine sequence, there will usually be a clear change in the diatom stratigraphy from freshwater, to brackish and finally to marine forms. The reverse will, of course, obtain in a regressive sequence. Diatom analyses of littoral sediments are often accompanied by pollen analytical work, and have been used most effectively to unravel the complex history of sea-level changes during the Flandrian around the North Sea basin and on the shores of the Irish Sea along the Cumbrian and Lancastrian coasts (Huddart et al. 1977; Tooley 1978). At Downholland Moss, Lancashire, du Saar (1978) was able to isolate three phases of marine to brackish water conditions, separated by two periods of significantly-reduced marine influence within a 2.5 m section of clays and peats (Fig. 4.6). Three marine transgressions and two significant regressions could, therefore, be established, and during one of those regressive phases, the locality was almost completely isolated from the sea. The great advantage of this type of analysis is that, in association with pollen data, it enables a detailed picture to be reconstructed of the changing position of the shoreline through time. Transgressive and regressive contacts can be located precisely and these can then be dated by radiocarbon.

Diatom remains in shallow water marine and lake sediments, therefore, provide a range of data on local habitat changes which, when carefully analysed along with other fossil evidence, can make a significant contribution to our understanding of both local and regional environmental changes during the Quaternary period. More recently, however, increasing attention has been focused on diatom remains in deep-ocean sediments, and the application of the study of these assemblages in the fields of marine biostratigraphy and palaeoecology. This aspect of diatom analysis is considered in more detail in the last section of this chapter.

PLANT MACROFOSSIL ANALYSIS

An alternative approach to the reconstruction of former vegetation patterns, and hence of palaeoenvironments, is the study of plant macrofossils. Indeed, the investigation of fossil plant remains can justifiably be regarded as one of the earliest branches of Quaternary studies (Dickson 1970), for research was underway into the Quaternary floras of the British Isles as long ago as the 1840s. Similar studies were being undertaken in Denmark and Germany and the results of many of these investigations were synthesised in Clement Reid's remarkable volume *The Origin of the British Flora* published in 1899. This book contained the first clear statement of Quaternary vegetation changes in western Europe, and appeared almost twenty years before the development of pollen analysis as a palaeoenvironmental technique.

The nature of plant macrofossils

Plant macrofossils range in size from minute fragments of plant tissue to pieces of wood that can be measured in cubic metres. They include fruits, seeds, stamens, leaves, megaspores, buds, cuticle fragments and, very occasionally, flowers. They are found in a variety of depositional environments, but most commonly in lacustrine and riverine sediments (especially in fine alluvium), and in acid peats. Occasionally, rich assemblages of plant remains are recovered from soils or sediments at archaeological sites where the fossil remains will include cultivated plants, weeds of cultivation and uncultivated species collected for food, while at the other end of the scale, large tree stumps which are the remnants of former woodland are found in degraded Flandrian blanket peats (Fig. 4.7) and exposed in sections in earlier deposits (Fig. 7.8).

The preservation of fossil plant material is very variable. Wood may

Fig. 4.7: Pine stump (*Pinus sylvestris*) of mid-Flandrian age exposed in degraded blanket peat, Rannoch Moor, Scotland.

survive in recognisable condition for many thousands of years, either in waterlogged sites or in very dry soils in arid environments (Western 1969), but in other situations decomposition may be extremely rapid. Seeds and fruits will survive in most deposits, their resistance to decay reflecting adaptation to withstand periods of dormancy. Some of the very small seeds, however, such as those produced by orchids and by certain members of the heather family (Ericaceae) are rarely preserved. Seeds of the grasses (Gramineae) are also seldom found in the fossil form (Birks and Birks 1980). Near sites of former human occupance, the soils will often contain carbonised remains of fruits and seeds and these will, in general, tend to be highly resistant to decomposition. The leaves of deciduous trees with their delicate structure are highly vulnerable to mechanical breakdown and decomposition and in lake sediments are rarely preserved except as very small fragments. Perfect specimens have occasionally been found, however, in sites where rapid burial in fine-grained alluvial deposits in still back-water situations has ensured perfect preservation (Watts 1978). By contrast, the robust needles from coniferous trees are often abundant as macrofossils, occurring in a variety of depositional situations. Also, leaves of dwarf shrubs from tundra environments such as willow (*Salix*) and bilberry (*Vaccinium*) have been found in a good state of preservation in lake sediments of Lateglacial age at sites in North America (Watts 1967) and Scotland (Birks and Mathewes 1978). Of the lower plants, mosses preserve very well in the macrofossil form, but lichen and liverwort remains are seldom found. In the great majority of cases therefore, Quaternary macrofossil analysis is concerned only with the study of wood, seeds, fruits and mosses, augmented by information provided by a limited number of easily identifiable plant remains such as conifer needles and certain leaves (Watts 1978).

Field and laboratory work

The larger plant macrofossils can be collected in the field from exposed sections or from sediment cores, but in most cases extraction takes place in the laboratory. A variety of techniques is available for the removal of the fossil remains from the sediment matrix, but the majority involve disaggregation of the material (particularly fibrous peats) with either nitric acid or sodium hydroxide solution, followed by careful sieving. Finer lake sediments can usually be broken down simply with the aid of a strong jet of water. During disaggregation, some fruits, seeds or leaves will rise to the surface of the liquid and these can be picked off with a fine paintbrush, while macrofossils can be removed in a similar fashion from the mesh of the sieves. Some fossils such as fruitstones can be kept dry, but others need to be stored in alcohol, formalin or other preserving fluids. The more delicate structures such as translucent leaves and seeds are best kept mounted on microscope slides. Wood, except in the case of some very obvious species (e.g. birch) must either be macerated or prepared in thin section for identification. Keys for the identification of wood sections can be found in Godwin (1956) and Jane (1970).

Most plant macrofossil remains can be examined on a white plate under binocular scanners or low-powered stereo-microscopes. On occasions, however, high-powered microscopy is required, and in recent years, the electron microscope has been employed for the differentiation of closely similar taxa (Connolly 1976). As with pollen analysis, identifications are based on a reference collection of seeds, fruits, leaves, wood, etc from the present flora, on atlases of macroscopic plant remains and, in certain cases, on keys of particular plant families. An excellent review of extraction procedures and identification techniques can be found in Dickson (1970).

Data presentation

The results of plant macrofossil analysis can be presented in a number of different ways. At many sites, particularly where archaeological investigations are being carried out or where a single stratum is being investigated, a simple species list is compiled of all taxa discovered. Where several levels are being examined, a tabular format may be adopted in which the presence or absence of particular plant remains are recorded. Alternatively, the data may be expressed as estimates of abundance using such descriptive terms as rare, occasional, frequent, etc. In this case the results may be depicted graphically (Fig. 4.8), and thus an impression can be gained of change in frequency of taxa through time. In many types of plant macrofossil work, one or all of the above methods of data presentation will be adequate, for the quantitative occurrence of plant macrofossils is often not as significant as it is for pollen, and the greatest

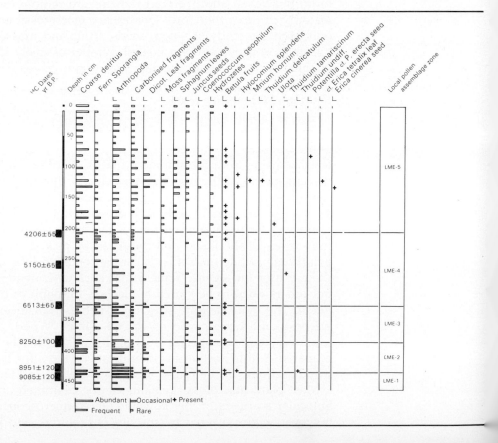

Fig. 4.8: Plant macrofossil diagram spanning the Flandrian from Loch Maree, Wester Ross, Scotland (after Birks 1972).

interest often lies in knowing presence or absence of a species at a particular stratigraphic horizon (Watts 1978).

Certain types of study, however, demand the presentation of plant macrofossil data in quantitative form. One approach is to construct a diagram which is essentially similar to the percentage pollen diagram discussed above in which the plant remains from different levels in a profile are expressed as a percentage of the total number of macrofossils identified from each sample, and these are plotted on a vertical time-sequence. The diagram can then be divided into zones on the same basis as a pollen diagram. A major problem with this type of diagram, however, is that certain macrofossil types (such as seeds of aquatic plants in limnic sediments) will be over-represented and thus the curves for other taxa will be heavily suppressed. Equally, it may be difficult to arrive at a satisfactory macrofossil sum, particularly when a range of different types of plant macrofossil material is present. An alternative approach is to construct a concentration diagram showing the occurrence of total numbers of plant macrofossils at different levels in the profile, and this may be converted to a macrofossil influx diagram showing accumulation per year if a dating framework can be established (e.g. Birks and Mathewes 1978). As with pollen concentration diagrams, however, fluctuations in sedimentation rates and an insufficiently sensitive time-scale can pose serious interpretative problems and, as yet, very few influx diagrams have been constructed for plant macrofossil data.

The interpretation of plant macrofossil data

The majority of plant macrofossils found in a sediment body are derived locally (autochthonous) and are therefore of limited value in the reconstruction of regional vegetational patterns. They do, however, provide data on the composition of former plant communities growing in and around the site of deposition. Important information can often be gained therefore on local hydroseral developments and also on changes in the trophic status of lake waters (see, e.g. Birks et al. 1976). Interpretation of the record is aided by the fact that, unlike pollen grains, a very large number of plant macrofossil remains can be identified to the species level, and therefore many of the problems arising from taxonomic imprecision which often prove so frustrating in pollen analysis are not encountered in the study of plant macrofossils. Moreover, there are some plants such as the rushes (Juncaceae) and poplar (Populus) whose pollen seldom survives in the fossil form, and which are therefore only occasionally represented in the pollen record. The former presence of these types may, however, be detected by their macrofossil remains. The same applies equally to those plants that are very low pollen producers.

Unfortunately, however, the occurrence of plant macrofossils in Quaternary sediments is sporadic. Many deposits, while rich in fossil pollen, are entirely devoid of recognisable plant remains. Other sediments may contain plant macrofossils, but a large quantity of material may be needed in order to produce relatively few fragments of fossil vegetable matter. Moreover, although identification is frequently possible to the specific level, the abundance of diagnostic detail in different fossil remains is very variable, and while there are seeds and fruits that are relatively easy to recognise, the identification of small fragments of achene or epidermis, for example, may require a great deal of work. For these reasons, plant macrofossils may only be worth studying if they are abundant, well-preserved and easily extracted from the sediments in which they occur. Under certain circumstances, however, it may be worthwhile examining the sediments of a site where macrofossils are sparse, if the results will help solve a problem which has been encountered in pollen analysis.

As with pollen analysis, a proper understanding of the origins of a plant macrofossil assemblage is required before palaeoecological inferences can be attempted. In fen or bog sites, macroremains tend to be almost entirely of local origin, apart from a few with very good wind dispersal such as fruits of birch (*Betula*) or seeds of sycamore (*Acer*). Hence, these records will usually be dominated by the stems, leaves and seeds of the peat-forming plants such as *Sphagnum* moss and the cotton grass (*Eriophorum vaginatum*), accompanied by the remains of species such as the bog myrtle (*Myrica gale*) and ling (*Calluna vulgaris*) which are often found growing on bog surfaces. In lake sediments, however, the plant macrofossil assemblage is more diverse, for although locally-derived fossils (particularly those of aquatic plants) will tend to predominate, exotic elements from outside the lacustrine ecosystem may also be present. The proportion of autochthonous to allochthonous fossil material will be determined by such factors as production and dispersal of seeds, fruits, leaves, etc, and mode of sedimentation within the lake basin.

Vegetational productivity varies considerably between species, and even between individuals of the same species, depending upon reproductive strategies (in the case of fruits and seeds) and vegetational response to environmental conditions. Watts and Winter (1966) noted, for example, that woodland shrubs and herbs tended to produce relatively few seeds, while certain trees and annual weedy plants, especially those found on the mud surfaces of lakes with markedly fluctuating water levels, may have a very high seed production. Other factors will also come into play. Many seeds and fruits will not find their way into the fossil record because they are taken for food by birds and animals, while others may be subject to attack by fungal parasites. Thus the seed population available for dispersal is what remains after such predation (Watts 1978).

Plant macrofossil material arrives at the lake surface by a variety of pathways. Wind clearly plays a major role in moving seeds and leaves, but as with the dispersal of pollen, the nature of the vegetation surrounding the lake may have an important limiting effect. Dispersal through a woodland stand is, for example, less efficient than wind transport over bare or open ground. Sheetwash, solifluction and snowbed melts on slopes around the basin catchment may be important processes (Glaser 1981), particularly during periods of reduced vegetation cover and soil instability. Birds and animals may be instrumental in transporting plant material to the lake surface either on or within their bodies. Transport of vegetative remains by inflowing streams may also be important, especially following periods of heavy rain, and will be particularly instrumental in carrying into the lake the seeds and fruits of such waterside species as alder (*Alnus glutinosa*) which are largely dependent on running water for their dispersal (McVean 1953). Finally, there will be the input from vegetation growing around the basin littoral, and from both floating and submerged aquatics within the lake itself.

Once the plant material arrives at the lake surface, its incorporation into the lacustrine sediments is not necessarily straightforward. Some fruits and seeds will become waterlogged and sink immediately, while others may remain afloat for a considerable period of time. Seeds of aquatic species are particularly noted for their flotation characteristics (H. H. Birks 1973) with certain types deriving their buoyancy from a covering of corky tissue, while others such as the white water lily (*Nymphaea alba*) have an aril consisting of a thin cellular bag containing many air bubbles. Those seeds which float for long periods may be washed up on the lake shore and thus disappear from the fossil record. Even those that sink to the lake floor do not necessarily become incorporated into the sediments immediately, but may be moved by turbulence or by bottom-living creatures some distance from their original point of deposition before becoming incorporated into the lake floor.

Watts (1978), in an extremely useful review of the problems of interpretation of plant macrofossil assemblages, has proposed the following general model for the recruitment of vegetative material to lake sediments (Fig. 4.9). The model is designed for seeds, but also has application to other types of plant macroremains.

1. After predation and parasitism has been accounted for, the surviving seed population is dispersed.
2. Wind, particularly strong or violent winds, carries seeds to lake surfaces. This process is most effective in the case of trees where seeds are launched from a height. Alternatively, some seeds are carried to lakes by streams and surface runoff during periods of high rainfall.

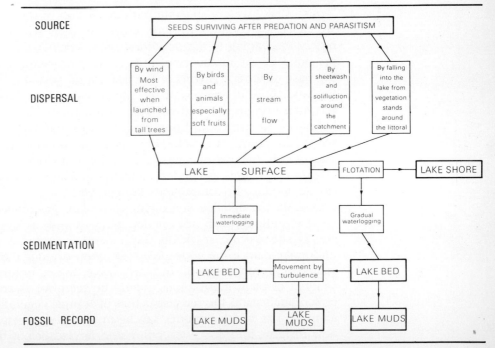

Fig. 4.9: General model of seed recruitment to lake sediments (modified after Watts 1978).

Birds and animals play a minor role in transporting soft fruits.
3. Seeds that are not waterlogged remain afloat and are carried into shallow water by wind and wave action.
4. Seeds of some species become waterlogged preferentially and sink, while others succeed in resisting waterlogging altogether and are washed ashore.
5. The waterlogged seeds are moved along the lake bed by turbulence until they settle out by gravity in the coarser marginal sediments or are trapped near the margin by submerged plants.

The model is, at present, speculative and is designed to stimulate experimental work involving trapping such as that carried out in a recently deglaciated area of Norway (Ryvarden 1971). One of the very few studies on modern macrofossil assemblages in lake sediments by H. H. Birks (1973) demonstrated that macroremains were deposited predominantly within the zone of rooted aquatics, and that the fossils consisted almost entirely of plant material from the aquatics themselves. The occurrence of upland plants from the basin catchment was both accidental and unpredictable. Clearly, further studies of this type are required if we are to gain a proper appreciation of the origins of the macrofossil assemblages in Quaternary lake sediments.

Although the foregoing discussion has been concerned with lake sites, many of the points apply equally to plant macroremains in riverine sediments. In fluvial deposits, however, a further complicating factor is the often frequent occurrence of plant macrofossil material from different time periods that has been incorporated into a single assemblage. This is a problem common to all fossil assemblages and has already been considered with reference to reworked or secondary pollen. Mixed assemblages of plant fossil material are occasionally encountered in lake sediments where, for example, there have been marked fluctuations in the levels of lake waters and intermittent erosion of the shoreline, but they are especially common in river terrace deposits where they can often be highly misleading to the unwitting palaeoecologist. A good example is provided by the discovery of nuts of the hornbeam (*Carpinus betulus*) at the base of terrace gravels of the last glacial period in the Fens of eastern England. Careful analysis of the site revealed that the fossils were of secondary origin and had been derived from sediments deposited at the height of the preceding Interglacial (Sparks and West 1972). Unlike secondary pollen, it is not always possible to detect reworked plant macrofossil material by signs of physical and chemical deterioration, and often a meticulous evaluation of the assemblage is necessary in order to isolate such exotic components (Spicer 1981).

Overall, the detailed investigation of plant macrofossils requires many more man-hours than does pollen analysis and, as the results tend to be largely site-specific, markedly fewer plant macrofossil studies have been undertaken since the widespread adoption of pollen analysis as a palaeoenvironmental tool. In many ways, this is unfortunate, for the analysis of plant macrofossils should be seen not as an alternative to pollen analysis in the reconstruction of former vegetation, but as an adjunct. The two techniques can be mutually supportive and can provide new insights into flora, vegetation and landscapes of the Quaternary, as well as providing information on local site developments. When used in conjunction therefore, plant macrofossil analysis and pollen analysis offer a more secure basis for palaeoecological inference than either technique used in isolation.

ANALYSIS OF SMALL ANIMAL REMAINS

Fossil insect analysis

Fossil insects are often abundant in a wide range of Quaternary deposits. Typically these include sediments that accumulated in ponds or near lake margins, in backwaters of rivers, in peats or indeed in any depositional environment where conditions were suitable for the preservation of plant debris. The insect remains are usually striking because of their brilliant and often irridescent colouring of blues, greens and golds, and consequently their presence in Quaternary sediments has long been a source of fascination to laymen, naturalists and entomologists alike. Over the last twenty years, however, largely through the work of Professor Shotton, Dr Coppe and their colleagues at the University of Birmingham, the study of fossil insects has provided an exciting new method of investigating environmental changes during the Quaternary period.

Insect remains in Quaternary sediments

Many different orders of insects have been found in Quaternary deposits. These include bugs (Hemiptera), two-winged flies (Diptera), caddis flies (Trichoptera), bees, ichneumons etc (Hymenoptera), dragon flies (Odonata) and beetles (Coleoptera), while members of the ant family (Formicidae) have also been investigated in the fossil form. Although some of these have provided useful palaeoenvironmental information (Osborne 1972; Buckland 1976), many are of limited value due to the delicate nature and poor preservation of certain fossils (e.g. Diptera), to the relative scarcity of fossils of orders such as Odonata, to the difficulties encountered in identifying some groups such as the midges (Chironomidae) beyond the generic level, and finally to the imperfect state of knowledge on the modern ecological affinities of many insect types. To a very great extent, these difficulties do not arise with the Coleoptera, and this particular insect order has tended to dominate the interests of Quaternary entomologists. As a consequence, the remainder of this section is concerned very largely with fossil beetles as palaeoenvironmental indicators.

Coleopteran remains are often abundant in Quaternary deposits, usually outnumbering all other insect types. The chitinous exoskeletons are highly robust and contain sufficient structural detail to permit many fossil types to be identified to the species level. They have been collected and studied by entomologists in many parts of the world and there is now a considerable body of accumulated knowledge in the form of atlases and monographs upon their distribution and ecological associations. The order Coleoptera is one of the largest in the insect kingdom and within Britain alone over 3800 named species are known to occur (Coope 1977a). They occupy a very wide range of habitats, having colonised almost every terrestrial and freshwater situation, some even having been found within the inter-tidal zone (Coope 1977b). Yet, many species are distinctly *stenotypic* in that they show a very clearly marked preference for particular environments, and it is this characteristic above all others which makes the Coleoptera such valuable palaeoecological indicators.

Laboratory methods

The extraction of fossil insect remains from the sediment matrix invariably takes place in the laboratory. Occasionally, the insect fragments can be removed by hand where, for example, they occur on marked bedding planes in clays or felted peats. More frequently, however, flotation techniques are required (Coope 1968a). The most commonly used method involves disaggregation of the sediment using water or, where necessary, a caustic soda solution, followed by careful sieving. The residues remaining on the sieves are then mixed with paraffin (kerosene) and the insect remains along with some plant macrofossils float to the surface. The floating fraction is decanted, washed and sorted under a low-power binocular microscope. The insect fossils may range in size from a few millimetres to several centimetres and very careful examination may be necessary at this stage in order to ensure that very small specimens are not overlooked and that the subsequent collection is not biased heavily in favour of the more conspicuous species (Coope 1961). The fossil remains are then gummed onto cards or stored in alcohol and examined under a microscope. In recent years, as in other fields, increasing use has been made of the electron microscope where ultra-detailed study has been required. Since the entire insect fossil is rarely recovered from the sediment body, keys to identification are of lesser value, and therefore reference must always be made to a collection of modern species types, preferably those displaying component parts of the complete carapace as discrete entities. Some parts of the fossil possess few diagnostic features, but in many cases, the heads, thoraces, elytra (wing covers), and often the genitalia (particularly in the male) display a wealth of useful characters that enable specific determinations to be made. A review of the techniques and problems encountered in identification is provided by Coope (1970a).

Data from fossil insect analyses are usually presented in the form of a species list showing numbers of individuals occurring within a particular sample. Occasionally, further information is provided on the parts of insects that have been recovered, for example heads and elytra. The data are not usually presented in diagrammatic form.

Coleopteran analysis and Quaternary environments

Coleoptera exhibit a number of characteristics which make them one of the most valuable components of the terrestrial biota in the reconstruction of Quaternary environments. Not only are they relatively abundant in a wide range of deposits, but they appear to combine both evolutionary and physiological stability with a sensitivity to climatic change that is seldom found in the plant or animal kingdoms. In addition, they frequently display a marked preference for very restricted environmental conditions.

The fact that speciation does not seem to have taken place to any great extent during the Quaternary period is extremely important. So far, the only unequivocal evidence for evolutionary change in Coleoptera has been found in fossils of lower Pleistocene age from Alaska (Matthews 1974), and even in these cases, the differences are so small as to be of a comparable order of magnitude to present-day racial differences between individual species (Coope 1977b). Indeed, evidence from Arctic Canada suggests that the last major phase of evolution at the species level took place as long ago as the upper Miocene period (Matthews 1976). In Britain, the very close similarity between skeletal elements of living Coleoptera and the very large number of skeletal remains that have so far been examined have led to the conclusion that this particular insect order has maintained a remarkable morphological stability throughout at least the upper Quaternary, and most probably during the entire period. It would seem 'therefore' that there are good grounds for believing that the British Quaternary fossil Coleoptera represent exactly the same species as defined by living assemblages (Coope 1977a).

Equally significant is the fact that the ecological affinities of most coleopteran species do not seem to have changed to any great extent during the Quaternary. This is obviously more difficult to establish, but the available evidence tends to suggest that, in the great majority of cases, species of beetles are found in similar associations in both fossil and living assemblages. Moreover, independent evidence of a palaeobotanical or geological nature indicates that most fossil species were associated with similar types of environment to those that they occupy today. It does seem, therefore, that physiological stability in Coleoptera accompanied morphological constancy throughout the greater part of the last two million years. There are, however, exceptions to the rule. Coope (1977a) notes, for example, that *Hypnoidus rivularis* (Gyll) is a species which now has its southern limit of distribution across Fennoscandia at about 60 °N, and yet is found at a number of sites in Britain, including the Lateglacial Interstadial of the Windermere profile, in association with species of a more temperate aspect. On the other hand, *Timarcha goettingensis* (L) is today widely-distributed in Europe south of latitude 60 °N, and yet is encountered in Middle Devensian deposits at Upton Warren in central England in association with many species now found in tundra environments. In both cases, the beetle species may well have changed their ecological tolerances, and could therefore be misleading as palaeoenvironmental indicators if found in isolation. In most Quaternary insect assemblages, however, careful examination of the total assemblage will usually identify these aberrant species.

On the regional scale, the fundamental factor that governs the distribution of all insect types is climate, of which thermal conditions are by far the most important. Modern distribution maps (Fig. 4.10) show that the geographical range of many species of beetle corresponds with definable climatic zones, and especially with summer temperature parameters. Those insect species whose distributions are narrowly

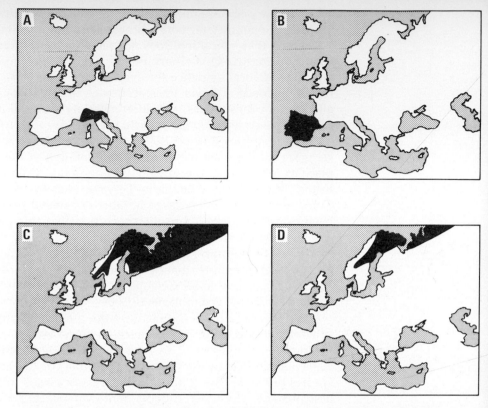

Fig. 4.10: Present day distributions of Coleoptera found in Lateglacial deposits at Glanllynnau, north Wales. A. modern world distribution of *Asaphidion cyanicorne* Pand.; B. modern world distribution of *Bembidion ibericum* Pioch.; C. modern European distribution of *Helophorus sibericus* Mot.; D. modern European distribution of *Boreaphilus henningianus* Sahlb. (after Coope and Brophy 1972).

restricted are termed *stenotherms* while those with a broader climatic range are considered as *eurytherms* and are of lesser importance in palaeoclimatic reconstructions. Clearly there are problems in utilising this type of data as the basis for inferences about former climatic conditions, particularly as it can never be proved that an insect species has colonised the entire climatic range to which it is suited. On the other hand, Coope (1977b) has argued that the enormous scale of the changes in geographical distribution of species in response to climatic change during the Quaternary, and the rapidity of many of these changes, suggests that Coleoptera would have been able to colonise a very wide range of habitats relatively quickly. In the majority of cases, therefore, there would seem to be good grounds for assuming that where a coleopteran distribution coincides with a climatic boundary, that border marks the geographical range of a species.

Two examples will serve to demonstrate how this approach can be used to make inferences about Quaternary climatic conditions. A number of insect assemblages of interglacial age have been described from southern Britain, including that from the type site of the last (Ipswichian) interglacial at Bobbitshole in Suffolk (Coope 1974a). That assemblage contained twenty-one named species of Coleoptera, all of which now have a markedly southern European distribution, with six now being extinct from the British Isles. The ecological affinities of the insect fauna suggest a climate some 3 °C warmer than that of the present day in south-east England, with average July temperatures of around 20 °C. This temperature estimate corroborates the qualitative statements of West (1957) and Phillips (1974), both of whom suggested, on the basis of pollen and plant macrofossil evidence, that summer temperatures during the Ipswichian Interglacial were rather warmer than those prevailing at present. In this case, therefore, the palaeobotanical inferences and palaeoentomological evidence are in agreement.

An instance where such close agreement is not found occurs at the beginning of the Lateglacial in Britain. The Lateglacial period spans the time interval from *ca.* 14 000 to 10 000 BP and has traditionally been divided by palynologists into pollen zones I, II and III, the first and last being interpreted as cold phases with a flora of steppe and tundra conditions. The intervening zone II was believed to represent a warmer episode as indicated by the expansion of woody plants, notably juniper (*Juniperus*) and birch (*Betula*). Coleopteran analyses from a large number of sites have shown, however, that the most thermophilous fauna (which frequently includes species whose range falls well to the south of Britain at the present day) was always found in zone I, while pollen zone II was characterised by progressively less warmth-loving assemblages. Climatic deterioration continued into zone III as reflected by the appearance of many northern species that are now extinct in the British Isles (Coope 1970b). This apparent discrepancy between the botanical and entomological evidence (Fig. 4.11) seems to be explicable largely in terms of different rates of response of plants and insects to rapid climatic amelioration (Coope *et al.* 1971). The more mobile thermophilous Coleoptera were able to respond immediately to a thermal improvement at the beginning of the Lateglacial, but by the time birch-woods had become established in southern Britain, the 'climatic optimum' of the warm phase had already passed, and climatic deterioration was once again beginning to force the treeline southwards. During late pollen zone II and pollen zone III, both plants and insects present a consistent picture of a climate of arctic severity (Ch. 7).

Although climate is the dominant factor governing regional distributions

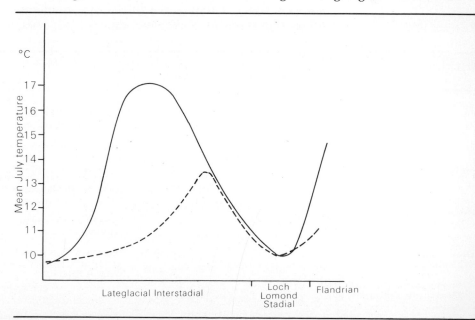

Fig. 4.11: Diagrammatic illustration of the variations in mean July temperature in lowland Britain during the Lateglacial, inferred from fossil Coleoptera (solid line) and fossil pollen data (dashed line) (after Coope 1970b).

of insects, any insect assemblage will contain species from a variety of different local habitats. Botanical factors, soil conditions, micro-climatic variations, hydrological and chemical fluctuations will all restrict the distribution of insects on the local scale (Coope 1967). To a palaeoecologist who is interested in the environmental history of a particular site, it is therefore important that each of these habitats is identified and, as far as possible, quantified. This is especially true in the case of the archaeologist who may be reliant on fossil insect remains to interpret structures such as grain stores, tanneries, sewers, refuse pits, stables etc. Of considerable importance here is the need to establish the source of the insect remains comprising the assemblage. Kenward (1975a, 1976) has discussed the difficulties involved with this type of work and has noted that, although most assemblages contain a large proportion of insect remains that originated from near the point of deposition, they often contain an element of '**background fauna**' composed of insects that have flown to the site, or that have been derived from the regurgitated pellets or faeces of birds (Table 4.5). As a consequence, single species or faunas consisting of a group of individual species may provide misleading information about former local environments. Kenward has therefore suggested that, because of the complexities of some insect death assemblages, reliable palaeoecological inferences can only be made on the basis of large and diverse insect faunas, with fifty or so species as a minimum, that form associations characteristic of particular habitats. Although these problems are perhaps not quite so acute in those aspects of Quaternary entomological work where the principal concern is with macroscale climatic changes, they do provide a salutary warning about the injudicious use of single indicator species in palaeoecological studies.

A further problem in archaeological work arises where man has created a range of wholly artificial environments, for some ecological changes may have occurred in certain insect species as they adapt to these new biotopes. Many natural and artificial environments, while being very different in terms of macroenvironment, may contain microclimatic niches

Table 4.5: The abundance of Coleoptera and Heteroptera of various habitats in an assemblage from a modern drain sump, and the availability of the habitats in the immediate surrounding area (after Kenward 1976).

Ecological grouping	Number of individuals	Percentage of total fauna	Availability of habitat
Aquatic and aquatic marginal	9	3.5	Not recorded within 250 m
Open ground	12	4.5	Some habitat for most species within 10 m
At roots of low vegetation	34	13.0	Scattered isolated plants present within 30 m
Phytophages	28	11.0	Hosts of some recorded within 100 m, but rare
Rotting plant matter			
Primarily	38	15.0	Certainly absent within 10 m, probably some accumulations within 250 m
Facultative	72	28.0	
Herbivore dung			
Primarily	4	1.5	Absent within 250 m; probably very rare within 1 km
Often in dung	21	8.0	
Dead wood	22	8.5	Present within 2 m
Synanthropic	6	2.5	Study area entirely dominated by man – no natural habitats
Often in association with man	57	22.0	
Often in areas disturbed by man	259	100.0	

that are very similar, and in some cases that are identical (Buckland 1976). Moreover, it would be expected that, in those insects that are markedly *synanthropic* (living in close association with man) such as the grain weevil (*Sitophilus granaricus*), slight physiological changes may have occurred to adapt to these artificial habitats. Although in most cases the general archaeological interpretation will not be affected, care clearly needs to be exercised in the palaeoecological inferences that are drawn from archaeological sites based on the known ecological affinities of modern insect species.

These problems notwithstanding, fossil insect assemblages are one of the most valuable sources of evidence at our disposal for inferring former ecological conditions. Few elements within the Quaternary biota possess the flexibility to provide, on the one hand, palaeotemperature estimates for the interglacial stages, while at the other end of the scale, furnish information on the extent of infestation of furniture beetles in Roman building timbers (Osborne 1971)!

Molluscan analysis

Mollusc shells, particularly those from freshwater or terrestrial deposits, are some of the most common fossil remains in Quaternary sediments, and consequently have a long history of investigation. As in other branches of palaeontology, much of the early work in the eighteenth and nineteenth centuries was concerned with taxonomy-and, by comparison, little consideration was given to the palaeoecology of molluscan assemblages. In the late nineteenth and early twentieth centuries, however, workers in Britain such as A. S. Kennard, B. B. Woodward and F. W. Harmer began to utilise molluscs as palaeoclimatic indicators, and also as a means of dating geological events, tasks for which we now know them to be manifestly unsuited (Kerney 1977a). The increasing use of pollen analysis as a palaeoenvironmental technique led to a gradual decline in interest in molluscan studies in the period before and after the Second World War, and the modern phase of molluscan investigations did not begin until the 1950s when B. W. Sparks of the University of Cambridge finally placed molluscan analysis on a secure foundation by adopting, for the first time, a quantitative approach to the study of shell-bearing deposits (Sparks 1957, 1961; Sparks and West 1959). This approach, developed initially for the investigation of non-marine Mollusca, has since been applied to the study of marine shells (e.g. Norton 1967), and has enabled a considerable body of data to be assembled over the last twenty years on local environmental and climatic changes during the Quaternary period.

The nature and distribution of molluscs

Molluscs are invertebrates in which the soft parts of the body are generally enclosed within calcareous shells. The two principal groups as far as the palaeoecologist is concerned are the **Gastropoda** (snails) or **univalves** which usually possess a single spiral or conical shell (although in the case of the slugs the shell is reduced to an internal remnant), and the **Bivalvia** (mussels and clams) in which the animal possesses two valves. Other groups include the **Scaphopoda** (tusk-shells) and the **Cephalopoda**. The latter are the most highly-organised of the Mollusca and range from *Nautilus* which has a large calcareous, external, coiled shell, through the squids which have only a thin horny vestige of a shell embedded in the mantle, to the octopuses which have no shell at all (Buchsbaum 1948). Some molluscs breathe by gills (**Prosobranchs**) and are mainly aquatic, while others such as the snails and slugs breathe by a rudimentary lung (**Pulmonates**) and although they can live in water, are primarily terrestrial. The crystalline form of calcium carbonate in the shells of most molluscs is pure aragonite, although the internal shells of certain slugs, are composed of calcite. In both cases, the shells are subject to little change either in their crystalline or chemical composition following the death of the organism, and are therefore usually referred to as **subfossil** rather than fossil (Evans 1972).

Land and freshwater Mollusca are extremely useful palaeoecological tools because of their wide distribution and preservation in a variety of deposits. They show a marked preference for habitats that possess sufficient lime for building their shells, although they are found not only in chalk and limestone regions, but also in calcareous drifts in colluvial deposits, in cave earths, in loess and on coastal dunes and beaches. Molluscs do occur in non-calcareous environments for some species (e.g. *Zonitoides excavatus*) are strongly calcifuge, and are found only in non-calcareous regions. In such areas, however, the number of species are more restricted, shells are often thinner and less well-preserved, and weathering and leaching in acid environments is more likely to lead to shell dissolution. In general, the richer the base status of the locality, whether it be land or freshwater, the richer is the fauna, and molluscan remains would be expected to be discovered in a wide range of river, marsh, lake, woodland and open-land sediments (Sparks 1961). Terrestrial molluscs are also found in a variety of archaeological deposits including

Fig. 4.12: The 'Red Crag' deposits of eastern Suffolk and Essex, England. These Early Pleistocene marine deposits consist largely of shell fragments (here both gastropods and bivalves are shown) in well-formed ripple bottomsets and foresets.

soils, ditch, pit and well sediments, ploughwash and other colluvial deposits, and in occupation horizons and building debris (Evans 1972).

Marine molluscs occupy a great range of ecological niches from pools and rock outcrops in the inter-tidal zone to the deeper waters off the edge of the continental shelf, although they are seldom found at depths greater than 1 km. The gastropods, scaphopods and bivalves are principally benthonic and sedentary in life and, upon the death of the individual, often become fossilised *in situ* to form autochthonous death assemblages. On occasions, however, shells may be exhumed and moved away to form a transported or allochthonous death assemblage (Norton 1977). Marine Mollusca are often well-preserved in Quaternary sediments (Fig. 4.12), as they are less frequently exposed to sub-aerial weathering agencies than their freshwater or terrestrial counterparts, although those from the inter-tidal zone may be heavily fragmented by wave action.

Field and laboratory work

Although molluscan remains can be collected from open sections in the field by hand, they are best extracted under laboratory conditions. Bulk samples from sections or from cores are air dried and immersed in water, a small quantity of H_2O_2 being added if there is organic material present. The froth, which includes the snails, is then decanted through a 0.5 mm sieve. The process is repeated several times until no more snails are present in the froth, and the residual slurry is then poured into a second 0.5 mm sieve. Both sieves are dried and the dried residues passed through another set of sieves (1 mm, 710 μm, 2411 μm) for ease of sorting. Molluscan remains can either be removed by hand, or with the aid of a moistened brush under low-powered binocular microscope or scanner. Identifications are always based on type collections of modern shell material, and on the numerous molluscan reference works that are now available. In many respects, identification of molluscan remains is not as difficult as in other branches of palaeontology. This is particularly true of land and freshwater molluscs, for which not only is there an extensive taxonomic literature, but in Britain at least, the total number of Late Quaternary species both living and extinct does not exceed 200 (Evans 1972). Nevertheless, identifications may be complicated by the fact that many molluscan subfossils have been damaged by mechanical abrasion, others have their diagnostic characteristics masked by carbonate overgrowths, while others may be very highly comminuted. In certain cases, specialised techniques allow even small fragments to be identified. For example, the marine genera *Mytilus*, *Modiolus* and *Pinna* have shells which possess a characteristic crystal structure that can be recognised with a high-powered microscope (Norton 1977). Similarly, differences in shell microsculpture are often diagnostic of land Mollusca, and specific determination of fragmentary remains can be made using scanning electron photomicrographs (Preece 1981).

The results of molluscan analyses can be presented in a variety of ways. Some workers prefer the use of species lists, perhaps using symbols to depict the frequency of occurrence (+ = presence; * = common; 0 = abundant, etc). This type of tabular format is usually adopted where a limited number of samples has been taken from a profile and where a full

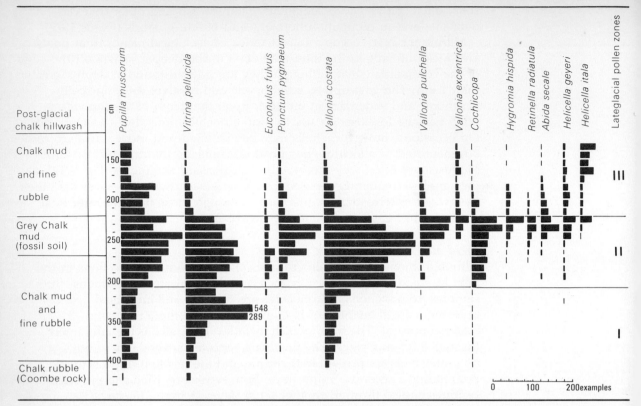

Fig. 4.13: Lateglacial molluscan histogram from Oxted, Surrey, showing the absolute abundance of species per 2 kg sample (after Kerney 1963).

quantitative assessment of the change in species composition over time is required. It is often employed in studies of marine Mollusca (e.g. Peacock *et al.* 1977, 1978). More frequently, however, the data are depicted graphically, either in histogram form for single samples or, where a sequence of sediments is being investigated, on a diagram which has the vertical axis showing the depth below ground datum, and the horizontal axis the number of species plotted. The results can be presented in terms of relative abundance or absolute numbers (Fig. 4.13), in both cases the histogram bars being drawn proportional to the thickness of the sampled horizons. Once constructed, the diagram can be divided into **molluscan assemblage zones** which allow further palaeoecological inferences to be made. These zones will initially be of local significance, but may be extended (as in pollen analysis) to form a zonation scheme that has regional applicability, as has been demonstrated by Kerney (1977b) and Kerney *et al.* (1980) for southern Britain (Table 4.6).

Molluscan analyses and Quaternary environments

Non-marine Mollusca.   Mollusca possess a number of advantages over other fossil groups that have been used in the reconstruction of Quaternary environments. First, specimens can nearly always be identified to species level, and therefore more meaningful palaeoenvironmental conclusions can be drawn. Secondly, shells are present in oxidised sediments (e.g. slopewashes, loess, tufa) which generally lack other fossil remains such as pollen or Coleoptera. Thirdly, specimens are large enough to be recognised in the field (see Kerney and Cameron 1979) and

^{14}C dates BP for mollusc zones (Kent)	Mollusc zones (Kent)	Lateglacial and postglacial Mollusc Assemblage Zones (Kent)	Approximate pollen zone equivalents
	f	Open ground fauna. As (e), but with appearance of *Helix aspersa*	VIIb–VIII
	e	Open ground fauna. Decline of shade-demanding species (e.g. *Discus rotundatus*). Re-expansion of *Vallonia*	
4 540±105(BM-254)	d²	Closed woodland fauna. Expansion of *Spermodea, Leiostyla, Acicula*	VIc–VIIa
7 500±100(ST-3410)	d¹	Closed woodland fauna. Expansion of *Oxychilus cellarius*	
8 470±190(Q-1425)	c	Closed woodland fauna. Expansion of *Discus rotundatus*	V–VIb
8 980±100(St-3411)	b	Open woodland fauna. Expansion of *Carychium* and *Aegopinella*. *Discus ruderatus* characteristic	
9 305±115(St-3395)			
9 960±170(Q-1508)	a	As (z), but with decline of bare soil species (notably *Pupilla*) and corresponding expansion of catholic species. Appearance of *Carychium, Vitrea, Aegopinella*	IV
11 900±160(Q-618)	z	Open ground fauna. Restricted periglacial assemblage with *Pupilla, Vallonia, Abida, Trichia*	II–III
11 934±210(Q-463)			
13 180±230(Q-473)	y	Open ground fauna. Restricted periglacial assemblage dominated by *Pupilla* and *Vallonia*	I

Table 4.6: Regional zonation scheme for the Lateglacial and Flandrian periods in southern Britain, based on molluscan faunas (after Kerney 1977b).

a very good idea of the general palaeoecology can generally be gained. This may help to determine where samples should be taken for other fossil groups. Fourthly, much is known about their present-day ecology and geographical distributions (see e.g. Ellis 1978; Kerney 1976).

Mollusca can be used in a number of ways to make inferences about former local habitats and about environmental change through time. One approach is to divide species into groups as a basis for palaeoecological reconstructions. Sparks (1961), for example, recognised four groups of freshwater mollusca:

1. A **'slum' group** composed of individuals tolerant of poor water conditions, ephemeral or stagnant pools with considerable changes in water temperature, for example, the water snails *Lymnaea truncatula* and the small bivalve *Pisidium casertanum*.

2. A **'catholic' group** comprising Mollusca that will tolerate a wide range of habitats except the worst slums, for example, *Lymnaea peregra* and *Pisidium milium*.

3. A **ditch group** that includes species often found in ditches with clean or slowly-moving water and abundant growth of aquatic plants, for example, *Valvata cristata* and *Planorbis planorbis*.

4. A **moving water group** composed of molluscs commonly found in slightly larger bodies of water such as moving streams and larger ponds where the water is stirred by currents and wind. Typical species are *Valvata piscinalis* and *Lymnaea stagnalis*, together with the larger freshwater bivalves.

Land mollusca can similarly be divided into four groups:

(a) **Marsh** and associated species, for example, *Vallonia pulchella*;

(b) **Dry land** species, for example, *Vallonia costata* and *Pupilla muscorum*;

(c) *Vallonia* spp. which may sometimes be separated as indicating open unwooded conditions;

(d) **Woodland** species.

Fig. 4.14: Local habitat changes shown by ecological groups of molluscan faunas in the Ipswichian Interglacial deposits at Histon Road, Cambridge. The diagram at the left-hand side shows the ratio of freshwater to land species. The changes most probably reflect a meandering stream shifting across the floodplain (after Sparks 1961).

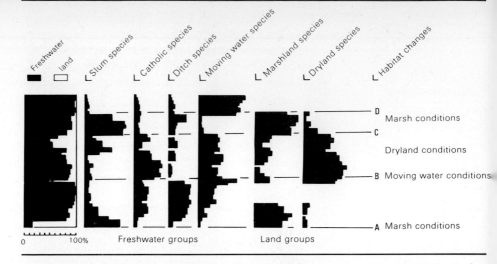

Fig. 4.14 shows how this type of subdivision can be employed to reconstruct local environmental changes, in this case in the Ipswichian Interglacial deposits at Histon Road, Cambridge.

Groups of Mollusca have also been used to infer former climatic conditions, and Sparks (1961, 1964, 1969) has suggested four climatic or distributional groupings based mainly on the northern limits in Sweden and Finland:

(a) Species reaching to, or almost to, the Arctic Circle;

(b) species reaching to 63 °N;

(c) species reaching 60–61 °N, *i.e.* approximately the oak limit at present;

(d) species reaching only the very south of Sweden or confined to the European mainland.

These broad groupings have been used to provide an impression of climatic trends during the later part of the last interglacial at sites in southern and eastern England (e.g. Sparks and West 1959).

An alternative approach to palaeoenvironmental reconstruction based on molluscan evidence is to use single indicator species. West and Sparks (1960), for example, have shown how a change in frequency of *Hydrobia ventrosa* in the Ipswichian Interglacial deposits at Selsey in Sussex can be used to infer changes in water quality, in this case reflecting an increasingly brackish influence as sea-level rose. Species of Mollusca have also been used to infer regional climatic changes. Kerney (1968) has noted that three particular species of snails, *Pomatias elegans*, *Lauria cylindracea* and *Ena montana* were all common in Britain during the Flandrian climatic optimum (*ca.* 8000 to 5000 BP), yet all now have a more southerly distribution in Europe and are restricted to southern and western Britain. *Ena montana* requires summer warmth, while *Pomatias elegans* and *Lauria cylindracea* are essentially oceanic and cannot tolerate low winter temperatures. Kerney therefore suggested that, on this evidence, both summer and winter temperatures in Britain are now lower than they were during the climatic optimum of the Flandrian. Similarly, Sparks and West (1970) have used the occurrence of molluscan remains of *Vallonia enniensis*, *Clausilia pumila* and *Corbicula fluminalis* in the Ipswichian deposits at Wretton in Norfolk to infer a more southern and continental climatic régime during the last interglacial in Britain, with summer temperatures perhaps 2 °C higher than those of the present day.

Fossil Mollusca have been widely employed as indicators of regional landscape change, particularly of the nature and extent of the former vegetation cover. In Britain, studies have been carried out on both interglacial (e.g. Kerney 1971a) and cold climate deposits (e.g. Briggs and Gilbertson 1980), but most attention has been focused on molluscan assemblages in sediments of Lateglacial and Flandrian age (e.g. Kerney *et al.* 1964; Evans *et al.* 1978; Kerney *et al.* 1980). Both land and freshwater snails have also proved to be extremely useful in archaeological investigations, allowing inferences to be made about former occupation sites (Evans 1972), man-induced vegetation changes (Spencer 1975) and agricultural activities (Preece 1980). In this respect, Mollusca have proved to be particularly valuable in chalkland areas where archaeological evidence is often abundant, but where there is little scope for pollen analysis due to the general absence of peats or limnic sediments. The Pink Hill site near Princes Risborough to the west of London provides a good example of how changing frequencies of molluscan faunas with distinctive ecological affinities can be used to reconstruct land-use changes (Fig. 4.15). A section through a series of loams and colluvial deposits revealed evidence of an initial pre-Iron Age woodland clearance, followed by two 'open-country' phases of Iron Age and Romano-British agricultural activity, and finally the establishment of open grassland (Evans 1972).

Although these examples serve to demonstrate the various ways in

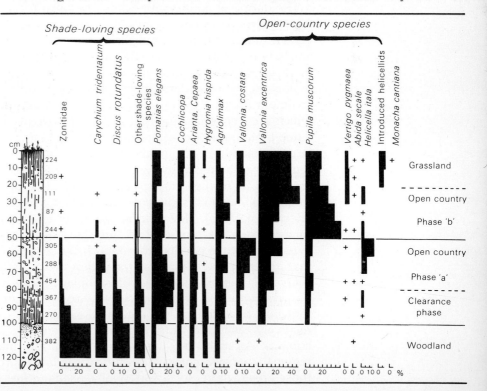

Fig. 4.15: Mollusan assemblages of the Pink Hill (near Princes Risborough, Buckinghamshire, England) dry valley fill (after Evans 1972).

which non-marine Mollusca can be used in Quaternary studies, as in other aspects of palaeoecology, problems are encountered in the interpretation of the fossil evidence. Many molluscan assemblages are mixed and contain fauna derived from a variety of ecological niches, and possibly also from strata of different ages. Such mixed assemblages clearly pose problems in environmental reconstruction. These difficulties may be further compounded by the fact that differential destruction of mollusc shells, either by natural or human agencies, can bias assemblages towards dominance by the more robust forms. In many deposits (e.g. in fluvial sediments that accumulated during cold stages) molluscan assemblages may be dominated by species whose environmental ranges are not strongly defined, in which case the fossil remains are of little value in palaeoecological work. In other circumstances, it may be difficult to establish the extent to which a change in the molluscan fauna reflects local habitat changes or regional climatic fluctuations. On the macroscale, most molluscan distributions are limited by climatic parameters, but Mollusca will often survive in favoured habitats where the microclimate allows, for example on sheltered south-facing slopes, for some time after the regional climate has deteriorated. Moreover, Mollusca migrate relatively slowly so that there may well be a time-lag in the record between regional climatic amelioration and molluscan migration – very similar, in fact, to that which is often noted in the pollen record. That being the case, it is clear that molluscan evidence can only provide fairly generalised information on former climatic conditions, and cannot be relied upon as a source of precise information on former temperature and precipitation variations.

The dating of molluscan assemblages also presents difficulties. Radiometric dates have been obtained from the relatively robust shells of certain marine Mollusca, but these dating methods are more difficult to apply to the generally more delicate shells of terrestrial and freshwater molluscs. Hence, in order to develop a chronology of non-marine molluscan changes, dates have usually been derived from the associated sediment matrix. However, by comparison with peats and lake sediments which are frequently highly organic throughout and can therefore be dated by radiocarbon, most sediments containing terrestrial Mollusca are largely devoid of such datable material, although occasionally organic remains can be recovered where waterlogging has prevented full oxidation. A further problem arises over the validity of radiocarbon dates from carbonate-rich sediments (such as lake marls) due to the 'hard-water effect' (see Ch. 5). However, recent work on the radiocarbon dating of tufa deposits has produced encouraging results (Thorpe 1980), and it may eventually prove possible to establish time-scales for molluscan changes that are comparable to those for pollen and Coleoptera.

Overall, subfossil molluscan assemblages offer a valuable data source for the reconstruction of Quaternary environments. In many areas, molluscan remains are abundant in Quaternary sediments, they are easily extracted, relatively easy to identify, and are representative of a wide range of habitats. As such, they can provide useful information on both local and regional changes, particularly during the later part of the Quaternary in those areas where, for a variety of reasons, other palaeobotanical or palaeozoological evidence is absent. Equally, however, they also serve in a useful corroborative capacity where, for example, plant macrofossils, pollen or Coleopteran remains are found.

Marine Mollusca.   The shells of Quaternary marine Mollusca have been found in a range of deposits in coastal areas. They often occur in beach gravels and estuarine clays now lying some distance above sea level, having been raised isostatically following the wastage of the last ice sheets. Molluscan assemblages have been recovered from boreholes both onshore and from the sea bed of the continental shelf, and shell remains, often highly fragmented, are found at localities inland having been stripped from a former sea bed and transported to their present site by glacier ice. Although perhaps less widely used in Quaternary studies than their freshwater counterparts (largely because they are less frequently encountered) marine Mollusca are nevertheless an additional source of palaeoenvironmental information, and they can also be a useful medium for the dating of Quaternary events.

The approaches to marine molluscan analyses are essentially the same as those employed in the study of terrestrial and freshwater forms. Some workers have divided the fauna into groups on the basis of present-day ecology, the most important determining factors being salinity, water depth and, above all, water temperature. It is common, therefore, to recognise marine Mollusca with, for example, lusitanic, boreal, subarctic or high-arctic affinities. In his discussion of the shell-bearing marine clays of Lateglacial age around the coasts of Scotland, Sissons (1967a) distinguishes between deposits from the 'Arctic seas' which are thicker and more widespread on the east coast of Scotland, and which are characterised by shells of such species as *Leda arctica, Pecten groenlandicus, Cardium groenlandicum* and *Thracia myopsis,* and sediments from the 'sub-arctic seas' which are more typical of the sea lochs on the west coast, and which contain shells of such species as *Tellina calcarea, Astarte borealis, Pecten islandicus, Cyprina islandica* and *Littorina littorea.* The former group was believed to have been living in the seas around Scotland while glacier ice still covered much of the mainland, while the latter group was characteristic of the succeeding period when ice cover was markedly less extensive.

Alternatively, indicator species may be used to infer former water temperatures. Mangerud (1977) used the variation in occurrence through a sequence of marine deposits at Ågotnes in Norway of three particular molluscs (*Modiolus modiolus, Littorina littorea* and *Mytilus edulis*), whose present-day distribution and ecology are fairly well-known, to plot the palaeo-positions of the North Atlantic Polar Front[1] off the west coast of Norway during the Lateglacial period. The evidence suggested that the warm waters of the Atlantic reached the south-west coast of Norway prior to 12 600 BP, at which time the Polar Front lay some distance to the north. During the Younger Dryas (*ca.* 11 000 to 10 000 BP), colder waters moved southward and possibly excluded the warmer Atlantic waters from the Norwegian coast. Similarly, in East Greenland, Hjort and Funder (1974) used the occurrence of the common mussel (*Mytilus edulis*) in deposits dated approximately between 8000 and 5500 BP to show that sea temperatures may have been somewhat higher around those coasts during the Holocene Climatic Optimum by comparison with the present day.

Once again, however, both of these approaches must be used with care. Although a considerable amount is now know about a number of molluscan species, relatively little is known about the ecology and distribution of many others. Precise data on the thermal tolerances of a great many marine Mollusca are, unfortunately, scanty (Norton 1977), and therefore reliable palaeotemperature estimates of sea waters are not always possible. Moreover, even divisions into broad groupings such as those mentioned above may not always be undertaken with certainty. Detailed work on the molluscan assemblages obtained from the deep borehole through the early Quaternary deposits at Ludham in East Anglia, for example, showed that admixtures tended to occur between supposedly distinctive ecological groups (Norton 1967). Further, changes in water temperature as indicated by foraminiferal and other fossil evidence were not always reflected in the molluscan record. It would seem that many Mollusca can tolerate a very wide range of sea-water temperature and, moreover, the possibility seems likely that several species have changed their thermal requirements, perhaps in the very recent past. Clearly, therefore, inferences about former sea-water temperatures using the uniformitarian approach can only be of a generalised nature, although it may be possible to derive more precise palaeotemperature estimates from marine Mollusca by using the relative abundance of oxygen isotopes in the calcium carbonate of their shells (Shackleton 1969a).

Inferences about former water depths may be obtained from marine molluscan assemblages. On the basis of the known depth range of present-day species, Mollusca can be divided into littoral, infra-littoral or sub-littoral forms (Norton 1977), this type of grouping having proved to be particularly valuable in the interpretation of marine facies in the 'pre-glacial' Quaternary sequence in East Anglia (Norton 1967; West and Norton 1974; West et al. 1980). Again, however, although it may be possible to recognise assemblages characteristic of, for example, inter-tidal areas or estuarine systems, it may not be possible to derive precise figures for former water depths. Because of the relative ease by which mollusc shells may be transported from their original point of deposition, there will always be a proportion of allochthonous shells within any assemblage which may make it difficult to draw firm conclusions about the nature of former habitats.

One final aspect of marine molluscan analysis that has yet to be mentioned concerns their application in the dating of Quaternary events, particularly changes in land-and sea-level, and advances of glacier ice. Marine shells on raised shorelines and in marine clays have been dated radiometrically in many parts of the world in order to provide a chronology of isostatic and eustatic events (Broecker and Bender 1972). These range from the dating of Flandrian marine transgressions around the coasts of north-west England (Tooley 1978) to the dating of sea-levels of interglacial age in Barbados and New Guinea (Mesolella et al. 1969; Bloom et al. 1974). More recently, marine shells have been employed in 'amino-acid dating' (Ch. 5). In Britain, several localities exist where shell-bearing marine clays have been incorporated into glacial drifts, and radiocarbon dates on marine shells provide a minimum age for the respective ice advances. Particularly important sites include the shelly drift on the Cheshire Plain dated at *ca* 28 000 BP (Boulton and Worsley 1965) which demonstrates a Late Devensian age for the uppermost tills in that part of western England, and the dates of 11 800 ± 170 BP on glacially-transported marine shells in the Menteith Moraine at the head of the Forth Valley (Sissons 1967b) and 11 300 ± 170 BP on similar materials from the Kinlochspelve Moraine on eastern Mull (Gray and Brooks 1972), which together provide clear evidence for the glacial phase in Scotland known as the Loch Lomond Readvance.

Overall, although perhaps more limited in their application than freshwater and terrestrial Mollusca, marine Mollusca can still make a positive contribution to the reconstruction of Quaternary environments. This is especially true in the field of oxygen isotope work, in dating, and in the study of land and sea-level changes, and it is in these areas more than any other that subfossil marine Mollusca are likely to prove of greatest value.

Ostracod analysis

Ostracods are small, laterally-compressed arthropods whose time range extends from the Cambrian to Recent. The fossil record is extremely well-documented, the first fossil ostracod having been described as long ago as 1813. They occupy a wide range of environments, but many have highly restricted ecological preferences. They are generally regarded by palaeontologists and stratigraphers as the most useful group of Crustacea in the geological sciences.

The nature and distribution of ostracods

The majority of ostracods are between 0.15 and 15 mm in adult length. They consist of an outer shell or **carapace** which contains the soft body parts of the living organism (Fig. 4.16). The carapace is usually ovate,

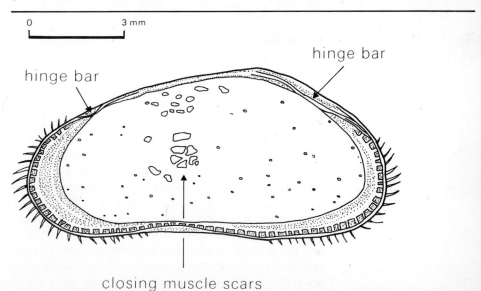

0   3 mm

hinge bar

hinge bar

closing muscle scars

Fig. 4.16: Drawing of the inner face of the female right valve of the freshwater ostracod *Candona suburbana* Hoff. (after Kesling 1965).

kidney-shaped or bean-shaped and consists of two chitinous or calcareous valves that hinge above the dorsal region of the body. The biological classification of Recent ostracods rests very largely on the characteristics of the soft parts, but as these features are very rarely preserved in the fossil form, the palaeonotologist is forced to base his taxonomic assessments on the nature of the carapace which fossilises relatively easily.

Ostracods occur in fresh, brackish, saline and even hypersaline waters. Some freshwater ostracods are even found in terrestrial niches, living, for example, in the moist humus of forest soils or in the aerial part of the freshwater floating plant accumulations (Pokorny 1978). Ostracods probably originated, however, in marine environments and the largest number of species still occupies the oceans from the shoreline down to the abyssal depths. The majority of marine ostracods are bottom-dwelling forms, and only a small number occupy the planktonic realm. Moreover, pelagic species usually possess weakly-calcified shells and are therefore relatively rare in fossil assemblages.

The distribution patterns of living ostracod communities are governed by a wide range of factors (Neale 1964). These include physical parameters such as water temperature, salinity and nature of the substrate, and biological factors such as food chains and natural associations (Robinson 1980). It is, however, difficult to cite any one control as universally dominant, for while many workers feel that, in the case of marine ostracods in particular, water temperature is the most important, others would argue that salinity is more fundamental, while in the freshwater situation, the nature of the substrate may be the overriding influence (Delorme 1969). Nevertheless, where autecological studies have been able to establish the major limiting factors that govern ostracod distributions, those species may be of considerable value in palaeoenvironmental reconstruction.

Excellent reviews of ostracod structure and ecology can be found in Neale (1964, 1969), Löffler and Danielpol (1977), Bate and Robinson (1978) and Pokorny (1978).

Collection and identification

Ostracods are often collected along with Foraminifera from lacustrine and marine sediments. The deposits are usually disaggregated in water (although occasionally hydrogen peroxide may be required), sieved and then dried. The ostracod carapaces and valves can be picked out by hand, using a very fine brush. The use of a low-powered binocular scanner of $\times 40$ or $\times 60$ magnification allows the majority of determinable remains to be collected. Individual ostracods are usually mounted on a slide and examined under a high-powered microscope. The carapaces possess a considerable range of morphological features that aid in identification including extensive ornamentation of frills and spines, and internal details such as muscle scars, pore canals and duplicature. Ostracods are usually studied under reflected light, but transmitted light may be necessary to see the internal features. Identifications are based on modern type collections, stereoscan photographs and descriptions in micropalaeontological manuals, and the data are expressed either as species lists or in diagrammatic form showing the change in frequency of occurrence through time. Further details on collection, preparation and study can be found in Brasier (1980).

Ostracods in Quaternary studies

Certain very rapidly evolving ostracod lineages are useful markers in marine stratigraphic sequences, especially where Foraminifera are absent. However, as they lack planktonic larvae, many shallow and warm water species cannot cross physical barriers and are therefore restricted to particular geographical areas. Moreover, some of the problems already considered in the interpretation of terrestrial fossil assemblages are also found in ostracod analysis. At the generic level, the poor state of taxonomy often precludes the comparison of fossil and recent forms. Also there is evidence to suggest that several ostracod species that are now benthonic in character developed from shallow water ancestors. Fortunately, it appears that migration in the opposite sense, in other words from deep and cold to shallow and warm water, seems unlikely to have occurred. Further, there are indications that the dominant elements in certain benthonic fossil assemblages may be due less to environmental factors than to selective preservation of the more thick-shelled species. Providing that these difficulties can be overcome, however, marine ostracods can be extremely useful as palaeotemperature and palaeosalinity indicators, and they may also provide valuable data on palaeobathymetry.

Around the coasts of the British Isles, the Clyde Beds of the Glasgow area and the Errol Clay of Tayside both contain a rich ostracod fauna characterised by species that today are found no further south than the waters of eastern Norway in the first case, and the Barents Sea or the fiords of east Greenland north of latitude 76° in the case of the Tayside clays. These deposits are believed to be of early Lateglacial age (pre–13 000 BP) and are indicative of very cold climatic conditions off the coasts of Scotland at that time (Robinson 1980). Similarly, the presence of some north European Atlantic ostracod species at particular levels in marine sediments of the Italian Quaternary sequence has provided a method for the recognition of cold climate episodes (Ruggieri 1971).

As ostracods can be good indicators of salinity, they are often useful in studies of sea-level change where changes in the character of ostracod assemblages can provide evidence for transgressive or regressive sequences. In the Somerset Levels of south-west England, Kidson *et al.* (1978) used ostracod evidence, along with foraminiferal, molluscan and plant macrofossil remains to show that within the Burtle Beds, an interglacial marine transgression reached a height of up to 12 m above the present (OD), at which time temperatures appeared to be similar to those of today. Moreover, biometrical analysis of the ostracod *Cyprideis torosa* found within the deposits suggested that the marine transgression may well have been of last (Ipswichian) interglacial age. In a similar integrated study, Haynes *et al.* (1977) traced the course of the Flandrian marine transgression along the coast of Cardigan Bay in west Wales, where once again, the transition from brackish to salt-water conditions was clearly represented in the ostracod record at a large number of sites.

Freshwater ostracods have in recent years been less extensively studied than their marine counterparts, due largely to a difficult taxonomy, and a certain lack of interest on the part of research workers. In typical organic muds ostracods can be rare (Frey 1964), but in marl or other shallow water sediments with a high calcium, sodium and magnesium content, they can be abundant. In general, they possess thinner carapaces than the marine ostracods and can be easily destroyed by mechanical breakdown and chemical corrosion. Because of their largely calcitic shell, the fossil remains are easily leached out leaving, in extreme cases, a complete gap in the sedimentary record (Delorme 1969). Fossil assemblages can be further biased by the incorporation into lacustrine deposits in particular of fragments of older or younger material although *in situ* species can often be recognised by the presence of moult-stages which cover the life history of an indigenous fauna.

In Europe, ecological associations of freshwater ostracod species have been determined by Diebel and Pietrzeniuk (1977) for spring tufas and loess spreads of Interglacial, Lateglacial and Holocene ages, while Absolon (1973) has used assemblages identified as 'spring', 'small pond', 'lake' and 'large lake' to reconstruct environmental changes in both Lateglacial and earlier Pleistocene sediment sequences. In North America, the use of freshwater ostracods in palaeoclimatic investigations has been elegantly demonstrated by L. D. Delorme. In detailed study Delorme (1971) derived ecological tolerance limits for over 100 species of recent freshwater ostracods from the Canadian Prairies, and then applied this information to fossil ostracod assemblages obtained from the marl deposits in the Stringer Valley of northern Saskatchewan. Using these data, it proved possible to identify ten environmental zones covering the mid-Holocene period, during which time five 'humid' and five 'arid' periods appear to have occurred, with one of the 'arid' phases culminating in a severe drought (Table 4.7). In a similar investigation in the Mackenzie River area of north-west Canada (Delorme *et al.* 1976), palaeoclimatic interpretations based on shelled invertebrates (ostracods and molluscs) indicated that the mean annual temperature during the time interval between 14 410 and 6820 year BP was between 8.2 and 11.6 °C higher than at the present day, while the annual precipitation was between 55 and 234 mm greater than at the present time.

These examples serve to demonstrate how both marine and freshwater ostracods can be used effectively as palaeoenvironmental indicators, either by themselves or, more fruitfully, in conjunction with other organisms

Table 4.7: Mid-Holocene environmental zones based on ostracod data from Sturgeon Lake, Saskatchewan, Canada (after Delorme 1971).

Zone	Maximum summer air temperature (°C)	Maximum bottom water temperature (°C)	Average mean salinity ppm	Terrestrial vegetation
X	28	25	1000	forest
IX	33	30	6000–19 000	grass
VIII	28	23	1500	forest
VII	Drought			grass
VI	33	30	25 000	grass
V	28	24	1 000	forest
IV	33	30	13 000	grass
III	28	23	650	forest
II	33	30	13 000	grass
I	28	23	600	forest

such as molluscs and Foraminifera. They are valuable in that they can provide information not only on local habitat parameters such as water temperature and salinity but, once the ecologically-limiting factors are established, they can also form a basis for palaeoclimatic inference. Their full potential in Quaternary palaeoclimatic reconstruction has, perhaps, not yet been fully realised.

Foraminiferal analysis

Foraminifera are protozoans that possess a hard calcareous shell often distinctively coiled to resemble that of a gastropod or cephalopod. They were first described and illustrated in the sixteenth century, but were not studied systematically until the mid-nineteenth century following the remarkable voyage of HMS *Challenger* which began in 1872. The discovery during that expedition of living Foraminifera in deep-sea waters, and fossil remains in sediments that were dredged from the sea floor, revolutionised marine micropalaeontology. Since then Foraminifera have become invaluable tools in palaeoenvironmental analysis, in palaeo-climatic reconstruction and in Quaternary stratigraphy.

The nature and distribution of Foraminifera

Foraminifera consist of a soft body (protoplasm) enclosed within a shell or **test** secreted by the organism which is variously composed of organic matter, minerals (calcite, aragonite or silica) or agglutinated particles. The tests may be single chambered, but more frequently consist of a series of chambers connected by openings known as **foramina**, from which the group derives its name. In many common species, the chambers are added in a spiral pattern, producing a coiled shell, while others develop far more complicated structures (Fig. 4.17). Foraminifera range in size from less than 0.40 mm (the planktonic forms) to some of the benthonic species which may measure up to 10 cm in width (so-called large Foraminifera). They are tolerant of a range of salinity and temperature, being found in salt marshes, shallow brackish water in estuaries, on the continental shelf and in the waters of the deep oceans of the world. Most Foraminifera are marine and benthonic, although a few genera are pelagic, while a very small number of species (**thecamoebids**) are adapted to

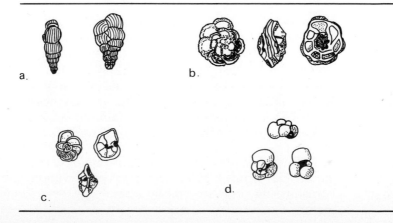

Fig. 4.17: Drawings of Foraminifera, suborder Rotaliina, superfamily Globigerinacea. a. *Heterohelix* (×97); b. *Globorotruncana* (×36.5); c. *Globorotalia* (×15.5); d. *Globigerina* (×18.5) (after Brasier 1980: a. and c. based on Loeblich and Tappan 1964; b. based on Glaessner 1945; d. based on Morley Davies, 1971).

freshwater environments. The open ocean planktonic forms have proved to be particularly useful in global correlation and climatic reconstruction and these will be considered in more detail in the last section of this chapter. The present discussion will be concerned principally with foraminiferal remains in shelf-seas and inshore waters.

Further discussion on the nature and distribution of Foraminifera can be found in Brasier (1980), and in the various contributions in Funnell and Riedel (1971) and Haq and Boersma (1978).

Collection and identification

Foraminifera can be extracted from sediments obtained from surface samples or from cores. The matrix is usually crushed and disaggregated using either water or hydrogen peroxide. The samples are then washed through sieves and the residues dried; the retained Foraminifera can then be picked out by hand with the aid of a binocular microscope. Where large numbers of sand grains are present (in some shelf sediments, for example) the foraminiferal remains can be concentrated using a heavy liquid such as ethylene bromide/absolute alcohol solutions (Knudsen 1977) the tests being 'floated' from the grain sand. The smaller Foraminifera are examined under a high-powered microscope using reflected light. Occasionally staining (with e.g. malachite green or a similar food dye) is required to bring out the surface structures more clearly. Larger Foraminifera are often studied in thin section where wall and growth plan may be better seen under transmitted light. As with ostracods, identifications are based on type collections, descriptions and stereoscan photographs in foraminiferal manuals. The data can be expressed simply as species lists or, more commonly, either in percentage form or as abundance per unit volume of sediment. Further information on collection and study of Foraminifera can be found in Douglass (1965), Todd *et al.* (1965) and Brasier (1980).

Foraminifera in Quaternary inshore and shelf sediments

Foraminiferal remains in sediments of most inshore waters and the shelf seas are dominated by benthonic forms in contrast with the deep-sea sediments in which planktonic Foraminifera are particularly abundant. Palaeoenvironmental inferences from these bottom water assemblages are constrained by a number of factors (Lord 1980) and include:

Statistical reliability.   A complete assemblage of at least 300–400 individuals is required for statistical comparability between samples. The assemblage may still be biased, however, as the agglutinated Foraminifera are more susceptible to post-mortem disintegration than the calcareous or siliceous tests. Care is also required in the use of heavy liquids for the concentration of foraminiferal tests, and the heavy residues must always be picked over carefully for the foraminiferal remains left behind, because of pyritisation or permineralisation of some tests.

Reworking.   Reworking and redeposition of tests is a frequent occurrence, and while such factors as wear, poor preservation, and unusual population structure may indicate a mixed assemblage, errors in interpretation may still occur.

Sampling.   Isolated samples can be misleading, and several samples from sediment sequences should be taken whenever possible.

Taxonomic comparability.   Although most of the Foraminifera of north-west Europe are well known, not all are easy to recognise and confusion over specific differentiation of, for example, *Elphidium* types and related forms continues to pose a problem. *Elphidium clavatum* is a commonly cited species with cool-water/sub-arctic connotations, but it is often mis-identified.

Correlation.   Some foraminiferal species have limited geographical ranges, and as very few sedimentary sequences containing foraminiferal remains are adequately dated, inter-area correlation of successions can therefore often be difficult to achieve.

These problems notwithstanding, foraminiferal data have provided much useful information in a number of different fields of Quaternary investigation. The work of Haynes *et al.* (1977) and Kidson *et al.* (1978) discussed above has demonstrated how integrated microfaunal and microfloral studies, including the use of Foraminifera, have contributed to our understanding of sea-level changes around the coasts of the British Isles during both the Flandrian and Ipswichian Interglacials. The use of Foraminifera in inferring former water temperatures is illustrated by the work of Peacock *et al.* (1978) on the Lateglacial marine clays at Ardyne in Loch Fyne on the west coast of Scotland. The faunal remains (which also included molluscs and ostracods) indicate that in the period from before 12 000 BP to *ca.* 10 000 BP, bottom water temperatures were generally lower than in the seas around those coasts at the present day, although faunas with a distinctly temperate aspect are recorded between the horizons dated at 11 159 ± 47 BP and 10 801 ± 67 BP, perhaps corresponding with the later part of the Lateglacial Interstadial in the terrestrial record (Fig. 4.18). Significantly, the Foraminifera from the horizons dated 10 412

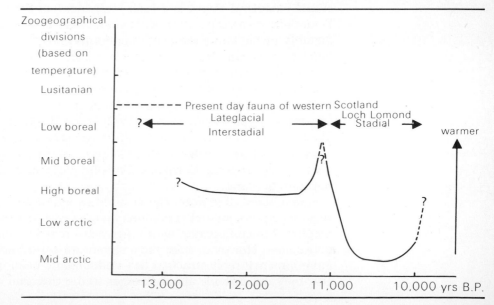

Fig. 4.18: Tentative curve, based partly on foraminiferal evidence, showing marine temperature changes in the Clyde sea area, western Scotland, during Lateglacial times (after Peacock *et al.* 1978).

± 136 to 10 195 ± 117 BP are dominated by arctic and sub-arctic forms, and clearly relate to the later part of the cold Loch Lomond Stadial.

Lateglacial marine sequences on and around the Norwegian and Danish coasts have also yielded foraminiferal assemblages which provide data on both water temperatures and salinity (Feyling-Hanssen 1964; Feyling-Hanssen *et al.* 1971; Nagy and Ofstad 1980). The earliest faunas are arctic in character reflecting both low water temperatures (much lower than the present day) and low salinity values, the latter resulting from the high meltwater influx from the wasting Late Weichselian ice sheet. The Foraminifera are primarily benthonic, although a few arctic planktonic forms occur. Early Flandrian foraminiferal remains are still primarily benthonic but contain fewer high-arctic forms and indicate increasing water temperatures and a decreasing freshwater influence. Middle Flandrian faunas are primarily boreal in character, and contain a high percentage of planktonic forms typical of the Foraminifera found in the north-east Atlantic waters at the present day. Both temperature and salinity characteristics of Norwegian coastal waters, therefore, appear to have changed little from the Middle Flandrian to the present.

One of the most useful applications of foraminiferal analysis has been the support it has provided for pollen-stratigraphic and molluscan investigations of the lower Quaternary deposits in East Anglia in south-east England (Funnell 1961; Funnell and West 1962, 1977; West 1980). The basal Quaternary sediments in that area are known as **crags**, are primarily marine in origin, and accumulated in a shallow basin of deposition to the east of the Chalk escarpment (Fig. 4.12). Pollen analysis of cores obtained from the Royal Society Borehole at Ludham (West 1961) showed a sequence of alternating warm and cold phases that preceded the deposition of the earliest glacial tills in East Anglia. Foraminiferal analysis from an adjacent borehole and from other sites in Norfolk and Suffolk substantially confirmed the palaeoclimatic inferences based upon the pollen sequence, with thermophilous assemblages coinciding with the closed woodland stages and high frequencies of boreal and arctic Foraminifera corresponding with more open terrestrial conditions. The majority of the fauna are benthonic forms and provide useful data on local habitats, for example, estuarine or brackish water bay-head environments, as well as on water depth. As such they give some indication of the changing position of the early Quaternary coastline in this part of Britain. Occasionally, planktonic forms are found as in the Pre-Ludhamian at Stradbroke in Suffolk (Beck *et al.* 1972) where the occurrence of planktonic Foraminifera including *Globigerina bulloides* and *G. pachyderma*, both of which are northern forms, suggests an increased injection of ocean waters into the North Sea, probably from around the coast of northern Scotland.

Foraminifera, therefore, can be used as indicators of both local and regional environmental conditions and, as with the ostracods, are best employed in conjunction with other fossil remains, particularly molluscs and pollen. However, their most significant contribution to Quaternary environmental reconstruction has undoubtedly been in the information that they can provide about changes in the character and movement of ocean waters, a topic which is considered in more detail in the final section of this chapter.

ANALYSIS OF LARGE ANIMAL REMAINS

Fossil animal bones constitute what are arguably the most spectacular remains of the Quaternary biota, yet until comparatively recently little systematic work had been undertaken on the analysis of fossil animal bones or on the palaeoecology of the Quaternary vertebrate fauna. Traditionally, animal bones have been collectors' items and museums in many parts of the world are full of skulls, femurs, and antlers of Pleistocene mammals that have been placed there by enthusiastic amateurs and professionals alike. A very large number of these bones were removed from exposures in cliffs and in river valleys during the heyday of collecting towards the end of the last century by well-meaning Victorian enthusiasts who, unfortunately, often paid scant regard to the stratigraphic context within which the bone was lying, or indeed to the less spectacular but equally important collecting of smaller animal remains which together formed the total assemblage of the stratum. Thus, although the fossil remains indicate that now extinct animals such as the mammoth (*Elephas primigenius*) and woolly rhinoceros (*Coelodonta antiquitatis*) were to be found in many areas of the Northern Hemisphere during the Quaternary, and also that creatures such as the hippopotamus (*Hippopotamus amphibius*) and musk-ox (*Ovibus moschatus*) formerly occupied ranges which are very different from those of the present-day species, surprisingly little is known about the smaller vertebrates and their Quaternary distributions. It is only in the last few decades that any real progress has been made in our understanding of the spatial and temporal variations in animal populations during the Quaternary, and their relationships to vegetational and climatic changes. For these reasons, rather less attention has been paid to animal bones in the reconstruction of Quaternary environments than to other lines of biological evidence.

The structure of teeth and bones

Vertebrate remains consist principally of teeth and bones. Teeth are extremely complex in structure but in most mammals consist of three distinct substances of differing hardness: the hard brittle outer casing (**enamel**), the softer **dentine** of which the greater part of the tooth is composed, and **cement** which covers the dentine of the roots and occasionally the valleys and folds of the main tooth body (Cornwall 1974). In the fossil, the enamel provides the most durable element except where burning has affected the original dental material, in which case the dentine of the tooth roots may prove to be the most resistant. Teeth are of considerable importance in palaeoenvironmental work for not only do they provide data on the age (years of life) of the animal, but they also give an indication of its dietary preferences (*i.e.* herbivore or carnivore). In more recent sediments, they tend to be outnumbered by fragmentary remains of bone, but in older Quaternary deposits, such as those of Cromerian age at West Runton in Norfolk (Stuart 1975) and Sugworth in Oxfordshire (Stuart 1980), teeth are often proportionately more strongly represented than elements of the post-cranial skeleton.

Fresh animal bone consists of both organic material and inorganic material in the approximate ratio by weight of 30:70. The organic fraction is contained within the shafts of long bones (e.g. femurs, tibias, vertebrae) and comprises cell tissue (fats etc) and a fibrous protein called **collagen** (Chaplin 1971). The collagen is very resistant to decay and may survive for thousands of years following the death of the animal (Garlick 1969), while the remaining organic matter undergoes autolysis after death and is rapidly decomposed. Surrounding the collagen fibre is bone mineral material, the principal component of which is a phosphate of calcium, **hydroxyapatite** ($Ca_{10}OH(PO_4)_6$). The structure and composition of animal bones are of considerable interest to the palaeoenvironmentalist as they affect the way in which fossilisation takes place, and the chemical structure of bones in particular provides a means of dating the fossil material (see Ch. 5).

Fossilisation of bone material

Quaternary vertebrate remains have been recovered from a wide range of deposits. These include cave and fissure sediments, lacustrine and marine deposits, riverine sediments (especially river terraces), peat bogs, soils and a variety of situations associated with man and his activities such as middens, cess-pits and burial chambers. At some sites, whole skeletons have been found, but more frequently the fossil assemblages consist of disarticulated skeletons and an admixture of bones of varying sizes and in differing states of preservation. Animal bones are perhaps more susceptible to physical and chemical changes than any other biological remains encountered in Quaternary deposits. Chaplin (1971) and Cornwall (1974) provide useful reviews of the factors affecting the preservation of bones in different depositional environments.

As soon as a bone becomes incorporated into a body of sediment, it begins to undergo chemical changes that vary considerably in nature and degree with the chemistry of the surrounding matrix. In most deposits where air is freely circulating, the mineral parts of the bone will tend to be more resistant to decay, while the organic substances will break down rapidly into simple compounds such as ammonia, carbon dioxide and water. Mineral salts in solution in the surrounding sediment, particularly calcium and iron, will be deposited in the vacant pore spaces and the bone may eventually become completely **permineralised**. As the process usually proceeds fairly slowly, a high degree of mineralisation will generally indicate considerable age. On the other hand, if a soil is acid and depleted in bases, both the organic and mineral fractions will decompose and the bone will disappear completely leaving no trace of its former existence. This is one reason why deposits in caves in limestone regions are often comparatively rich in mammalian remains. Similarly, prehistoric burials on chalklands in areas such as southern England have often yielded well-preserved bones, while those in adjacent regions where porous, sandy soils are found contain few bone remains.

In waterlogged situations, a completely different set of reactions occurs. In deep lakes where the substrate is of limestone and where bases are in plentiful supply, bones are not only well preserved, but are often extremely hard. In some cases even the organic elements have been converted into a stable wax-like substance known as **adipocere** (Cornwall 1974). At the other extreme, in peat bogs or in oligotrophic lakes, the anaerobic nature of the depositional environment often results in the organic portions of the bones being preserved while attack by humic acids leads to complete destruction of the mineral fraction. Skeletal remains will, therefore, be found in a soft or pulpy state in advanced stages of decalcification. Occasionally, other organic components such as hair and skin will be found in such deposits, the most remarkable example being the discovery of the almost perfectly preserved corpse of an Iron Age man at Tollund in Denmark (Glob 1969). Other sediments in which the organic parts of animal remains may be preserved more or less intact include those in areas of high aridity where the faunal remains become desiccated, and permafrost regions, such as interior Siberia, where deep-frozen carcasses of mammoths have been discovered that have been almost unaffected by the passage of time. Clearly, however, these are all extreme situations, and a very great range of intermediate stages of decalcification, mineralisation and loss of organic content will be encountered, depending on local environmental conditions, and climatic and geomorphological factors that have affected the site.

Finally, there are the effects of burial on bone that are purely physical. The seasonal drying of clay soils, for example, will result in the fissuring and eventual destruction of even the strongest bones. Bones may be similarly shattered by frost-heaving and by the action of ground ice. Soil creep, solifluction and mechanical abrasion in river gravels will have similar damaging effects and will result in the progressive fragmentation of bone remains.

Field and laboratory techniques

Because bones can be found in such a variety of conditions, great care must be exercised in the excavation of bone-bearing deposits. Mapping and surveying augmented by field description, sketches and photographs should precede the removal of bone fragments from the sediment matrix. In some cases, it may be possible to remove the larger bones by hand. These can be left to dry out and then cleaned with a brush or by gentle agitation in water. Many bone remains, however, even if heavily mineralised are quite brittle and it may be necessary to treat these with a highly-fluid penetrative plastic solution (e.g. polyvinyl acetate in toluene) before removal from the matrix can be attempted. If the bones are wet, especially those that are markedly decalcified, an emulsion of the plastic solution in water may be needed in order for the strengthening material to penetrate the bone fibres. Bones that are so treated, however, are effectively useless for any subsequent chemical analysis or for radiocarbon dating. The presence of very small bones or teeth (of rodents, for example), can usually only be detected by sieving the matrix following the removal of the larger faunal remains by hand.

Identification of bone remains is usually carried out in the laboratory and often proceeds in two stages. As most identifications are based upon fragmentary evidence, the first step is to identify the bone of which the fragment is a part. This is usually achieved by comparing the fragment with fresh skeletal material from a range of animals of different sizes. The second, and more difficult stage, is to track down the animal from which the unknown bone was derived. Here a reference collection of type material is essential, although the development of a type collection for the Quaternary vertebrates involves many more difficulties (and considerably more expense) than are encountered in the construction of a reference collection for Quaternary pollen grains or coleopteran remains. The Mammalia, for example, include a proportion of taxa that are now extinct, while evolution and speciation during the Quaternary pose additional complications. Moreover, as relatively few sites have so far been properly investigated, it is scarcely surprising that very few museums in the country possess a good or reliable reference collection suitable for identification purposes. In spite of these difficulties, however, positive identifications of Quaternary vertebrate remains are steadily increasing and it is now proving possible to construct fairly detailed faunal lists for the major stages of the Quaternary (Stuart 1976, 1982).

The interpretation of fossil vertebrate assemblages

The first stage in the interpretation of fossil bone is to establish how a particular grouping of vertebrate bones came to be associated together. The various factors that can result in a more or less distorted picture of the living community as represented by the fossil assemblage are reviewed by Stuart (1974). Three different depositional environments will serve to demonstrate the potential complexities of fossil vertebrate assemblages.

Cave deposits.   Some of the richest vertebrate assemblages in the world are those found in cave sediments, particularly in limestone regions, yet the ecological history of cave faunas is frequently very difficult to interpret because of the multiple origins of the fossil material. Some bones, for example, may have been washed into the caves or fissures by streamflow and are therefore allochthonous to the site. Caves were often occupied during the Quaternary by carnivores including cave bear (*Ursus*), wolf (*Canis lupus*), red fox (*Vulpes vulpes*), sabre-toothed cats (e.g. *Smilodon*) and spotted hyaena (*Crocuta crocuta*), and hence many cave assemblages will be biased towards the prey of these animals. Small vertebrate remains could, for example, have been derived almost entirely from pellets dropped by owls roosting in the cave roofs, and could include either woodland or open-country rodents depending on the species of the owl involved. Many of the large vertebrate bones will have been dragged into the cave by predators so that the resulting assemblage will give at least some indication of the original large vertebrate fauna of the vicinity. However, the cave assemblages will inevitably be biased towards the predators themselves as many would have eventually died in the caves and thus contributed their bones to the assemblage. In some instances, the caves seem to have acted as natural pitfall traps with animals having fallen in through holes in the cave roof, a good example being Joint

Mitnor Cave in Devon (Sutcliffe 1960). These assemblages will be partly biased towards the scavenging animals such as the hyaena which would have been attracted to the cave by dead and dying animals. The difficulties of interpretation of vertebrate assemblages in cave sites are further exacerbated by the often complex stratigraphy of cave sediments (Ch. 3).

Lacustrine sediments.   Lake sediments often contain whole or partial skeletons of large mammals, amphibians and fish. Remains of large mammals such as elk (*Alces alces*), reindeer (*Rangifer tarandus*) and mammoths (e.g. *Elephas primigenius*) found in lake deposits probably represent animals that either died by drowning after breaking through thin ice or becoming trapped in the soft mud on the lake floors. Many human occupation sites, such as the early Mesolithic hunting settlement at Star Carr in Yorkshire (Clark 1954), were by lakes and rivers and therefore a proportion of the remains of animals that were hunted also found their way into the lake. Fish and amphibians clearly reflect the former presence of these animals in the lake waters, but again man may have been responsible for the concentration of faunal remains in the littoral sediments.

Fluviatile sediments.   The origins of the vertebrate remains in river sediments may be almost as diverse as those found in cave deposits. Large vertebrate remains become incorporated in riverine deposits in similar ways to those outlined above for lacustrine situations. From an analysis of the assemblage, it may be possible to gain some impression of relative population densities, of the lifespan of particular taxa, and of the distance of their habitats from the site of deposition. Fish and amphibian remains will be locally derived and will tend to be over-represented in the assemblage. The small animal remains are, however, more difficult to interpret. Some may be the remains of waterside creatures such as voles and rats, but Mayhew (1977) has emphasised the role of predatory birds as being responsible for the concentration of small animal bones in Quaternary fluviatile sediments. At West Runton in Norfolk, he was able to match corroded teeth and bone in river sediments of Cromerian age with specimens from the regurgitated pellets of modern kestrels and buzzards, and he therefore suggested that a major proportion of the fossil small mammal material from that site was transported to the point of deposition by avian predators. Stuart (1980) has suggested that carnivore droppings would provide an additional source of small mammal remains. Ironically, due to the extreme fragility of avian bones and their tendency to float, bird remains are seldom preserved in lacustrine or fluvial deposits. Consequently, the Quaternary history of the avian fauna is poorly understood.

A further difficulty that frequently arises over the interpretation of fluviatile fossil assemblages is that vertebrate remains of different ages are easily incorporated into the sediments as the river banks are eroded. Because similar (but not necessarily identical) animal populations existed during successive warm and cold stages of the Quaternary, the likelihood of erroneous ecological interpretations from mixed bone assemblages is very real. This problem can be tackled in two ways. On occasions it may be possible to recognise bones of different ages on the basis of the varying degrees of physical deterioration, although this method is essentially subjective and, therefore, not altogether reliable. An alternative approach is to establish the relative ages of the bones by chemical means, the most widely used techniques being the Fluorine, Uranium and Nitrogen methods described in Chapter 5.

Under certain circumstances it may be possible to use the chemical composition of bones to infer climatic change. Such a technique was described by Röttlander (1976) using bones from a sequence of sediments in a limestone cave in south-west Germany. Detailed chemical analyses on both the organic and inorganic fractions of bones from successive horizons showed that at some levels the bone material was in a better state of preservation than in others. The cultural remains within the cave earths enabled the sequence to be dated to the Lateglacial period, and this was confirmed by palynological and pedological evidence from nearby sites. The levels of good bone preservation appeared to equate with the Older and Younger Dryas (cold phases) while the three levels of relatively poor

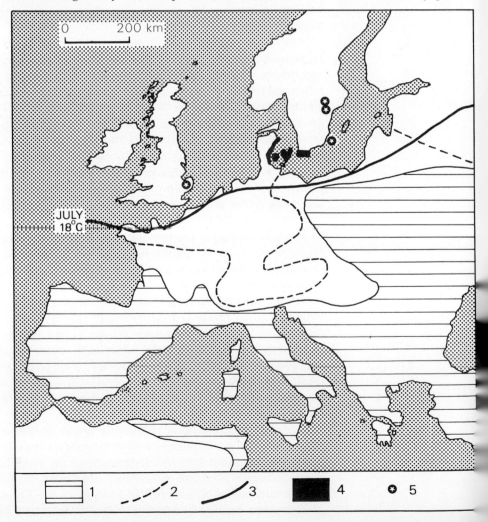

Fig. 4.19: Present day distribution of the freshwater pond tortoise (*Emys orbicularis* L.) west of about 30 °E, and its northward extension during the Flandrian. 1. Present day breeding range. 2. Northern limit of breeding and non-breeding individuals. 3. July 18 °C isotherm. 4. Areas with numerous Flandrian records. 5. Isolated Flandrian localities in Sweden and East Anglia (after Stuart 1979).

bone preservation were tentatively correlated with the Bölling, Alleröd and Early Holocene periods (warm phases). These results remain to be substantiated by work at other sites, but if proved reliable the technique may be valuable not only in palaeoclimatic work, but also in the relative dating of cave sediments in which bone is preserved.

The more traditional approach to the reconstruction of former climatic conditions on the basis of faunal evidence utilises the known ecological preferences of present-day species as a basis for inferences about past climates. During the last interglacial, for example, hippopotamus (*Hippopotamus amphibius*), the pond tortoise (*Emys orbicularis*) and the lesser white-toothed shrew (*Crocidura* cf. *suaveolens*) were all found in southern Britain (Stuart 1976). The hippopotamus is now confined to tropical Africa, the pond tortoise is now found only in the Mediterranean and in south-east Europe (Fig. 4.19) with its northern breeding range apparently limited by the 18 °C July isotherm (Stuart 1979), while the lesser white-toothed shrew is also essentially southern European in its distribution. These data are, therefore, strongly suggestive of warmer summers and milder winters in the British Isles during the Ipswichian Interglacial by comparison with the present day, a hypothesis supported by both palynological and coleopteran evidence. However, a straightforward relationship between animal distribution and climatic parameters may not always be correct, for it may well be that the primary adaptation of the Quaternary vertebrates was to vegetation and only secondarily to climate (Zeuner 1959). Climatic inferences may be particularly suspect during interglacials when both closed woodland and open grassland conditions often occurred in close proximity. In some Ipswichian Interglacial deposits in Britain, the two mammoths *Elephas primigenius* and *Elephas antiquus* are found shortly after the thermal maximum, yet the traditional interpretation is that the latter was the warm species while the former is indicative of cold conditions. A more likely explanation is that *E. antiquus* was adapted to woodland while *E. primigenius* was a native of open vegetation, perhaps akin to the warm loess steppe of Siberia (Sparks and West 1972).

A further problem with the 'indicator species' approach to the reconstruction of both climate and vegetation is that certain species appear to have changed their ecological affinities during the course of the Quaternary (Stuart 1974). The hamster (*Cricetus cricetus*), for example, is now an obligate steppe creature, yet it was present in the mixed oak woodland of the Cromerian Interglacial in southern Britain; similarly, the musk-ox (*Ovibus moschatus*) which is now a tundra animal, occurred in open but hardly sub-arctic conditions at the end of the last interglacial, and was also present in the Early and Middle Devensian steppe faunas of central England. The fossil record of *Cervus elaphus* is even more intriguing, for red deer remains have been found associated with a variety of environmental conditions ranging from Cromerian Interglacial oak-woods, to the interstadial or perhaps even full-glacial tundra environments of the Devensian cold stage. These examples serve to reinforce the point that, in the absence of corroborating palaeobotanical data, assemblages are more reliable indicators of former environmental conditions than a single taxon.

While fossil vertebrate assemblages should be employed wherever possible in palaeoenvironmental reconstruction, this aim is frequently difficult to achieve in practice. Taxonomic imprecision as a consequence of the poor state of preservation of fossil material, or simply the lack of type material, often places constraints upon the reconstruction of whole animal communities from fossil assemblages. Further limitations are imposed by the manifold problems associated with evolution and extinction, and by the fact that many fossil animal assemblages appear to have no modern equivalents. The frequency of climatic and environmental change during the Quaternary may have resulted in an acceleration in the rates of evolution of morphological characteristics compared with preceding geological periods. Kurtén (1968) has suggested that, for the Pleistocene as a whole, the average 'turnover rate' was 11 per cent per 200 000 years and that the mean longevity of species was *ca*. 3 million years. However, from the penultimate interglacial to the close of the last glacial, a 9 per cent turnover rate per 75 000 years and a mean species longevity of 1.6 million years has been estimated. Clearly, therefore, the older the fossil assemblage under investigation, the more acute the difficulties of interpretation become, for not only does the proportion of extinct species increase with the age of the assemblage, but also the degree of phyletic relation to the living form decreases (Lundelius 1976).

Finally, difficulties in correlation between individual fossiliferous sites are a recurrent problem. Many vertebrate assemblages cannot be dated precisely and the pollen-stratigraphic control is often not sufficiently sensitive to allow reliable correlations to be made between sites that are widely separated. A particularly frustrating problem has been the continued inability of Quaternary researchers to effect correlations between the often rich cave assemblages and open sites such as river terraces or lacustrine situations where fossil remains, by comparison, are relatively sparse. Paradoxically, the stratigraphy of the open sites is usually more clearly defined and the deposits often contain pollen, although in recent years some success has been achieved in the extraction of pollen from cave earths (Campbell 1977).

Many of the interpretative problems outlined above are common to the investigation of all fossil assemblages, and a number will be resolved in time with improvements in identification techniques, an increase in the amount and quality of type material and refinements in dating methods. The rates of evolutionary change and the high numbers of extinctions, however, are unique to the vertebrates, and impose considerable constraints upon palaeoecological interpretation. In some respects, therefore, vertebrate fossils are less valuable palaeoenvironmental indicators than pollen grains or coleopteran remains. On the other hand, the judicious application of the indicator species approach and, where possible, the vertebrate assemblage as far as it is known, can provide useful corroborative data for other sources of palaeoenvironmental inference. A major element in Quaternary investigations is the reconstruction of former landscapes, and for much of the Quaternary, vertebrates both great and small formed significant components of the landscape. A proper appreciation of the ecology of Quaternary environments can, therefore, only be achieved by the inclusion of an analysis of vertebrate remains.

MICROPALAEONTOLOGY OF DEEP-SEA SEDIMENTS

It has already been shown (Ch. 3) how the proportions of oxygen isotopes in fossil remains can provide a record of the expansion and contraction of ice sheets during the Quaternary, and the applications of this technique in Quaternary stratigraphy will be considered in a later chapter. The microfaunal and microfloral remains in ocean sediments, however, can also provide valuable data for palaeoenvironmental reconstruction as the fossil assemblages retain a record, at least in part, of former water temperatures that will, in turn, be a reflection of former climatic conditions. From an analysis of the biostratigraphy of the deep-sea sediments, therefore, it is possible to make reasoned inferences about climatic régimes that prevailed over large areas of the world's oceans at different times during the Quaternary period.

Some of the marine organisms found in the deposits of the deep-ocean floors that are employed in palaeoclimatic research have been discussed above in relation to fresh or brackish water situations, or to relatively shallow shelf seas. These include Foraminifera, diatoms and ostracods. Of the remaining marine organisms, the most valuable in terms of their application to palaeoclimatic research have proved to be the Radiolaria and coccoliths.

The nature of Radiolaria and coccoliths

Radiolaria are amoebic protozoans that secrete elaborate skeletons composed largely of amorphous (opaline) silica, and this is extracted from sea water in the same way that Foraminifera extract calcium carbonate. The skeleton, which consists of a complex network of elements, is contained within the living protoplasm and thus the hard parts forming the fossil are never subject to dissolution in sea water until the creature dies. The single-celled radiolarians are usually circular in shape and average between 100 and 2000 μm in diameter (Fig. 4.20). They are found in all ocean waters from the tropics to the sub-polar seas, and occur from surface waters down to depths of over 4 km. As with the Foraminifera, they are tolerant of a range of temperature and salinity conditions.

Coccolithophores are the most common members of a group of unicellular autotrophic marine algae known as **calcareous nannoplankton**. They are generally spherical or oval in shape, and are mostly less than 20 μm in diameter. The cell secretes a skeleton of minute calcareous

Fig. 4.20: Radiolarians and Coccolithophores. A. Acantharian radiolarian *Zygacantha* skeleton (×160); B. Polycystine radiolarian *Actinomma* (scale unknown); C. recent coccolithophore *Cyclococcolithinia* with coccolith shields (×2870) (after Brasier 1980: C. based on Campbell 1954).

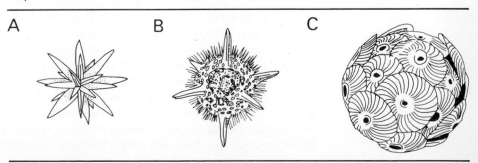

A   B   C

shields, and these may envelop the cell completely to form a hollow sphere (**coccosphere**) which eventually disintegrates, and falls to the ocean bed. The individual button-shaped remnants ranging in size from 1 μm to 15 μm are known as **coccoliths** and are usually all that remain of the former living creature. Although a few species are adapted to either fresh or brackish water, the majority of present-day coccolithophores are marine. Coccolithophores can tolerate a range of salinities. Being photosynthetic, they are confined to the photic zone of the water column and are rarely encountered below 200 m depth.

The distribution of Radiolaria and coccolithophores, as well as Foraminifera and diatoms is partly determined by nutrient requirements. The planktonic forms are all found in abundance in zones of upwelling for example, or pronounced vertical mixing, where mineral nutrients are readily available. For this reason, large numbers of these micro-organisms are frequently encountered just seaward of the continental slope. In most cases, however, the fundamental determinant is water temperature, and detailed ecological studies have shown that many species are associated with water masses that possess distinctive thermal characteristics. From the present distribution of marine plankton, and allowing for current circulation and coriolis effects, it is possible to recognise distinct equatorial, tropical, sub-arctic and arctic provinces. Hence, the analysis of marine microfossil assemblages can provide a unique source of information on ocean palaeotemperatures and, by implication, on former climatic conditions.

Further details on the structure and ecology of Radiolaria and Coccolithophores can be found in Buchsbaum (1948), Brasier (1980), and in the relevant chapters in Funnell and Riedel (1971), Ramsay (1977), Swain (1977) and Haq and Boersma (1978).

Microfossils in ocean sediments

Planktonic Foraminifera are the major contributors to deep-sea sediments and, along with coccoliths, account for more than 80 per cent of modern carbonate deposition in seas and oceans (Brasier 1980). Most of the shells now being deposited are from planktonic species of *Globigerina* and it has been estimated that about 30 per cent of the present ocean floor (over 60 million km^2) are covered by the grey mud known as *Globigerina* ooze. These oozes are forming at all depths up to 5 km in ocean waters between 50 °N and 50 °S. Coccolith oozes form principally in the tropical and subtropical regions, where the remains may average up to 30 per cent by weight of the sediments. In Arctic regions, by comparison, the values may be as low as 1 per cent (McIntyre and McIntyre 1971). By contrast with Foraminifera, however, coccolith remains settle much more slowly, and are therefore more susceptible to carbonate dissolution. Although some coccoliths may settle out more rapidly if they are contained within the faecal pellets of planktonic grazers (Haq 1978), it has been estimated that less than 25 per cent of all coccolith species are actually preserved in the fossils of ocean sediments (Ruddiman and McIntyre 1976). Below 3–5 km, the calcium carbonate compensation depth, nearly all $CaCO_3$ enters into solution, and thus only the most resistant calcareous fossils will be found.

The sediments there will be dominated by silicious remains, predominantly Radiolaria.

Radiolaria accumulate in abundance in equatorial sediments where productivity is high in the water column above. However, as the productivity of calcareous organisms is also high, the radiolarian remains are often masked by foraminiferal and coccolith fragments. Only in areas such as the tropical northern Pacific, where large areas of the sea floor lie below the carbonate compensation depth are radiolarian oozes encountered (King 1978). Diatomaceous oozes are also found in the abyssal depths of the Pacific and Indian Oceans. Silicious oozes are most common in the high latitude areas of the north Pacific and around Antarctica. In these regions, calcareous fossils are rare and both radiolarian and diatom remains are abundant. As with carbonates, however, silica is soluble in sea water, dissolution being especially rapid in the upper levels of the water column. Only those radiolarian species with a solid opaline skeleton reach the sea floor and, overall, it has been estimated that perhaps as few as 10 per cent of all Radiolaria find their way into the fossil record. Similar low values have been suggested for diatoms. Both Radiolaria and diatoms are very prone to exhumation and reburial in younger sediments and this poses further problems of interpretation for the marine biostratigrapher.

Laboratory work

Faunal and floral remains are extracted from deep-ocean cores in the laboratory by disaggregation of the sediment in water or, where necessary, in hydrogen peroxide (Foraminifera, Radiolaria), nitric acid or hydrochloric acid (Radiolaria), or sodium hexametaphosphate (Calgon) in the case of coccoliths. The larger fossils can be hand-picked from the meshes of sieves, while for the smaller remains, particularly the coccoliths, it is necessary to concentrate the microfossils into a liquid which can then be mounted on a microscope slide. High-powered microscopy (up to ×1600 may be necessary for ultra-detailed study) using transmitted, reflected and polarised light is required and, as in other micropalaeontological work, increasing use is being made of the electron microscope. Identifications may be made difficult by the solution of diagnostic elements, by the mechanical wear of the skeletal remains, and by the tendency, especially in the case of carbonate fossils, for calcite overgrowth and recrystallisation to obscure the morphology of the surface features. Further details of extraction and identification procedures can be found in Imbrie and Kipp (1971), Brasier (1980), and in the sections by Hay, Burma and Todd et al. in Kummel and Raup (1965).

Palaeooceanography and palaeoclimatology

The initial approach to Quaternary temperature investigations using data from deep-ocean cores was based simply on the presence or absence of certain key species in fossil assemblages. The early work by Schott (1935) demonstrated that the abundance of the planktonic foraminiferal species *Globorotalia menardii* could be used to infer climatic change. This idea was

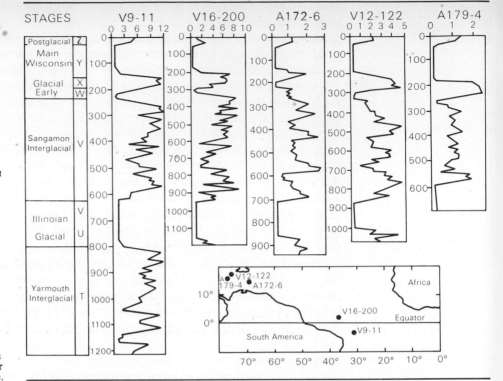

Fig. 4.21: Frequency curves for the *Globorotalia menardii* complex in five deep-sea sediment cores from the Caribbean and Equatorial Atlantic (inset map). The scales at the tops of the columns are the ratios of the number of shells of the *G. menardii* complex to the total population of foraminifera in the samples. The climatic zones indicated by the curves have been correlated with the Late Quaternary glacials and interglacials of the North American sequence (after Ericson and Wollin 1968).

developed by Ericson and his colleagues (Ericson *et al.* 1956; Ericson and Wollin 1968; Wollin *et al* 1971), and they constructed a series of palaeotemperature curves based on the abundance of *G. menardii* in sediments from the floors of the Caribbean and subtropical Atlantic (Fig. 4.21). High percentages of *G. menardii* were interpreted as indicating warmer, possibly interglacial periods, while reduced frequencies reflected cold, glacial periods. Further north, McIntyre *et al.* (1972) and McIntyre and Ruddiman (1972) and Kellogg (1976) used certain faunal indicators, particularly the markedly polar foraminifer *Globorotalia pachyderma*, along with the absence of coccolith remains at certain levels in cores from the north Atlantic to record the migration of the Polar Front over the past 225 000 years (Fig. 4.22).

   Although work of this nature provided valuable new insights into Quaternary climatic changes, the evidence was not always easy to interpret. It has already been shown that only a small proportion of the planktonic ocean fauna and flora actually reach the sea floor and enter the fossil record. The death assemblage in a body of ocean sediment, therefore, very rarely reflects accurately the former living assemblage in the water column above. For the same reason, however, indicator species in these inherently biased fossil assemblages are not always a reliable index of palaeotemperature change. Moreover, although temperatures are generally believed to be the major determinant in the distribution of planktonic fauna and flora, other factors need to be considered such as salinity variations and seasonal temperature fluctuations.

   In an attempt to overcome some of these problems, Imbrie and Kipp

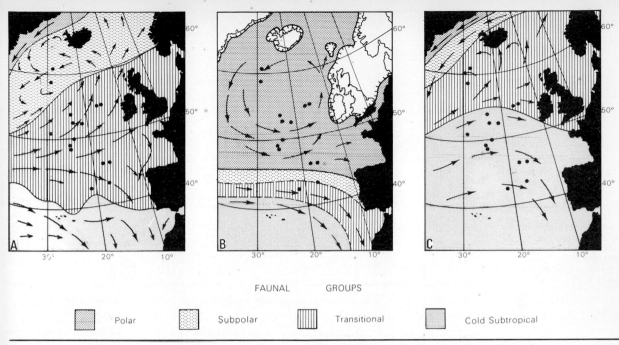

FAUNAL   GROUPS

Polar   Subpolar   Transitional   Cold Subtropical

Fig. 4.22: Oceanographic and palaeooceanographic maps of the north-east Atlantic Ocean depicting inferred surface currents and ecological water masses. A. today; B. maximum glaciation *ca* 18 000 BP; C. maximum interglaciation *ca.* 120 000 BP (after Ruddiman and McIntyre 1976).

(1971) developed a method of palaeontological analysis based on a complicated series of ecological equations known as **transfer functions**. The relative abundance of some twenty-six species of planktonic Foraminifera was calculated for sixty-one core-top samples distributed throughout the Atlantic basin. Factor analysis distinguished five distinct ecological assemblages–tropical, subtropical, polar, subpolar and gyre margin, and these could be related to observed oceanographic parameters including summer and winter temperatures and salinity values. The palaeoecological equations or transfer functions based on least-squares regression were then used to relate these to fossil assemblages at different depths in the cores from the Caribbean. The method has been further refined by Imbrie *et al.* (1973) and Kipp (1976) and now forms the basis for many palaeooceanographic investigations.

Recently members of the CLIMAP[2] team (e.g. Cline and Hays 1976), have amassed a wealth of impressive data on the Quaternary history of the world's oceans and atmosphere. Within the north Atlantic, for example, Ruddiman and McIntyre (1976) and Kellogg (1976) were able to construct ecological water masses characterised by specific temperature and salinity characteristics for different periods during the Late Quaternary based on the occurrence of distinctive foraminiferal and coccolith assemblages in the deep-ocean cores (Fig. 4.23).

These water masses appear to have migrated across more than 20° of latitude (some 2000 km) at rates of up to 200 m per year. From these data it would appear that glacial surface water temperatures in the area were up to 12.5 °C lower in winter and 13.0 °C lower in summer than at present, and within the last 600 000 years alone, at least eleven separate southward movements of polar water have occurred. Using a transfer function approach, Prell and Hays (1976) demonstrated that glacial water temperatures in the Caribbean were only 2–3 °C lower than at the present

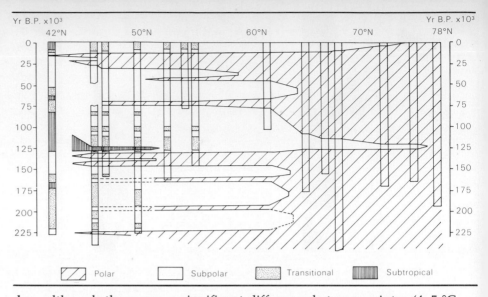

Fig. 4.23: Changes in ecological water masses in the Norwegian Sea and the northern North Atlantic over the past 225 000 years (after McIntyre and Ruddiman 1972; Kellogg 1976).

day, although there was a significant difference between winter (4–5 °C lower) and summer (1–2 °C lower). The evidence also indicated that the Intertropical Convergence Zone was located over South America more during the glacial stages than at present. In the south Atlantic transfer function analysis of radiolarian assemblages (Morley and Hays 1979) showed that at the height of the last glacial, at about 18 000 BP, sea surface temperatures were 2–5 °C cooler than today, and that there was a northward latitudinal shift of the Polar Front, the Subtropical Convergence and the Equatorial Divergence. Although the inferred temperature differences are small compared with the large changes (exceeding 10 °C in some areas) reported by McIntyre et al. (1976) for the north Atlantic, they do nevertheless reflect major oceanographic and climatic deviations from present-day conditions in the south Atlantic.

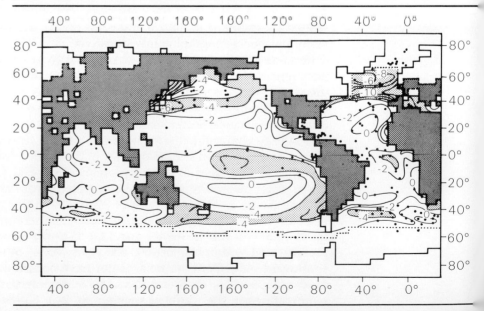

Fig. 4.24: Differences between global August sea-surface temperatures 18 000 years ago and modern values. Contour interval is 2 °C. Areas where the temperature change was greater than 4 °C are shown in light stippling. Ice-free areas are shown in darker stippling. Dashed lines on the continents are ice margins; dotted lines indicate sea-ice margins (after CLIMAP 1976).

The above examples refer only to research carried out in the Atlantic, but the CLIMAP scientists have also undertaken work in the Pacific and Indian Oceans. Their most spectacular contribution to date has probably been the maps they produced of the surface of the earth at 18 000 BP (CLIMAP 1976). These show the extent of glacier ice and vegetation zones of the continental areas and, most impressively, the differences in sea-surface temperatures between then and the present day (Fig. 4.24). The latter map brings out very clearly the extremely low temperatures that prevailed in the north Atlantic at that time, a point that will be considered in more detail in Chapter 7.

CONCLUSIONS

Despite the very considerable advances that have been made in recent years in physical and chemical methods of investigation, biological evidence in the form of fossil fauna and flora still remains probably the most effective means at our disposal for inferring past environmental conditions. The analysis of all forms of biological evidence, however, is time-consuming, often costly, and requires a very high level of specialisation, and these factors must be set against the type of information and the level of sophistication in the data that are being sought. No single technique can provide all the evidence that we need in order to understand fully the nature of Quaternary environments. The data sources outlined above each offer a slightly different perspective, and the point has been made repeatedly that the most fruitful lines of enquiry are frequently those in which two or more techniques are employed in conjunction, or where biological evidence is supported by geomorphological, sedimentological or geochemical data. In these circumstances, the tools are available to enable the Quaternary scientist to attempt a reconstruction not only of environments at specific periods in the past, but also a picture of changing environmental conditions through time. Before such a step can be taken, however, two further aspects of Quaternary research merit consideration, namely the development of a time-scale for environmental change, and the means whereby Quaternary sequences at widely separated localities can be correlated both spatially and temporally. These form the subject matter of the following two chapters.

NOTES

1. **Polar Front**: a prominent hydrographic/oceanographic boundary in the North Atlantic, termed the Subarctic Convergence, separates warm water of high salinity flowing northwards from cold low-salinity water flowing south from the Arctic. This boundary is informally termed the 'Polar Front' and controls weather patterns in the region.

2. CLIMAP: Climate/Long Range Investigation Mapping and Prediction. A team of earth and ocean scientists who came together in 1971 to carry out integrated research into the climatic history of the Quaternary. Their administrative base is the Lamont-Doherty Geological Observatory at Columbia University, New York.

CHAPTER 5

Dating methods

INTRODUCTION

Dating techniques in the Quaternary time range fall into three broad categories:

Methods that provide age estimates.   There are two types of dating technique that enable the age of fossils, sediments or rocks to be established in years BP. These are **radiometric methods** which rely on radioactive decay or related phenomena, and **incremental** methods which are based on the regular accumulation of biological or lithological materials through time.

Methods that establish age-equivalence.   These methods are concerned with the recognition of contemporaneous horizons in separate and often quite different stratigraphic successions. Certain distinctive stratigraphic markers are regionally and, in some cases, globally synchronous, and if these can be traced laterally between successions, they can be taken to represent time-planes within the different sedimentary sequences. If the age of the markers can be established in one locality by the application of any of the techniques in category (a), then equivalent horizons within other successions can be indirectly dated by correlation.

Relative age methods.   These techniques only establish the relative **order of antiquity** of fossils or stratigraphic units. The relative antiquity of geological materials is most obvious where superposition[1] can be established, but under certain circumstances, the relative age of Quaternary landforms and sedimentary units can be discerned from the degree of degradation or alteration resulting from the operation of chemical processes through time.

   In this chapter, the principal methods used in Quaternary dating are considered under these categories. The techniques and the applicable dating range of each method are shown in Fig. 5.1.

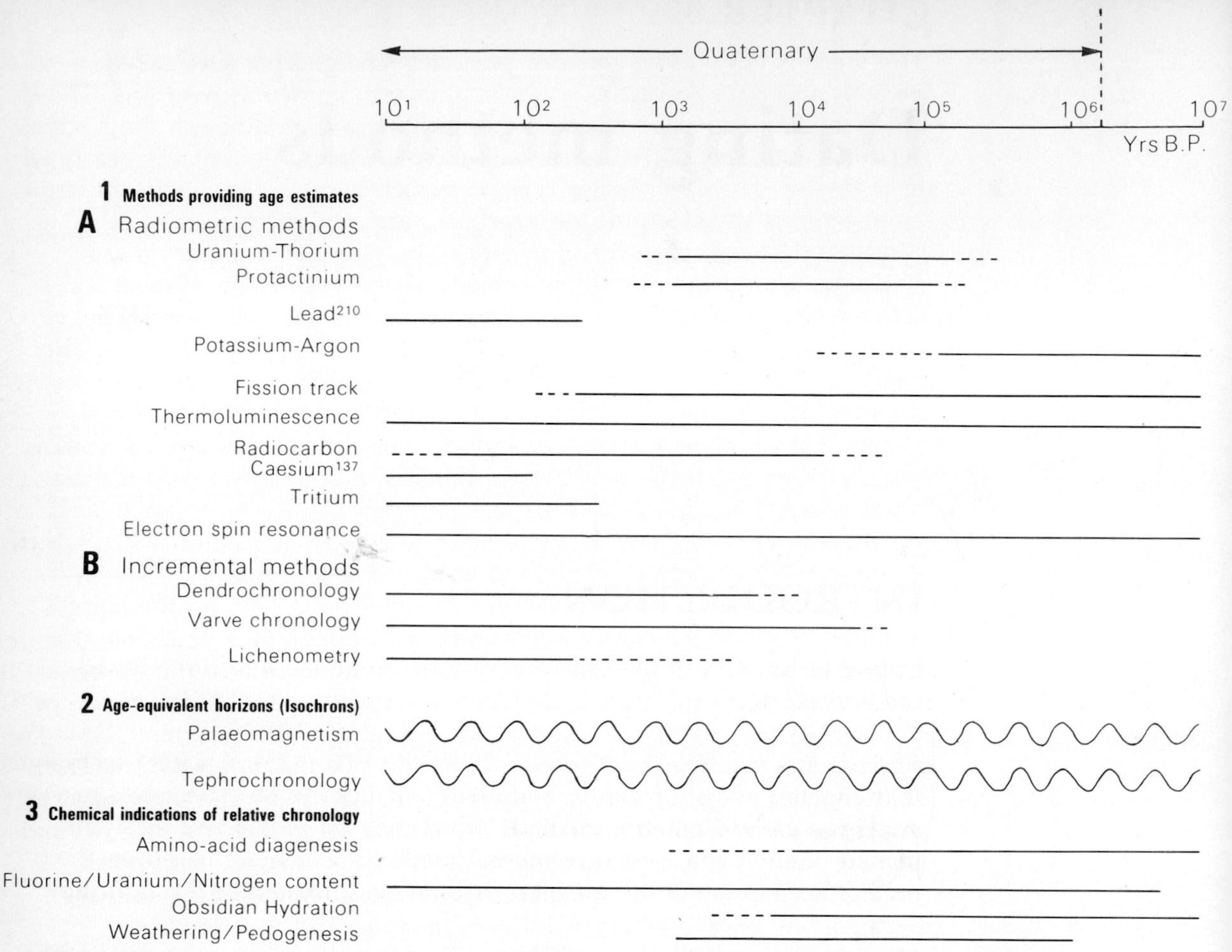

Fig. 5.1: Ranges of the various dating methods discussed in the text. Broken lines show possible extensions with improvements in techniques; wavy lines indicate that dating is limited to specific intervals within the Quaternary time-scale.

RADIOMETRIC DATING TECHNIQUES

Radiometric dating methods are based on the radioactive properties of certain unstable chemical elements, from which atomic particles are emitted in order to achieve a more stable atomic form. Some radioactive elements such as uranium occur naturally and are commonly found in rocks, sediments and fossils. **Radioactive decay** (atomic transformation) is **time-dependent**, and if the rate of decay is known, the age of the host rocks or fossils can be established. Rates of radioactive transformations are extremely variable; some elements decay in days or even seconds, whereas others transform gradually over millions of years. A number of radiometric dating methods have now been developed, but in this section only those techniques that are directly applicable to the Quaternary time-scale are discussed. Radiocarbon dating, however, is considered in more detail than the others for it is probably the most widely employed of all the radiometric dating methods.

The nucleus and radioactivity

The nucleus of an atom contains positively-charged particles called **protons** and particles with no electrical charge known as **neutrons**. These are extremely densely packed in the nucleus so that although the nucleus occupies only about 10^{-14} of the volume of an atom, it contains nearly all of the mass. The other major type of particle contributing to the structure of an atom is the **electron**. For practical purposes, electrons can be considered as tiny particles of negative charge, with negligible mass, spinning around the nucleus in orbitals. Electrons vary in number for different chemical elements, and they are arranged in electron shells, or orbitals, of different radial distance from the nucleus (e.g. Fig. 5.3). The analogy is often drawn between this arrangement and the planets orbiting the sun. Strictly speaking, this is not correct, as modern physics has shown that sub-atomic entities cannot be considered as discrete particles as such. However, for the purposes of this discussion, it will help if these nuclear units are regarded as nuclear particles.

Chemical elements are classified according to **atomic number** (Z), which is the number of protons contained in the nucleus. Hydrogen has an atomic number of 1, oxygen is 8 and uranium is 92. The **atomic mass number** (A) of an element is the number of protons plus neutrons; that of hydrogen is 1 and of oxygen is 16. It is convention to give the numerical value of A as a superscript and Z as a subscript on the left-hand side of the symbol for a chemical element, for example, $^{238}_{92}U$ (Uranium 238). The atomic mass number of elements can vary, since the number of neutrons in the nucleus is not always constant. Elements having the same number of protons, but a different number of neutrons (e.g. ^{16}O and ^{18}O; ^{12}C and ^{14}C) are known as **isotopes**. They have the same chemical properties, since Z remains constant for each element, but isotopes differ in mass. Each isotope of an element is called a **nuclide**. The particles that constitute the nucleus are bound together in a way that is not fully understood, but it is thought that if a nucleus contains too many neutrons, it becomes unbalanced and the repulsive forces between the similarly-charged particles overcome the binding forces keeping them together. This results in spontaneous emission of particles, which is the basis of radioactivity. Isotopes involved in such radioactive processes are known as **radioactive nuclides**.

Three types of particles are emitted during radioactive decay. **Alpha (α) particles** consist of two protons plus two neutrons and are the positively-charged nuclei of helium atoms. They collide with surrounding atoms and acquire electrons to form helium gas. Nuclides that emit alpha particles lose mass and positive charge. By this process, the atomic number changes, and thus one chemical element can be formed by the 'decomposition' of others. **Beta (β) particles** are negatively-charged electrons, and their emission does not alter mass, but changes atomic number. It is also possible for an electron to jump from one orbital into another, and in some rare cases they may even transfer into the nucleus. **Gamma rays (γ)** are not important in the formation of decay time constants, and are not discussed here.

The element that undergoes atomic transformation is termed the **parent nuclide** (or 'mother nuclide') and the product is the **daughter nuclide**. This single-stage transformation is known as **simple decay**. Many radioactive transformations, such as uranium series (see below), involve more complex pathways where the transformation of the nuclide with the highest atomic number to a stable nuclide involves the production of a number of intermediate unstable nuclides. This is known as **chain decay** (Fig. 5.2). Intermediate nuclides involved in such chains are both the product of previous transformations and the parents in subsequent radioactive decay, and such nuclides are termed **supported**. **Unsupported decay** involves the transformation of a parent nuclide that is not, in itself, the product of decay, or is separated from earlier nuclides in the chain as a result of biogenic or sedimentary processes.

Radioactive decay processes are governed by atomic constants or physical laws. The number of transformations per unit time is proportional to the number of atoms present, and for each nuclide there is a **decay constant** (λ) which represents the probability that an atom will decay in a given period of time. The transformation of an individual atom occurs spontaneously and unpredictably, but where a large number of atoms of a particular nuclide are considered, there is a predictable time rate at which overall disintegration proceeds. The law of radioactive decay is given by:

$$\frac{\delta N}{\delta t} = \lambda N$$

Fig. 5.2: Chain decay pathways and half-lives of intermediate nuclides during the decay of ^{238}U, ^{235}U and ^{232}Th to stable lead. The elements are arranged vertically according to atomic number. Loss of an α particle leads to a decrease in atomic number, whereas emission of a β- particle leads to an increase. Some of the extremely short-lived nuclides have been omitted (h = hours; m = minutes; s = seconds).

where N is the number of atoms, t is a time constant, and λ is the decay constant for that nuclide. For all nuclides the rate of decay is exponential (see Fig. 5.5), and is best considered in terms of the **half life ($t_{\frac{1}{2}}$)**. This is the period of time required to reduce a given quantity of a parent nuclide to one half. For example, if 1 gm of a parent nuclide is left to decay, after $t_{\frac{1}{2}}$ only 0.5 gm of that parent will remain. It will then take the same period of time to reduce that 0.5 gm to 0.25 gm, and to reduce the 0.25 gm to

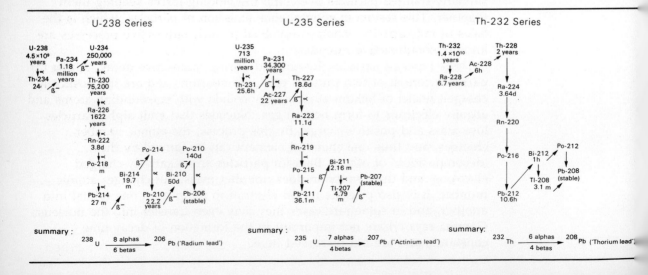

0.125 gm, and so on. The relation between the half-life and the decay constant is given as:

$$t_{\frac{1}{2}} = \frac{\log_e 2}{\lambda} = \frac{0.693}{\lambda}$$

The application of the principle of radioactivity to geological dating requires that certain fundamental conditions be met. If an event (such as the cooling of magma, the accumulation of a body of sediment, the death of an animal and the burial of its bones etc) is associated with the incorporation of a radioactive nuclide, then providing (a) that none of the daughter nuclides are present in the initial stages and (b) that no parent or daughter nuclides are added to or lost from the materials to be dated (*i.e.* the radioactive process has proceeded within a *'closed system'*), then an estimate of the age of that event can be obtained if the ratio between parent and daughter nuclides can be established, and if the decay rate is known. All estimates of time derived in this way are termed **radiometric clocks**; some methods are based on measurements of the progressive disappearance of nuclides during disintegration, while others ('accumulation clocks') measure the increasing quantity of a particular nuclide through time.

Uranium-series dating

[238]Uranium, [235]Uranium and [232]Thorium all decay to stable lead isotopes through complex decay series of intermediate nuclides with widely differing half-lives (Fig. 5.2). The helium gas formed by α particle emission may become trapped within host rocks, or may slowly diffuse out, ultimately to be liberated into the atmosphere. In theory, the age of a rock or mineral can be obtained from the amount of stable lead produced, or from the amount of helium liberated, but these measures are really only applicable to the dating of extremely old rocks (Jäger and Hunziker 1979). Within the limited time-scale of the Quaternary, only those intermediate nuclides with relatively short half-lives can be employed. However, nuclides with half-lives of only a few years or less are impractical for dating and even the intermediate nuclides with half-lives of hundreds or thousands of years cannot be used for dating of materials where radioactive disintegration has proceeded in an undisturbed system. This is because in most rocks, an equilibrium state has been achieved in which nuclides formed by decay are disintegrating at rates similar to their rate of production by the parent nuclide. If uranium decay is to form a basis for the dating of Quaternary events, the chain must be interrupted and some decay products selectively removed. When this occurs in nature, the techniques known as the **disequilibrium methods of uranium-series decay** can be employed.

Disequilibrium methods are based on the following geochemical principles. Uranium and weathering products containing uranium are highly-soluble, whereas other products of the U-series, such as thorium and protactinium, are readily absorbed or precipitated. Thus thorium and protactinium are co-precipitated with other salts and filter out with sediments to accumulate on the floors of lakes and on the sea bed, while uranium remains in solution. A selective separation, or **fractionation**,[2] of these decay products therefore occurs. Accumulating sediments will contain quantities of thorium and protactinium but will be deficient in uranium, whilst organisms that secrete carbonate direct from ocean waters will build a carbonate shell that contains uranium, but very little thorium or protactinium. In theory, therefore, age can be computed of lake or ocean floor sediments by measuring the rate of decay of thorium or protactinium within the sediments, while the age of carbonate fossils can be derived from the accumulation of decay products of uranium within the carbonate matrix. The former is based on the decay of *unsupported* intermediate nuclides, while the latter measures the accumulation of decay products of uranium which are *supported* (see Ku 1976).

The first of these methods has been used to establish rates of ocean sedimentation. If the disequilibrium system described above has proceeded without interruption, the ^{230}Th content of ocean floor deposits should decrease with depth from the present-day sediment surface. By measuring ^{230}Th content through a sediment column, therefore, the ages of individual horizons can be computed. The half-life of ^{230}Th is 75 200 years, and since quantities of ^{230}Th are normally measurable to about five half-lives, sediments up to *ca*. 350 000 years in age can be dated (Broecker 1963, 1965; Shotton 1967). Thorium was formerly named ionium and the method is still often referred to as the **Ionium method** or **Ionium-excess method**. In a similar way ^{231}Pa, with a half life of 34 000 years, can be used to date sediments back to *ca*. 150 000 years. An error source in both methods, however, is the occurrence of daughter nuclides derived from uranium that was already present in the sediments. Moreover, because of variations in the rate of precipitation of nuclides over time, non-linear plots of nuclide content with depth will often occur (Broecker and Ku 1969; Ku 1976). Hence specific horizons within the sediment column may be difficult to date accurately.

Of more limited application, but nevertheless of increasing importance in Quaternary dating, is the **Lead-210 method**. ^{210}Pb is a member of the ^{238}U decay series with a half-life of only 22 years. ^{226}Ra (radium) decays to yield the gas radon, which escapes into the atmosphere. Radon decays via a series of very short-lived nuclides to ^{210}Pb which is eventually removed from the atmosphere by precipitation or fall-out. This unsupported ^{210}Pb accumulates in lake and ocean sediments where it decays. Measurement of ^{210}Pb content can be used to establish rates of sedimentation in lakes over a time-scale of the past 150 years or so (Olsson 1979; Wise 1980).

The second disequilibrium method measures the accumulation of daughter products of uranium. Since ocean waters and lake waters contain uranium but negligible amounts of ^{231}Pa and ^{230}Th, the extent to which protactinium and thorium have reappeared in the skeletal, shell or inorganic carbonates precipitated in lake or ocean waters is a direct reflection of age. The ^{230}Th/^{234}U method in particular has been used to date a range of Quaternary materials, including the following:

Bone.   Following the death of an animal, uranium enters and is trapped within the bone apatite (see p. 206). The age of fossil bone can be determined using either the ^{230}Th/^{234}U or ^{231}Pa/^{235}U methods (Szabo and Rosholt 1969; Howell *et al.* 1972). Usually both ratios are measured, and the samples are accepted as having remained closed systems during isotopic changes only if the computed ages are concordant. One problem with dating bone, however, is that there is a time lag of *ca.* 2700 years between death of an animal and uranium uptake (Szabo 1980).

Speleothems.   A recent development in ^{230}Th/^{234}U dating is its application to cave calcite precipitates – stalagmites, stalactites, flow-stones, etc, the palaeoclimatic significance of which has already been considered (Ch. 3). ^{234}U is precipitated from karst waters during the formation of speleothem carbonate, and almost all of the ^{230}Th subsequently found appears to be authigenic, *i.e.* it has originated from decay of ^{234}U that forms part of the speleothem chemistry. Some detrital ^{230}Th may enter cave systems, however, although this can be corrected for by measuring ^{232}Th content, since ^{232}Th is invariably associated with detrital ^{230}Th (Ford *et al.* 1972).

Lacustrine deposits. ^{230}Th ages have been obtained from carbonate materials from the dried-up lake beds of ancient Lakes Lahontan and Bonneville in the south-west of the United States (Kaufman and Broecker 1965), and also from pluvial Searles Lake in California (Peng *et al.* 1978). The dates from Searles Lake range from 24 500 to 32 000 BP, and compare favourably with radiocarbon dates from the same site (see Ch. 3). One problem with dating lake limestones, however, is that they contain detrital uranium and thorium nuclides derived from the surrounding country rock and these may be difficult to isolate in dated samples (Kaufman 1971; Szabo and Butzer 1979).

Peat.   Peat, along with other organic material, takes up uranium from groundwaters and can become relatively enriched in the element (Vogel and Kronfeld 1980). Since peat has a high adsorption capacity, any percolating groundwater will transfer its uranium content to the upper peat surface layer, thus protecting the older layers from further acquisition of uranium. Initial U-series dating at several European sites has yielded encouraging results, although again, the incorporation of uranium-bearing inorganic detrital material into the peat deposits has proved to be a problem. Nevertheless, refinement of the method offers the exciting prospect of dating interglacial peats and other organic deposits that currently lie beyond the range of the radiocarbon technique.

Molluscs.   Thus far, the use of fossil molluscs has been relatively unsuccessful, due partly to the fact that they contain initially only very small amounts of uranium (one fiftieth of that contained in corals, for example), and also to the fact that they do not function as closed systems, for diagenetic uptake of uranium is very common following death of the organism (Broecker and Bender 1972). Hence, uranium-series dates on molluscs have tended to be regarded as unreliable (Kaufman *et al.* 1971). On the other hand, apparently meaningful dates have been obtained on, for example, Mollusca from interglacial high shorelines at a number of localities around the Mediterranean basin (Stearns and Thurber 1965; Butzer 1975). In view of the fact that many of the strandlines that need to be dated are not characterised by coral (a more suitable dating medium), but often contain abundant Mollusca, attempts are likely to continue to find a means of obtaining reliable uranium-series dates from fossil molluscs.

Corals.   Corals deposit aragonite skeletons containing 2–3 ppm uranium, but virtually no thorium. It is generally believed that after death, coral skeletons act as closed systems until the coral is dissolved or changes to calcite (Neumann and Moore 1975). Measurements of age must be based on unaltered coral, since recrystallisation to calcite can result in preferential loss of uranium relative to thorium and protactinium (Ku 1976), a process which has resulted in conflicting interpretations of uranium-series dates based on fossil coral (Marshall and Launay 1978). Nevertheless, uranium-series dating of coral reef complexes has produced valuable information on the chronology of changing sea-levels and rates of land uplift during the late Quaternary. For example, dates on both emergent and submerged reef complexes in the Bahamas (Neumann and Moore 1975), Bermuda (Harmon *et al.* 1981) and Florida (Broecker and Thurber 1965), all areas considered to be tectonically stable, show that sea-level during the last interglacial (*ca.* 125 000 BP) stood at +5 m relative to present, whereas by 95 000 BP it had fallen by some 20 m. On the other hand, U-series dates on the remarkable 'staircase' of raised coral reefs on the Huon Peninsula of New Guinea show that more than twenty reef complexes have emerged during the past 250 000 years, giving rates of uplift of between 1.0 and 2.5 m per 1000 years for different parts of the coastline (Bloom *et al.* 1974).

Two methods that allow materials older than 350 000 years to be dated are the $^{234}U/^{238}U$ method and the U-He method. The first measures the excess of ^{234}U over ^{238}U in natural waters, and may be useful for dating marine deposits up to one million years in age, although there are problems in dating pelagic deposits (Ku 1976). He^4/U ratios have already been applied to dating corals up to about 600 000 years in age (Bender *et al.* 1979), and appear to offer the possibility of dating throughout the Quaternary time-scale (Bender 1973).

Potassium-Argon ($^{40}K/^{40}Ar$) dating

Potassium-argon dating is a technique that allows the age of volcanic rocks to be established. ^{40}K is a radioactive nuclide that undergoes **branching decay**, leading to one of two daughter nuclides depending on the type of transformation (Fig. 5.3). Most ^{40}K decays by β emission to produce ^{40}Ca, each particle emitted from the nucleus resulting in the conversion of a neutron to a proton. The atomic number is therefore increased by one, resulting in an element of different chemical properties, but with a virtually unchanged atomic mass. Electron capture (from the surrounding electron shells) by the nucleus is the alternative radioactive process. This converts a proton into a neutron (Fig. 5.3) to reduce the atomic number by one and yield argon, a gas. Only one of these branches, the $^{40}K/^{40}Ar$ pathway, is useful for dating, for ^{40}Ca is so ubiquitous in nature that it is not possible to separate ^{40}Ca atoms produced by the decay of ^{40}K from those already present in rocks at the time of formation.

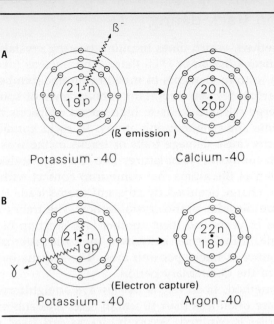

Fig. 5.3: Branching decay of ^{40}K. A: conversion of atoms of ^{40}K to ^{40}Ca through the emission of a β- particle from the nucleus. B: conversion of ^{40}K to ^{40}Ar through electron capture by the nucleus from one of the electron shells.

$^{40}K/^{40}Ar$ transformation is an accumulation radiometric clock. Dating by this method relies on an accurate measure of argon concentration that has accumulated through decay, the ratio of ^{40}K to ^{40}Ar reflecting the date of formation of the rock. Dates will only be valid therefore, for host rocks from which argon gas cannot escape. Fortunately, many mineral lattices retain argon, and it is only if rocks become melted, recrystallised or heated to a critical temperature that substantial loss of argon will occur. In the case of a volcanic rock that has been reheated or metamorphosed the method will determine the age at which the final phase of modification ceased and not the age of initial formation.

Radiometric dating by the $^{40}K/^{40}Ar$ method is largely restricted to volcanic and metamorphic rocks, since sedimentary rocks do not retain argon. Yet not all volcanic and metamorphic rocks are suitable, for sufficient potassium must be present to make dating possible. The nature of the mineral lattice is also an important factor, for not all minerals retain argon over long time periods, particularly when under stress. Orthoclase and microcline, for example, do not retain argon well. Biotite, on the other hand, is one of the most suitable minerals, for not only does it retain argon, but it is also rich in potassium. A further constraint in Quaternary research is that the method is only really effective in the time range >100 000 years, for younger samples tend to yield age determinations with standard errors so large as to make the date effectively meaningless. The principles and applications of this method are discussed in detail in Dalrymple and Lanphere (1969).

The principal contributions of $^{40}K/^{40}Ar$ dating in Quaternary research have been in the dating of lava flows in East Africa that are intercalated between sedimentary deposits containing remains of early hominids (Hay 1967, 1973; Fitch et al. 1976), thereby providing a time-scale for the early stages of human evolution, and also in the provision of a time-scale for the palaeomagnetic stratigraphic sequence (see below).

Fission track dating

This method, which dates uranium-bearing crystals, is based on the **spontaneous fission** of ^{238}U: that is the nucleus (of atomic number 92) divides to form elements of medium atomic number from about 30 to 65 (e.g. barium 56). An important consequence of spontaneous fission is that the energy released leads to high-speed collisions between fission fragments and neighbouring atoms. In rocks containing uranium, fission fragments cause damage trails or **tracks** in the wake of their movement through the host crystal lattice. The 'damage' induced is a result of ionisation of the atoms that come into contact with fission fragments. The positive charge acquired by adjacent atoms leads to mutual repulsion and therefore disorder in the crystal lattice. The tracks can be retained for millions of years and their number is a function of both uranium content and time. If the former is known, then the latter can be computed from the number of tracks per unit area. The technique is applicable to the whole of the Quaternary period.

The method, however, is not always straightforward. First, fission tracks are often less than 10 μm in size, and sub-microscopic examination of samples is required. Secondly, tracks can 'heal' or be erased (a process known as **annealing**) through the heating of the host materials or through spontaneous diffusion of ions. The longevity of fission tracks depends on temperature and rock type (Fleischer *et al.* 1965). Thirdly, since the density of fission tracks depends on uranium content and age, the method cannot be applied where the density of tracks is too low or, at the other extreme, so high that counting becomes impossible. Ironically, rocks with a rich uranium content are not suitable because of the very high density of tracks. Laboratory procedures are described and fission tracks illustrated in Fleischer *et al.* (1969), Green (1979) and Naeser (1979).

Materials that have been dated by this method range from man-made glass less than 1000 years old (in some cases as young as 20 years old), to some of the oldest rocks on earth. Geological applications have concentrated on the dating of volcanic rocks several million years old, where the age of mineral formation or the thermal history of the rocks is under investigation (Naeser 1979). The use of fission track dating in the Quaternary has so far been restricted largely to archaeological materials, or to the dating of volcanic rocks as a cross-check on other methods (e.g. Hurford *et al.* 1976). However, the technique has been employed with some success in the dating of microtektites in deep-sea sediments, where the quantities of available material were too small for $^{40}K/^{40}Ar$ dating (Gentner *et al.* 1970). Fission track dating may therefore have a wider application in Quaternary research than has hitherto been considered.

Thermoluminescence (TL) dating

Any material that contains uranium, thorium or potassium (many sediments and volcanic rocks contain all three), or lies in close proximity to other materials containing these radioactive substances, is subject to continuous bombardment by α, β and γ particles. This leads to ionisation in the host materials and the 'trapping' of metastable electrons within minerals. These electrons can be freed by heating, and under controlled conditions, a characteristic emission of light occurs which is proportional to the number of electrons trapped within the crystal lattice. This is termed **thermoluminescence**, and the light emitted is additional to the normal incandescent light which would result from heating. The latter will be emitted each time a sample is heated, whereas thermoluminescence, once liberated, can only reappear after further exposure to radiation. Thermoluminescent properties will accumulate progressively in a sample exposed to continuous radiation. Hence, thermoluminescent intensity is a product of **radiation dose rate** and time; if the former can be calculated then an age can be assigned to the onset of electron agitation and trapping.

Thermoluminescent data are commonly expressed as a plot of the intensity of light emitted by the sample as it is heated from room temperature through to temperatures in excess of 400 °C (Fig. 5.4). This is termed a **glow curve**, in which peaks are found at temperatures characteristic of the energies of the trapped electrons in the sample (Melcher 1981). A natural glow curve (curve a, Fig. 5.4) exhibits the thermoluminescent properties inherent in a geological sample. Before age can be calculated, two further measures need to be established: an **equivalent dose rate**, obtained by exposing the sample to a known dose from an artificial radioactive source and then measuring the induced glow curves (curves b and c, Fig 5.4), and an **annual dose rate**, which is computed from the radioactive content of the sample (*i.e.* amounts of uranium, thorium and potassium). Age is then given by:

$$\text{TL age} = \frac{\text{(natural TL)}}{\text{(equivalent dose rate)} \times \text{(annual dose rate)}}$$

(Aitken 1974)

Until recently, thermoluminescence was restricted largely to the dating of archaeological samples, especially pottery. TL properties are lost upon

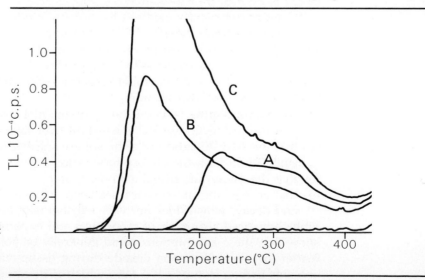

Fig. 5.4: Thermoluminescent glow curves for loess samples of last glacial age. For explanation see text (after Wintle 1981).

heating so that fired pottery will have zero TL at the time of cooling and intensity will build up over time after the initial cooling. More recently, some exciting new developments in TL-dating have been reported, for it now appears that exposure to sunlight is sufficient to release TL in sediments. It may be possible, therefore, to date deep-sea sediments (Wintle and Huntley 1979) and loess deposits (Wintle 1981), in which TL properties have built up after sedimentation, providing that TL has been reduced to zero in the sedimentary particles during transportation and exposure. The technique and its applications are discussed in detail in Dreimanis *et al.* (1979), Wagner (1979) and Wintle and Huntley (1982).

Radiocarbon dating

The principles of radiocarbon dating were first established by Libby (1955), who synthesised evidence from radiochemistry and nuclear physics to determine the effects of high energy cosmic radiation (the **cosmic-ray flux**) on the biosphere. Free neutrons resulting from nuclear reactions in the upper atmosphere collide with other atoms and molecules, and one effect is the displacement of protons from nitrogen atoms to produce carbon atoms:

$$\begin{array}{ccc} \text{7 neutrons} & & \text{8 neutrons} \\ & N + \text{neutron} \longrightarrow & C + \text{proton} \\ \text{7 protons} & & \text{6 protons} \end{array}$$

The carbon nucleus produced by this reaction, ^{14}C, is a radioactive isotope of carbon which eventually decays to form the stable element ^{14}N:

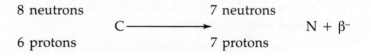

$$\begin{array}{ccc} \text{8 neutrons} & & \text{7 neutrons} \\ C \longrightarrow & & N + \beta^- \\ \text{6 protons} & & \text{7 protons} \end{array}$$

Decay is by beta transformation.

^{14}C atoms are rapidly oxidised to carbon dioxide and, along with other molecules of carbon dioxide ($^{12}CO_2$), become mixed throughout the atmosphere and absorbed by the oceans and by living organisms. In other words, ^{14}C, which is continually being produced in the upper atmosphere, becomes stored in various **global reservoirs**, the atmosphere, the biosphere and the hydrosphere.

Radiocarbon dating rests on four fundamental assumptions:
(a) that the production of ^{14}C is constant over time;
(b) that the ^{14}C: ^{12}C ratio in the biosphere and hydrosphere is in equilibrium with the atmospheric ratio;
(c) that the decay rate of ^{14}C is known; and
(d) that an equilibrium constant has been achieved between production and decay, which does not vary substantially over time.

Libby advanced a number of theoretical and experimental arguments to show that these assumptions could generally be accepted. All living matter that absorbs carbon dioxide during tissue-building, therefore, contains carbon isotopes in a ratio which is in equilibrium with atmospheric carbon dioxide. As long as the organic tissues continue to function, decay of ^{14}C is constantly being replenished by new ^{14}C. Upon death, ^{14}C within the organic tissues will continue to decay, but no replacement takes place. Hence, if the rate of decay of ^{14}C is known, date of death can be computed from the measured ^{14}C activity.

The activity of ^{14}C is approximately 15 dpm/gm (15 disintegrations per minute per gram), and this activity is halved every 5700 years or so (Fig. 5.5). The half-life of ^{14}C was originally calculated at 5568 ± 30 years (Libby 1955), but subsequently this has been more accurately determined as 5730 ± 40 years. Because a large number of ^{14}C dates were published prior to the calculation of the new half-life, it has been convention to base radiocarbon dates on the former half-life value. This avoids confusion as dates calculated using the same half-life, irrespective of value, are directly comparable (Godwin 1962). Conversion to the longer half-life can be made by multiplying radiocarbon ages based on the standard half-life by 1.03.

Measurement of ^{14}C activity

In order to detect ^{14}C decay in radioactive samples, extremely sensitive equipment is required. This is because samples supplied to laboratories are often small, and the natural occurrence of ^{14}C is such that for every one million million atoms of ^{12}C in a living organism there is only one atom of ^{14}C. The measurement procedure involves the detection and counting of emissions from ^{14}C atoms over a period of time to determine the rate of emissions and hence the activity of the sample. Beta transformation is a low energy process, and it is often difficult to distinguish beta particles from other particles of a similar energy value. Counters must therefore be shielded in aged (non-radioactive) lead, and surrounded by protective machinery that helps to detect and minimise background radiation from the surroundings or from materials of which the counter is constructed.

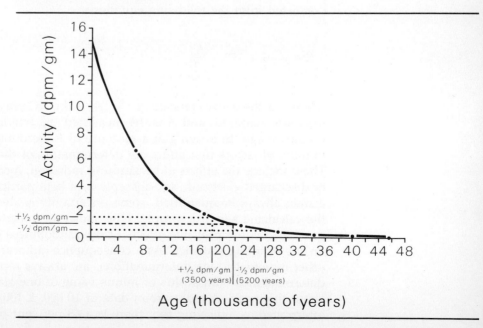

Fig. 5.5: Decay curve for radiocarbon.

Two principal methods are employed in conventional radiocarbon laboratories: **gas proportional counting** and **liquid scintillation counting**. In the former, a suitable gas (usually carbon dioxide, ethylene or methane) is prepared from the carbon in the sample and collected in a chamber, down the centre of which runs a wire with a positive charge. Discharged electrons are attracted to, and neutralised by, the positive charge and these can be counted by electronic means (Shotton 1977a). In liquid scintillation counting, samples are normally converted to benzene and this is mixed with an organic solvent and a 'scintillator'. The latter emits pulses of light (photons) as a result of radioactive disintegrations, the amount of light produced being proportional to the energy released during ^{14}C decay. The light pulses can be counted photoelectrically (Burleigh 1972).

From the decay curve (Fig. 5.5), it can be seen that material approximately 10 000 years old will have an activity of only 4 dpm/gm, and older samples correspondingly lower values. The limit of practical counting using conventional methods is eight half-lives (about 40 000 radiocarbon years), for beyond that age the curve becomes so flat and insensitive that it is difficult to separate samples of different activities with any statistical certainty. However, some techniques have been developed that allow an extension of the dating range of radiocarbon, and these are discussed later in this chapter.

In the calculation of radiocarbon dates, laboratories compare sample activities to a **modern reference standard**, namely 0.95 oxalic acid NBS (*i.e.* 0.95 times the activity of a bulk stock of oxalic acid held by the American Bureau of Standards). The reason for this is that there have been variations in ^{14}C production rates through time, and modern levels are artificially high (see below). Comparability between laboratories and between samples of different ages therefore requires reference to a standard. The time in years (R) since the death of an organism can be calculated from the following equation:

$$R = \frac{1}{\lambda} \log_e \frac{A_o}{A}$$

where λ is the decay constant of ^{14}C, A_o is the ^{14}C activity of the modern reference standard, and A is the measured ^{14}C activity of the sample of unknown age. In arriving at a measure of R, account has to be taken of a number of factors that affect the determination of the activity of a sample. These include the effects of background radiation, incorrect monitoring of discharged electrons, self-absorption of beta particles, etc. Because these cannot always be quantified, some uncertainty is always associated with the calculated age of a sample. One source of uncertainty that can be quantified is the probable effects of the randomness of radioactive decay on the counting statistics. As a consequence radiocarbon dates, along with other radiometric age determinations, are always reported as mean determinations with a plus or minus value of one standard deviation about the mean. A radiocarbon date of 10 000 ± 100 years should be interpreted as indicating that there is a 68 per cent probability that the true age of the sample lies between 9900 and 10 100 years BP. For a 95 per cent probability the range would have to be doubled. Even with two standard deviations on the 'date', there is still a one in twenty chance that the true age lies outside the range 9800 to 10 200 years BP.

It is important to remember, however, that age is not the factor that is being measured, but *activity of the sample* which, on the basis of a number of assumptions, is interpreted as indicating 'age'. The plus or minus refers to the uncertainty associated with determining activity, and this is why on older dates the plus value of the standard deviation is often quoted as being larger than the minus value. This can be seen in Fig. 5.5. by considering an activity of 1 dpm/gm ± 0.5 dpm/gm, which translated into 'age' gives a significantly higher plus than minus value. Because of the asymptotic decay curve, technically there is always a difference between the plus and minus value, but this is regarded as insignificant for younger material.

Sources of error in ^{14}C dating

Temporal variations in ^{14}C production.   Libby's assumption that ^{14}C levels had not varied significantly over time were first challenged by de Vries (1958). He showed that wood samples from the seventeenth century had a radiocarbon content some 2 per cent higher than that expected on the basis of comparison with nineteenth-century wood, and he demonstrated further discrepancies between radiocarbon and calendar dates in the period back to about 1500 AD. It was initially believed that this 'de Vries effect' was a phenomenon unique to the last 2000 years (Suess 1965), but evidence from tree-ring records now indicates that long-term fluctuations in ^{14}C activity have occurred (see below).

Suess (1970a) suggested three possible causes of secular variations in ^{14}C activity:

1. Variations in the geomagnetic flux, for a close geographical relationship has been found between magnetic field intensity and fluctuations in ^{14}C levels (Bucha 1970).
2. Modulation of cosmic-ray flux, where increased activity by the sun (solar flares) leads to attenuation of the cosmic flux, which in turn reduces the amount of neutron bombardment in the upper atmosphere.
3. The influence of the oceans, for the ocean reservoir is by far the dominant global reservoir, and any change in the rate of transfer of atmospheric ^{14}C into the oceans will affect the level of ^{14}C in the atmosphere, and ultimately in the biosphere.

In addition to natural variations, atmospheric ^{14}C levels have recently been affected by the activities of man. Over the past 100 years ^{14}C levels have been markedly diluted as a result of the combustion of fossil fuels, which has liberated large quantities of 'inert' ^{12}C into the atmosphere. In the last twenty-five years or so, however, this **industrial effect** has been offset by greatly increased production of ^{14}C resulting from the detonation of thermonuclear devices. The combined effects of industrial activity and atomic explosions means that modern organic samples are unsuitable as reference samples for radiocarbon activity. A value of 0.95 times the measured activity of the NBS standard is regarded as equivalent to the natural ^{14}C activity of AD 1890 wood (pre-industrial effect), and this is corrected to AD 1950, the reference year for all ages quoted in ^{14}C years BP. Temporal variations of atmospheric ^{14}C levels are discussed further by Damon *et al.* (1978).

Spatial variations in atmospheric ^{14}C activity. Despite the fact that atmospheric mixing is extremely rapid, there is evidence to suggest that, because of altitudinal or unusual local conditions, some areas may be characterised by atmospheric ^{14}C levels that depart significantly from the norm. Comparisons of ^{14}C levels in present-day and recent tree-ring sequences suggest considerable global variations, with higher levels in particular being recorded in North American material (Harkness and Burleigh 1974).

Isotopic fractionation.   There are three naturally-occurring isotopes of carbon and, of the total in circulation, about 98.9 per cent is ^{12}C, 1.1 per cent is ^{13}C and only 1.18×10^{-10} per cent is ^{14}C (Olsson 1968). In nature, however, a fractionation of this ratio commonly occurs (Craig 1953). Photosynthesis, for example, results in an enrichment of ^{12}C relative to the other isotopes in most plant tissues, whereas ocean waters preferentially absorb ^{14}C. These effects are small, but can significantly affect radiocarbon dates where measurement to less than $\pm$ 1 per cent error is required (Harkness 1979). In addition, fractionation can also occur in the laboratory induced by the conversion of sample carbon to the gas or liquid form. Most radiocarbon laboratories today make corrections for the probable effect of fractionation on samples. This is possible because there are good experimental and theoretical grounds for the general rule that, in nature, any isotopic adjustment of the ^{14}C:^{12}C ratio is about double that of the ^{13}C:^{12}C ratio. The latter can be measured in a small sub-sample of the material to be dated. The ratio is compared with a standard, which is PDB limestone (belemnite carbonate from the Cretaceous Peedee Formation of South Carolina), and values are published as deviations from this standard:

$$\delta^{13}C\text{‰} = \frac{^{13}C/^{12}C \text{ sample} - {}^{13}C/^{12}C \text{ standard}}{^{13}C/^{12}C \text{ standard}} \times 1\,000$$

Most samples have a negative δ^{13}C value compared to the PDB standard. For example, the normal range for wood, peat, gyttja and charcoal samples observed from European localities is -25 ± 5‰ (Oeschger *et al.* 1970), although lower values have also been noted, especially in aquatic plant tissues (Lerman 1972; Troughton 1972). Nevertheless, it is practice to 'normalise' ^{14}C activities during calculations of ^{14}C age, by treating each sample as if an average enrichment had taken place. The normal value is taken to be -25‰, which is the mean isotopic composition of wood. If the δ^{13}C value of a sample is found to be -25‰, then no adjustment is made. With a value of -30‰, however, a 5‰ depletion in the ^{13}C:^{12}C ratio is implied, which in turn indicates a probable 10‰ depletion in the

$^{14}C:^{12}C$ ratio. Thus, the measured ^{14}C activity would be increased by 10‰, which is equivalent to about 83 years (Harkness 1979).

Circulation of marine carbon.   Measurements of recent sea water samples have revealed that ^{14}C activity is often much lower than in contemporaneous atmospheric samples. Sea waters, therefore, have an **apparent age**. This amounts to about 300 to 520 years (if the reduced ^{14}C activity is translated literally) in samples from the surface waters of the North Atlantic, but discrepancies equivalent to 1000 years (reduced activity) have been found in deep-water samples (Broecker *et al.* 1960). This is due to the fact that ^{14}C is transferred only across the ocean-atmosphere interface, and there is extremely slow mixing of surface and deep waters. Cold water sinks in the northern latitudes of the Atlantic and flows very gradually southwards as North Atlantic Deep Water. While at depth, ^{14}C in the water decays in the absence of replenishment from the atmosphere. In the Pacific Ocean, **residence time** in deep water is apparently longer than in the Atlantic, for apparent ages of up to 2300 years have been recorded (Bien *et al.* 1963). Upwelling of deep water occurs near many coastlines, leading to reduced $^{14}C:^{12}C$ ratios in surface waters, and hence in molluscs and in other marine organisms inhabiting such areas. ^{14}C dates from marine fossils must therefore be adjusted to take account of the 'apparent age' of the fossils, different correction factors being required for different areas. Measures of apparent ages include 450 ± 40 years from the fiords of Norway, 550 ± 50 years for coastal localities of Iceland and south-west Greenland, 420 to 670 years for Spitsbergen (Mangerud 1972), about 750 years for Ellesmere Island, Arctic Canada (Mangerud and Gulliksen 1975), and 450 ± 50 years for estuarine waters of Scotland (Harkness 1979). The application of such correction factors assumes, of course, that oceanic circulation patterns have not changed significantly over the time period concerned.

Hard-water error.   In areas of carbonate rocks (limestones, chalk, coal measures etc), water is often rich in dissolved carbonate or particles of coal and other carbon-based compounds. Inert carbon is therefore found in unusually high proportions in such areas, diluting the local $^{14}C:^{12}C$ ratio. Sub-aquatic photosynthesis, water uptake in carbonate-rich groundwaters, and carbonate secretion by freshwater or offshore organisms may all be affected by such diluted ^{14}C levels. The resulting age error is termed **hard-water error** (Deevey *et al.* 1954; Shotton 1972). This problem may not be restricted only to areas of carbon-rich rocks or substrates, but may also affect sediments accumulating in newly-deglaciated areas. Rock flour in newly-deglaciated terrain may provide a supply of inert carbon that has been released as a result of comminution of igneous or metamorphic rock to yield silt and clay fractions that are easily transported in local groundwaters and filtered into lakes. In this way, inert carbon can become incorporated into lake sediments and biota (Sutherland 1980). The degree of hard-water error associated with different levels of contamination is shown in Table 5.1. Samples that are initially rich in organic carbon will only be seriously affected where the amount of

Contamination (%)	Years older than true age
5	400
10	850
20	1800
30	2650
40	4100

Table 5.1: The effect of contamination by inert carbon on the true radiocarbon age of samples (after Harkness 1975).

contaminant is high. Many organic sediments such as lake gyttjas, however, typically consist of only 4 to 5 per cent (and sometimes as low as 1%) organic carbon, and therefore relatively small amounts of inert carbon incorporated into the sediments can produce significant hard-water errors.

Contamination. Contamination of samples can occur because older or younger carbon has been added to the sample prior to, or during, collection and measurement. The hard-water error is an example of contamination by older carbon. Similar errors may result from recycling of older carbon which need not necessarily be inert, for example from older soils or sediments, and their eventual accumulation in younger sediments. This problem is commonly encountered where man-induced soil erosion has occurred around lake-basin catchments resulting in the inwash of older carbon and the subsequent incorporation of this allochthonous material into the lake sediments (Oldfield 1978; Dickson *et al.* 1978). Possible contamination by younger carbon can arise from root penetration to lower levels, infiltration by younger humic acids through older peat or soil horizons, and the downward movement of younger sediments through bioturbation. The effects of contamination by younger carbon are shown in Table 5.2. Because of the high activity of modern carbon in comparison to fossil materials, relatively small amounts of contaminant can result in major errors in radiocarbon dates.

Problems of contamination are often encountered where shell and bone are used for dating, since carbon exchange takes place more readily in carbonate structures. Contamination may be by older or younger carbon, depending on the dissolved contaminants introduced, and can result from the gradual accumulation of particulates or solutional carbon in the interstices of the carbonate matrix, or from the recrystallisation of the carbonate matrix and an **exchange** of sample carbonate with contaminant carbon. In the case of molluscs, it is thought that the former type can be avoided by choosing samples that exhibit a 'tight' shell matrix (Mangerud 1972).

Exchange usually affects the outer parts of a shell more than the inner

True age (years)	Measured age as a result of –		
	1%	5%	10% contamination
600	540	160	modern
1 000	910	545	160
5 000	4 870	4 230	3 630
10 000	9 730	8 710	7 620
infinitely old	36 600	24 000	18 400

Table 5.2: The effect of contamination by modern carbon on the true radiocarbon age of samples (after Harkness 1975).

layers, and it has therefore become practice to date outer and inner layers of mollusc carbonate separately. These are referred to as 'outer' and 'inner' dates, and where a noticeable difference in age occurs, the inner date is usually preferred. It has to be assumed that recrystallisation has not affected the inner layers, but this cannot be substantiated from just two data points. Olsson and Blake (1961) have therefore suggested that the outer layer should be removed first by acid digestion and then the *remainder* should be separated into outer and inner samples. If the difference between the dates from these two parts is small, then the level of uncertainty is reduced. The radiocarbon dating of marine shells is discussed in more detail by Mangerud (1972).

In the case of bone material, carbon exchange after death is always likely to have occurred and there are now serious doubts about the suitability of this type of material for radiocarbon dating. However, Burleigh (1972) and Harkness (1975) maintain that the problem of carbon exchange does not affect the protein (collagen) fraction, and recommend the extraction of bone protein for dating purposes. This is only possible where bones have been extremely well preserved, for collagen breaks down rapidly following death of the animal.

Radiocarbon laboratories have standard pre-treatment procedures to guard against the possible effects of contamination, but these cannot counteract nor indeed detect all forms of contamination. Samples are first checked for the presence of rootlets and other obvious signs of 'foreign' matter, and they are then usually cut or ground into small portions, examined and sieved to remove dust. This is followed by treatment with hot dilute acid to remove carbonates and with alkali solutions to remove humic acids that may coat the samples. Admixtures of chemically identical but non-contemporaneous materials cannot, however, be detected, and the onus is naturally upon the collector to ensure the utmost care in selection, handling and despatch of samples to the radiocarbon laboratory.

Radiocarbon dating of soil

One of the most difficult materials to date by the ^{14}C method is soil. All soils contain both organic and inorganic carbon and can, therefore, be dated by radiocarbon. However, soils are dynamic systems and receive organic matter over long time periods. Any radiocarbon date on a soil will therefore be heavily influenced by the **mean residence time** of the various organic fractions in the soil (Geyh *et al.* 1971). When a soil is buried, addition of organic matter ceases and a radiocarbon age will reflect both the mean residence time and the time that has elapsed since burial. The date at which pedogenesis commenced, which is of primary interest to the stratigrapher, will be almost impossible to establish. Further complications are added by the constant recycling that takes place within the soil profile, notably by humic acid filtration and root penetration (Campbell *et al.* 1967). Some meaningful dates have been obtained from buried palaeosols by comparing radiocarbon ages of the different organic fractions (e.g. Matthews 1980), or by using different materials in soils such as charcoal and humus (e.g. Vogel and Zagwijn 1967), but in general, palaeosols offer one of the least satisfactory media for the radiocarbon dating of organic material (Scharpenseel 1971; Scharpenseel and Schiffman 1977).

Of equal complexity is the dating of inorganic carbon in calcium carbonate enriched horizons of soils, due largely to the fact that as calcium carbonate is readily soluble, solution and reprecipitation can take place. Each time this occurs, new carbon is added to the system because of the carbon dioxide content of air (Birkeland 1974). In humid regions, radiocarbon ages of porous carbonates are often too young (Bowler and Polach 1971), while in arid regions the age of carbonates is sometimes greater than that of the soil from which they have been derived (Williams and Polach 1969).

Refinements in radiocarbon dating

Using conventional techniques, radiocarbon dating is normally restricted to an upper age limit of about 40 000 years, although counting under higher pressures with exceptionally low background radiation levels permits detection of sample activities equivalent to just over 50 000 ^{14}C years. Samples older than this possess such low levels of ^{14}C activity (Fig. 5.5) that either the levels of statistical uncertainty are so great as to render calculated dates meaningless, or the decay rates are so slow that it is only possible to obtain an 'indefinite' (*i.e.* 'greater than') age determination. Two refinements in the radiocarbon method have enabled the time range of ^{14}C dating to be extended, however: **isotopic enrichment** and the use of **accelerators**.

Isotopic enrichment.   Dates in excess of 60 000 ^{14}C years BP have been obtained by the use of elaborate enrichment methods, in which the amount of ^{14}C present in a sample is enhanced, usually by thermal diffusion, so that the frequency of decay can be more accurately determined by gas counters (Haring *et al.* 1958). More recently conventional detection methods have been further improved by the use of lasers to enrich the heavier isotopes of carbon (particularly ^{14}C) during the transformation of sample carbon to formaldehyde which acts as the counting medium (Hedges and Moore 1978).

Particle accelerator techniques.   In conventional radiocarbon dating the measurement of beta transformation is limited by the amount of raw carbon that is required for counting, the low activity of samples older than about four half-lives, and by the problems of eliminating background radiation, especially for older samples. A new approach has therefore been developed in which the *carbon atoms* are counted rather than the beta particles emitted during decay. For every atom of ^{14}C that decays there are between 10^9 and 10^{12} more radiocarbon atoms remaining in the sample, and therefore a much larger data base is available if the raw sample is examined rather than the tiny part that undergoes beta transformation during the count period. The concentration of ^{14}C atoms in samples can be detected using accelerators[3] as sensitive mass spectrometers (Muller 1977; Nelson *et al.* 1977; Doucas *et al.* 1978). Although still in the experimental stage, this new approach to radiocarbon dating may eventually prove superior to conventional methods since (a) counting statistics will be enormously improved; (b) age determinations will require much less time, involving hours of measurement instead of days; (c) background radiation should no longer prove to be a problem;

and (d) only very small samples (5 mg or less) will be required for dating purposes.

The benefits to the Quaternary scientist of this particular development in ^{14}C dating are likely to be considerable. Not only will the age range of the radiocarbon method be extended to approximately 100 000 BP, but the fact that only tiny quantities of sample carbon are required for dating purposes offers greater flexibility in applications of the method. For example, instead of dating bulk samples of peat, wood, organic detritus or soils, different fractions of a sample can be dated, even where the original sample is small in bulk. Tiny seeds or individual tree-rings could even be dated. If the experimental problems can be surmounted, there is little doubt that this technique promises the most significant advance in radiometric dating since the pioneering work of W. F. Libby over thirty years ago.

Other radiometric methods

In theory, a number of other radioactive reactions offer a basis for measuring time intervals within the Quaternary, but few have so far been developed as dating methods. Those that have been used include ^{10}Be:^{9}Be ratios (half-life 1.5 m years) and ^{36}Cl decay (half-life 300 000 years) (Muller 1977; Andrews and Miller 1980). Rapid progress is also being made in the development of techniques associated with short-lived nuclides. Caesium-137, for example, is an artificially-generated radioactive nuclide that has only been produced in significant quantities as a result of thermonuclear weapon testing, which commenced in 1954. ^{137}Cs has a half-life of 30 years, and has been used to investigate recent sedimentation rates and influx in lakes (McHenry et al. 1973; Pennington et al. 1973; Wise 1980). Similarly Tritium (^{3}H) with a half-life of 17.8 years has also been produced in abundance as a result of thermonuclear explosions. The sudden rise in ^{3}H levels in the mid-1950s can therefore be used as a marker horizon in modern sediments. However, ^{3}H is also produced naturally in the upper atmosphere as a result of cosmic-ray reactions, and it is hoped eventually to be able to use ^{3}H to date back to about 160 years ago (Muller 1977). Finally, over a much longer time-scale, the technique of **electron spin resonance** (ESR) of atoms may provide a means of dating speleothems and flint artefacts. The method is based on the relative amounts of induced energy states in electrons produced by the rupturing of electron bonding during radioactive decay. These reflect the incidence of radioactive emissions which will, in turn, be a reflection of the number of radioactive nuclides in the sample and the time over which decay has occurred. In theory, the method is applicable to the whole of the Quaternary time-scale, but is still in the early stages of development (Ikeya 1975; Garrison et al. 1981).

INCREMENTAL METHODS

Incremental dating methods are those based on regular additions of material to organic tissue or sedimentary sequences. Those which have been most widely used are **dendrochronology** (tree-ring dating), **varve chronology** and **lichenometry**. These techniques are restricted in application largely to the Holocene, although varve chronology has been used as a means of dating earlier Quaternary successions.

Dendrochronology

In most temperate trees, new water and food-conducting vessels are added to the outer perimeter of the trunk each growth season, following an inactive period in winter. The new vessels that commence growth in the spring tend to be larger and more thin-walled than those produced in late summer, as a result of heavier demands on water supply early in the growth season. Later in the year the vessels become gradually smaller and develop thicker walls. There is normally, therefore, a distinct 'line' between successive annual increments of wood growth and counting of these lines (**tree-rings**) allows the age of the tree to be established.

Tree-ring width is seldom uniform, for tree growth is influenced by a range of environmental factors, variations in which will produce different physiological responses within the trees (Fritts 1976). The most important determining factor for many trees is climate. Under conditions of stress, growth is retarded, and a narrow tree-ring will result; conversely, under more favourable conditions, growth rates are increased, and wider annual rings are produced. As a consequence, climatic variations over very short and precisely dated time-scales can often be inferred (**dendroclimatology**).

A greater variation in cellular structure is exhibited in hardwoods (deciduous trees) by comparison with softwoods (conifers). In wood anatomy the vessels in transverse section are often referred to as 'pores' (Jane 1970; Butterfield and Meylan 1980), and hardwoods can be divided into **ring-porous** types where the spring vessels are normally distinctly larger than those of the summer wood (e.g. oak, ash, elm) and **diffuse-porous** types in which the pores are more uniform in size (e.g. beech, birch, alder, lime). There is, therefore, a considerable variation in the appearance of tree rings between species, and not all trees show clearly defined annual bands. The most useful genera, or at least those most widely employed in dendrochronological work to date, are oak and pine.

Dendrochronological procedures

Measurement.   Sub-fossil and dead trees can be cut so that a complete cross-section can be examined, allowing comparisons of ring-width to be made in several radial directions. Living trees can be sampled with a metal increment corer which extracts small diameter cylinders of wood from the tree trunk. In the laboratory samples are dried, cleaned, polished and mounted prior to examination. Counting and measuring can be carried out by visual inspection of the rings under normal magnification, or on a moving stage under a binocular microscope. However, the use of scaling devices and automatic recorders allows speedier and more accurate determinations to be made (Pilcher 1973).

A recent advance in the measurement of tree-rings has been the development of **X-ray densitometry** techniques. Thin sections of wood are X-rayed and the negatives are then scanned by a beam of light and a photocell. The amount of light that is transmitted through the negative is proportional to the character of the negative image, which, in turn, reflects the density of the wood (Schweingrüber *et al.* 1978). This method has two advantages. First, ring counting is less likely to be affected by operator error or from errors arising out of limitations of the human eye, and secondly, it has been found that density variations are often more reliable than ring-widths as climatic indicators (see below).

Crossdating.   Ring patterns within trees from a limited geographical area can be matched through the technique of **crossdating**. Climatic variations within a particular area will be reflected in characteristic ring-width patterns in the trees in that area. Distinctive rings, or groups of rings, form markers and these can be used as a basis for cross-matching between trees of overlapping age range. In Fig. 5.6, for example, a sequence of tree-rings from a living tree (A) contains a group of much narrower rings, perhaps reflecting drought conditions around 1930. This group can also be identified in a record from a tree which was felled in 1954 to provide beams for a building (B), and a match has also been established between B and a ring series in a beam taken from an older building (C). In this way, the tree-ring record can be extended back in time. Where subfossil logs are available (preserved, for example, in peat bogs), crossdating allows a link to be established between these and the historical or living record, and tree-ring chronologies spanning several thousands of years can now be constructed (see below). A number of computer-aided techniques have now been

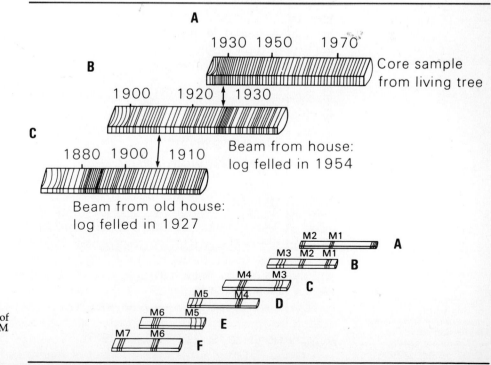

Fig. 5.6: Core matching of tree rings (crossdating). M = marker groups of three-rings. For explanation see text.

developed to aid in the process of crossdating older material (Baillie and Pilcher 1973; Cropper 1979).

Standardisation.   Trees grow more vigorously in youth than in old age, and as a result there is usually a reduction in ring-width with age. Crossdating can therefore be difficult where ring-width variations resulting from the effect of limiting environmental factors are complicated or even masked by ring-width variations due to age. A further problem is that the width of tree rings often varies with the height of the trunk. Each tree-ring series is therefore **standardised**, by transforming the measured ring-width values to **ring-width indices**. A regression line is fitted to the measured ring-width values, and this provides an indication of the general decline in tree ring-width with age (Fig. 5.7A). The measured ring-width value for each year is then divided by the value for that year obtained from the regression curve. The resulting ring-width indices (Fig. 5.7B) have been corrected for the ageing effect and therefore fluctuations in the tree-ring curves reflect the influence of environmental factors only. Standardisation indices are required from a number of trees in a locality if accurate crossdating is to be achieved.

Complacent and sensitive rings.   Local site conditions exert a major influence on tree growth and therefore also on tree-ring width. Some trees will experience more stress than others, depending on such factors as slope of the ground surface, water-retentive capacity of soils and sub-soil stratigraphy, relative amount of shade and exposure, genetic

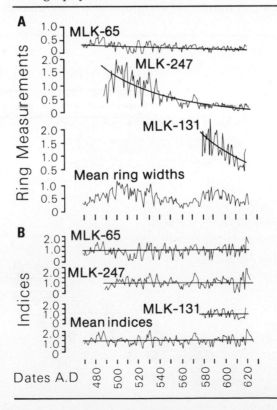

Fig. 5.7: Standardisation of ring-width measurements by using regression curves to approximate ageing effects in tree growth (after Fritts 1976).

characteristics and so on. If a tree exists in a situation where there is a constant and adequate supply of water, and where it is protected by shading from the extremes of temperature, then tree-ring widths in such a specimen may show little variation, reflecting the lack of interruption in physiological activities. A tree-ring series of this type is termed a **complacent series**, and is difficult to use in crossdating as distinctive markers cannot be identified. Dendrochronologists therefore prefer to select trees from stressed situations, where some climatic factor has been critical to growth. Wood production will be reduced at times of stress, and this is shown by a **sensitive series** of rings, reflecting a clear and immediate response to some limiting factor. Tree-ring chronologies have been largely derived, therefore, from semi-arid or arid sites where low moisture or high temperatures produce stress conditions, or from sites at higher altitudes and latitudes where low temperatures in particular have restricted growth. It is also necessary to compare records from single species, since different species have different responses to environmental factors. In living specimens it is relatively easy to check for comparable site conditions affecting individuals, but this is obviously more difficult to monitor where subfossil specimens are used, especially where these are not in the position of growth.

Missing and false rings.   Because trees are deliberately selected from stressed situations, there is always the possibility that during years of extreme climatic conditions a tree may fail to manufacture new vessels, or may only produce new material on restricted parts of the trunk. These are

Fig. 5.8: The White Mountains (foreground), northern California, home of the longest-living trees in the world. The White Mountains lie to the east and in the lee of the Sierra Nevada (background), and thus the bristlecone pines exist in a dry environment, despite the high altitude.

referred to as **missing** and **partial rings** respectively. Partial rings may be absent from a record where only part of the trunk can be examined, for example, in a wood core, a beam, or a partly-destroyed subfossil trunk. On the other hand, more than one growth layer can develop in a particular growth season where more severe conditions prevail for a short while during late spring or early summer, after the spring vessels have commenced growth. This may produce a change in structure that resembles an annual boundary, and such layers are referred to as **false rings** or **intra-annual growth bands**. These can usually be identified and corrected for by replication of a number of records, where missing and false rings in individual records will be identified by crossdating. In this way a master chronology can be established for a particular locality.

Dendrochronological records

It has long been known that bristlecone pines (*Pinus longaeva*, formerly *P. aristata*) growing in the White Mountains of eastern California (Figs 5.8 and 5.9) can achieve ages in excess of 4000 calendar years (Schulman 1954, 1958). Such long-lived trees are usually found on dry and rocky sites usually in a twisted and stunted form. The pines of the White Mountains have somehow adapted to an unusually dry environment, and in parts of the Sierra Nevada thrive at altitudes of 4000 metres. By crossdating between dead and living wood, Ferguson (1970) was able to extend Schulman's chronology back to 8200 years BP. This is the longest tree-ring chronology so far available. Other relatively long tree-ring chronologies have been established in North America on species of *Sequoia* and Douglas

Fig. 5.9: Bristlecone pines, White Mountains, California. Relatively-young living trees occupy the foreground, as well as dead individuals still in their growth positions. Elsewhere, trees that appear to be dead still support isolated patches of foliage, These tend to be the oldest living specimens, often with active vessels confined to one side of the trunk, providing a slender life-line between root system and foliage.

Fir (*Pseudotsuga*), but the longest of these, based on *Sequoia gigantea*, spans only 3100 years. Elsewhere in the world tree-ring research has lagged behind the pioneer developments in the USA. One of the principal reasons for this is that in few parts of the world are there trees that have an age range in any way comparable to the bristlecone pine, and long chronologies that have been established have involved the laborious crossdating of a large number of individual wood samples whose ring patterns span only 100 to 200 years (e.g. Baillie and Pilcher 1973).

In Europe, long chronologies have been developed only for a few areas where large numbers of subfossil timbers have been found. The two genera most often preserved, beneath blanket peats or rapidly accumulating river gravels (e.g. Becker and Schirmer 1977), are the oaks (*Quercus robur* and *Q. petraea*) and the Scots Pine (*Pinus sylvestris*). In north-east Ireland a record of cross-correlated tree rings stretching back over 8000 years has been derived from subfossil oaks (Pilcher *et al.* 1977). The record, however, is not continuous and a number of gaps remain where overlaps between local site records have yet to be established (Fig. 5.10). A series of tree-rings that is not tied directly to historically-dated or living wood is referred to as a **floating chronology**, for it is not possible, without resort to other dating means, to determine the duration of the gaps in the record. In Fig. 5.10 the ages of the floating sequence have been estimated on the basis of available radiocarbon dates.

Dendroclimatology

Dendroclimatology is the science of reconstructing past climatic conditions and histories from tree-rings and it is regarded as a branch of the general discipline of dendrochronology (Fritts 1976). The importance of dendroclimatological records is that palaeoclimatic information can be precisely dated, and inferences can even be made about seasonal

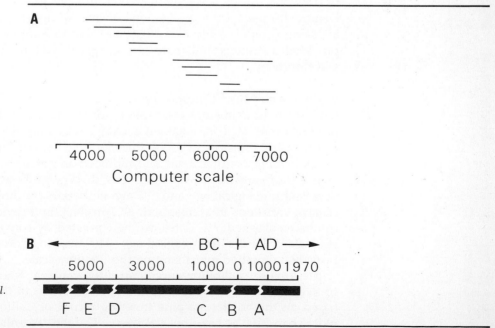

Fig. 5.10: A Overlapping tree-ring sequences used to reconstruct the dendrochronological record for north-east Ireland. B: The complete dendrochronological record for north-east Ireland (after Pilcher *et al.* 1977). A to F represent short gaps in the sequence.

variations. Further, as a proxy record of climate it enables climatic data to be extended beyond the limits of records from instrumental measurements.

Palaeoclimatic reconstructions based on tree-rings rest on assumed or demonstrable relationships between ring-width, or some other ring characteristic, and climatic parameters. These relationships are often complicated by lag effects between climatic inputs and tree response, for trees have the ability to store food reserves and water for several years, and this stored material may then be used in adverse years. Because the vigour of a tree in any single year is determined both by environmental influences during that year and during previous years, a sequence of annual rings is **autocorrelated**, although the degree of autocorrelation can often be discerned by statistical tests (Fritts 1976). Once the relationship between climatic inputs and tree-ring response has been established, previous climatic history can be reconstructed for the duration of the tree-ring chronology. Variations in weather patterns can be inferred from comparisons of a large number of dendroclimatological records from a region, and previous anomalies in climatic history may be revealed by contemporaneous changes in tree-ring characteristics from sites located throughout a region (see La Marche Jr and Fritts 1971; Blasing and Fritts 1976). Moreover, particular aspects of climatic control can sometimes be discerned, such as recurrent drought conditions (Stockton and Meko 1975) and occurrence of frost, reflected in damaged rings (La Marche Jr 1970).

More recently, it has been found that variations in wood density may provide clearer palaeoclimatic information than simple ring-width measurements. For example, X-ray densitometry investigations showed that in European temperate sites, maximum wood density could be directly correlated with summer temperature (Schweingrüber *et al.* 1978). On drier sites, however, a close relationship was detected between growth in latewood during the growth period and precipitation. Dendrochronological investigations on red spruce (*Picea rubens*) from sites in Maine (Conkey 1979) also suggested that maximum wood density provided a stronger and more season-specific climatic indicator than ring-width indices.

Calibration of the ^{14}C time-scale

One of the fundamental assumptions of ^{14}C dating is that atmospheric production of ^{14}C has remained constant over time. Dendrochronological records offer a means whereby this can be checked, for if the assumption is valid, then tree-ring dates and radiocarbon dates should show a broad measure of agreement over time. Any divergence between dendrochronological age and ^{14}C age is therefore a direct measure of secular variations in atmospheric ^{14}C production. Providing that a trend can be established, ^{14}C dates can be corrected by conversion to the equivalent dendrochronological age. The tree-ring chronology, therefore, provides a method for calibrating the ^{14}C time-scale.

The longest calibration curve currently available, showing a comparison of tree-ring and ^{14}C dates over the period 7100 BP to the present day, is based on the bristlecone pine tree-ring record of California (the tree-rings older than 7100 BP have not yet been ^{14}C dated as only a few have so far

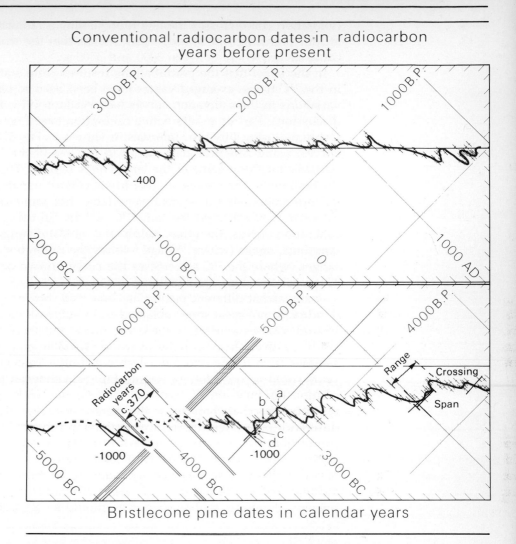

Conventional radiocarbon dates in radiocarbon years before present

Bristlecone pine dates in calendar years

Fig. 5.11: Calibration curve of radiocarbon dates based on dendrochronological records for the period approximately 6000 BP to the present day (based on Suess 1970b). For explanation see text.

been found). This curve was first published by Suess (1970b), and is reproduced in Fig. 5.11. The ^{14}C dates were obtained from groups of ten annual rings, each sample comprising approximately 20 gm of wood, and the curve is based on nearly 300 of these dated samples. For the period from 2000 BP to the present, radiocarbon years appear to be approximately equivalent to dendrochronological years, but beyond 2000 BP an increasing divergence is apparent so that a ^{14}C date of 5000 BP corresponds approximately to 5800 years on the dendrochronological time-scale. If the tree-rings are accepted as reflecting calendar years, then the implication is that ^{14}C levels were significantly lower at around 6000 BP by comparison with present-day values.

Although the Suess curve and subsequent calibration curves based on more data points (e.g. Michael and Ralph 1972; Damon *et al.* 1972) have led to a major reappraisal of the relationships of ^{14}C years to calendar ages, they are not easy to use in calibration because of the marked perturbations in certain parts of the curves. In places these result in ^{14}C dates having more than one calendar age. In Fig. 5.11, for example, the calibration curve crosses the line representing a radiocarbon age of 4400 BP at four points (a, b, c and d) which means that the true calendar age could lie anywhere between 3000 and 3500 BC.

In the belief that the perturbations resulted from statistical uncertainties in the ¹⁴C dates, averaged values have been used to smooth out the variations in the calibration curves (see Watkins 1975, for further discussion). Part of an alternative calibration based on an averaging technique using fifty-year intervals in shown in Fig. 5.12. Similar 'average' curves, using American data, result in the loss of the smaller perturbations, but some of the larger ones remain. This suggests that the fluctuations in the curves are due to short-term variations in atmospheric ¹⁴C production, and it seems unavoidable that some samples of different calendar ages will yield the same ¹⁴C activity. In using the American calibration curves, therefore, Ralph *et al.* (1973) distinguish between **crossings, spans** (where ¹⁴C age follows the calibration curve closely) and **ranges** (where the ¹⁴C age crosses the curve on two or more occasions). These relationships are illustrated in Fig. 5.11.

A somewhat different picture has emerged, however, from dendrochronological series obtained from subfossil oak wood in north-east Ireland. Oaks manufacture thick tree-rings, and therefore in the Irish study relatively large samples of wood (120–180 gm compared with 20 gm samples from bristlecone pine) each covering a time-span of only twenty years could be used for ¹⁴C dating. Fig. 5.13 shows a more linear relationship between ¹⁴C age and tree-ring age, with none of the marked perturbations that characterise the American curves. To what extent this difference reflects geographical variations in former fluctuations in ¹⁴C production, or errors associated with the data used in construction of the curves cannot yet be established. Clearly, however, the relationship between ¹⁴C age and tree-ring age needs to be monitored in many more localities before the reliability of calibration curves can be assessed. At present, therefore, calibration curves should be regarded as area-specific,

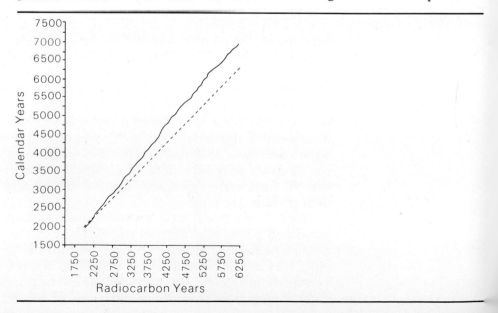

Fig. 5.12: Relationship between ¹⁴C and tree-ring years based on an averaging technique applied to North American data (after Switsur 1973).

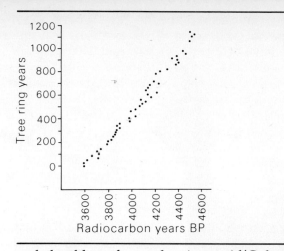

Fig. 5.13: Relationship between tree-ring years and ^{14}C years according to dendrochronological records from north-east Ireland (after Pearson *et al.* 1977).

and should not be used to 'correct' ^{14}C dates outside the region from which they were derived.

Since the bristlecone pine data indicate increasing divergence between ^{14}C years and tree-ring years through time (Figs 5.11 and 5.12), implying a progressive decrease of ^{14}C levels with age, extrapolation would suggest even greater discrepancies at earlier times. There are some indications, however, that this may not be the case. Dendrochronological studies in the Netherlands, for example, have shown that ^{14}C levels were more or less constant at around 7000 BP, but by 8000 BP may actually have been increasing (Vogel 1970). If this inferred trend is correct, therefore, it would suggest that beyond about 7000 BP, the curve in Fig. 5.11 should gradually swing back towards the straight-line diagonal, representing correspondence between ^{14}C years and tree-ring years. Evidence in support of this hypothesis has been obtained from work on varve chronology, which is considered in the following section.

Varve chronology

Rhythmic accumulations of sediments, forming laminae of fine sands, silts or clay, are common in the geological record. Often the laminae are arranged in couplets, with relatively coarse-grained layers alternating regularly with finer-grained bands (Figs 5.14 and 5.15). The general geological term applied to such sediments is **rhythmites**, except where the laminations arise because of annual variations in the supply of sediments, in which case they are termed **varves**. Because they are deposited annually, varves can be used as a means of dating, for time intervals can be calculated and a floating chronology established. Moreover, if a varve sequence can be tied to the calendar time-scale, then it may be possible to assign calendar dates to the varve record.

Glaciolacustrine and glaciomarine varves

Large supplies of sediment are deposited in proglacial lakes and shallow seas as a result of rapid spring and summer ice melt. The coarsest particles are deposited first on the lake or sea bed, leaving the finer clay particles in suspension. During winter, when the lake or sea adjacent to

Fig. 5.14: Late Wisconsinan varve sequence deposited in Late Pleistocene Glacial Lake Penticton near Kelowna, British Columbia, Canada.

Fig. 5.15: Close-up of the varve sequence shown in Fig. 5.14. The coarse-grained layers accumulated during the late spring and early summer months while the fine-grained units represent sediments that were deposited during the winter.

the shore is frozen, the suspended clay particles gradually settle out to produce a clay lamina that forms a marked contrast to the coarser summer layer. Glaciolacustrine and glaciomarine varves have usually formed the basis for varve chronologies.

The potential of such sediments as a means of dating was first recognised by the Swedish geologist de Geer who, as early as 1878, was investigating the exposed varve sequences in the Stockholm area. He discovered that as the last ice sheet wasted northwards across southern Sweden, it left behind a complex of moraines and moraine-dammed (formerly proglacial) lakes. In many places, overlaps of varve 'histories' had developed, the uppermost varves in one lake sequence being of the same age as the lower varves in a close neighbour. In a classic publication in 1912, de Geer presented measurements of individual varves and curves of relative thickness for each site investigated. Curves from two or more sites were compared and correlated on the basis of matching sections, using similar principles to those now employed in the matching of ring-width series in dendrochronology. By methodically extending his master chronology northwards and southwards across Sweden and into Denmark, de Geer compiled a continuous varve sequence extending back to about 15 000 years BP.

Other varved sediments

Studies of recent sedimentation in lakes and in the sea have shown that annual rhythmites are common in many temperate areas far removed from nearby glacier activity (Schlüchter 1979). As a general rule, both sedimentation and biomass production are affected by seasonal variations in lakes, and in certain temperate lakes chemical precipitation can also vary seasonally. In many lakes, reworking of sediments by currents or by bottom-dwelling organisms prevents the formation of laminations (**holomictic lakes**). In deep lakes with oxygen-deficient basal water, however, the numbers of bottom-dwelling fauna are restricted, vertical water circulation does not extend to the bottom of the water column (**meromictic lakes**), and fine laminations may therefore be preserved. Where seasonal variations in the accumulation of organic detritus occurs, couplets formed as a result are termed **organic varves** (Fig. 5.16). In some areas of calcareous bedrock light summer layers of $CaCO_3$ alternate with dark winter layers rich in organic humus (Welten 1944; Brunskill 1969). Diatom-rich summer layers in laminations have been observed in interglacial (Turner 1970) and modern (Saarnisto et al. 1977) sediments. Seasonal variation in iron oxide precipitation may also lead to the formation of laminations (Anthony 1977). In some lakes, variations in the accumulation of organic matter or of chemical precipitates are superimposed on particle-size variations, thereby emphasising contrasts between winter and summer layers (Saarnisto 1979). Organic varves have proved to be most useful as a means of verifying radiocarbon dates obtained from organic lake sediments (see below).

Sources of error in varve counting

Problems are encountered in the development of varve chronologies which are similar in many respects to those that are encountered in

Fig. 5.16: Organic varves of Hoxnian Interglacial age, Marks Tey, Essex, England. The lighter bands are diatom-rich summer deposits, while the darker bands are sediments that accumulated during the winter months and which have a lower microfossil content.

dendrochronology. Local site factors can lead to incorrect assessments of age, and also cause difficulties in correlation (cross-dating) between individual varve sequences. For example, adverse weather conditions in particular years can lead to a reduced input of sediment into a proglacial lake, or to reduced biomass or chemical precipitate in summer, so that an individual varve fails to develop, or is too thin to be recognised. An annual increment of sediment may also be absent because of the intermittent erosion of bottom layers at some sites. Alternatively, more than one set of laminae (**sub-laminations**) can develop within an annual increment of sediment. Sub-laminations can result from sudden seasonal changes, and in shallow lakes may be caused by redeposition of sediment by storm waves (Hansen 1940). Normally, sub-laminations are found in the coarse member, but sand layers can form in the winter silt and clay layers of glaciolacustrine varves through the operation of turbidity currents (Shaw and Archer 1978). Identification of varves on the basis of grain-size characteristics may not always be reliable, therefore, and proof of the annual nature of sediments must be sought from other evidence such as variations in microfossil content of the different laminations (Saarnisto 1979).

As in tree-ring chronology, crossdating of varve sequences involves the analysis of a number of records from each locality, in order to exclude or at least minimise the problems arising out of missing or false varves. Matching of relative varve thicknesses is only possible over limited geographical areas as climatic variations may produce quite different varve sequences even in adjacent regions. Because sections are not always conveniently located, long-distance correlations (**teleconnections**) have been attempted, and these have often led to errors in the subsequent chronologies. More recently, varves have been obtained from terrestrial, lake and shallow-water marine localities by coring. However, special coring devices that allow the sediments to be frozen *in situ* are usually required to ensure the recovery of undisturbed finely-laminated sediments especially from horizons close to the present lake or sea bottom (Swain 1973; Saarnisto *et al* 1977). Nevertheless, coring of unexposed varve sequences has introduced a much greater flexibility into sampling designs for work on varve chronologies. Finally, the clarity of varved sediments varies considerably and depends on the combined effects of a large number of factors, including basin morphometry, topographic situation, type and heterogeneity of the sediment supply, water depth, and relative contrast between winter and summer conditions (Saarnisto *et al.* 1977).

Application of varve chronologies

Patterns of regional deglaciation. The development of a regional varve

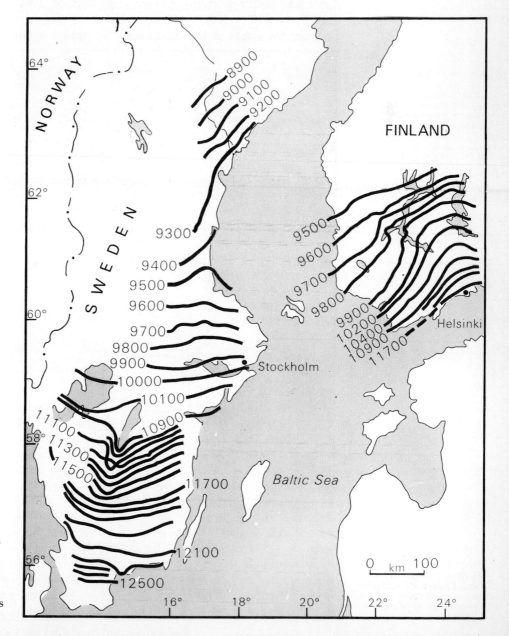

Fig. 5.17: Ice recession in varve years BP according to Tauber (1970). Tauber's BC varve dates have been converted to approximate BP ages by the addition of 2000 years to each date.

chronology spanning a long period of deglaciation is so far unique to Scandinavia. In North America, for example, despite the pioneering work of Antevs (1953) who developed a tentative varve chronology for the Great Lakes region, and in spite of the considerable amount of work that has been carried out in recent years on modern varves and varving mechanisms (e.g. Ashley 1979; Ludlam 1979), no extensive and regionally-applicable chronology has yet been established. The same is also true of the British Isles and most of western Europe. In Scandinavia, however, following on from the early work of de Geer (1912) and Sauramo (1918, 1923), glaciolacustrine varves have continued to attract interest as a means of dating the wastage of the last ice sheet. The master chronology of de Geer has been repeatedly revised (e.g. Nilsson 1968; Tauber 1970; Lundqvist 1975, 1980), and attempts have been made to quantify and account for errors in the chronology by applying correction factors to

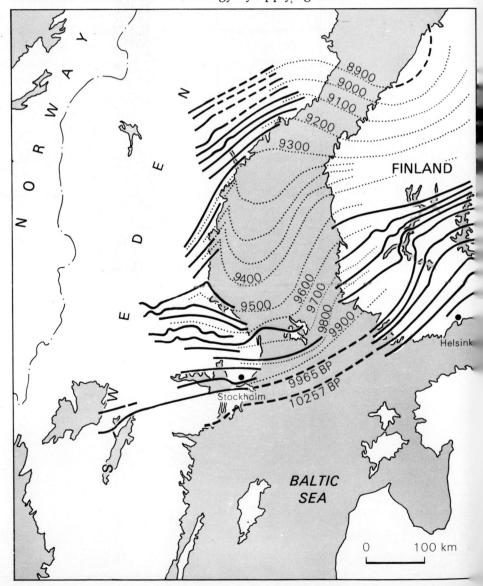

Fig. 5.18: The early Holocene deglaciation of Scandinavia according to the revised varve chronology of Mörner (1980b).

'varve years' (e.g. Fromm 1970). Maps have been produced showing successive positions of the ice front (e.g. Figs 5.17 and 5.18) and although these often differ in detail due, *inter alia*, to the application of correction factors and to uncertainties surrounding correlation of the ice limits in different parts of the Baltic region, a broad measure of agreement is evident over the general deglacial chronology of Scandinavia. Further refinement in the varve chronology, augmented by other dating methods, may eventually enable rates of retreat to be established for different parts of the ice front (e.g. Table 5.3.)

Calibration of the radiocarbon time-scale. Where sufficient organic material has accumulated in varved sediments to allow radiocarbon measurement, sequences of varves can be used as an independent means of calibrating the radiocarbon time-scale. One detailed radiocarbon-dated sequence that has been used in this way is the 10 000 years of varved sediments that have accumulated in Lake of the Clouds, Minnesota (Stuiver 1970). The relationship between ^{14}C and varve chronology at that site was considered over three time-periods. From 0 to 2500 calendar years it was found that one ^{14}C year equalled approximately one 'calendar' (*i.e.* varve) year; from 2500 to 5500 calendar years, one ^{14}C year was the equivalent of about 1.26 calendar years, and from 5500 to 10 000 years, one ^{14}C year equalled around 1.03 calendar years. These data imply that there was no further divergence between ^{14}C and calendar age beyong 5500 BP, a finding that is in broad agreement with the dendrochronological evidence discussed

Conventional varve years converted to years BP	Mean ice recession (km/100 yr)	Radiocarbon age (yr BP)
9 980–9800	29	
10 000–9900	24	
10 100–10 000	32	
		10 000
10 200–10 100	6	
10 300–10 200	7	
10 400–10 300	3	Younger
10 500–10 400	8	Dryas
10 600–10 500	5	Stadial
10 700–10 600	4	
10 800–10 700	2	
10 900–10 800	4	
		11 000
11 000–10 900	15	
11 100–11 000	12	
11 200–11 100	12	
11 300–11 200	15	
11 400–11 300	11	
11 500–11 400	11	
11 600–11 500	15	
11 700–11 600	16	
11 800–11 700	14	
11 900–11 800	14	
12 000–11 900	19	
12 100–12 000	35	
12 200–12 100	16	
12 300–12 200	7	
12 400–12 300	10	
12 500–12 400	8	

Table 5.3: Rates of ice recession in southern Sweden in km/100 years after Nilsson 1968; Tauber 1970). The varve dates have been converted to approximate P dates by the addition f 2000 years. (^{14}C dates rom Mangerud *et al.* 1974).

above. Only a few organic varve sequences have so far been investigated in this way, however, and relatively little information has been published on uncertainties associated with the dates obtained. As with dendrochronology, therefore, further detailed work is required before precise correlations can be made between varve years and ^{14}C years.

Lichenometry

Lichenometry as a dating technique was pioneered by Beschel (see Beschel 1973), and rests on the principle that there is a direct relationship between lichen size and age. Where a surface has been recently exposed to lichen colonisation, providing that (a) the growth patterns of the lichens are known and (b) that no major time lapse has occurred between surface exposure and lichen colonisation, an estimate of the age of the substrate can be made.

Lichenometry has been most widely employed in the dating of glacier recession (e.g. Andrews and Webber 1964; Benedict 1967; Mottershead and White 1972; Karlén 1973). The largest lichen size (Fig. 5.19) is first established for morainic surfaces of known age (dated, for example, by radiometric methods or by historical evidence such as old photographs) and a lichen growth-curve can then be constructed based on these 'fixed points' (Fig. 5.20). Surfaces of unknown age can then be dated by relating the largest size of lichen on those surfaces to the growth-rate curve and deriving a calendar age. In this way, a detailed deglacial chronology can be established for a particular area. The method has been used with some success in Scandinavia and North America to establish glacial chronologies for the Neoglacial and Little Ice Age periods (e.g. Denton and Karlén 1973b; Miller 1973), and it has also provided the time-scale for the measurement of rates of plant colonisation of newly-exposed substrates in proglacial areas (e.g. Matthews 1978).

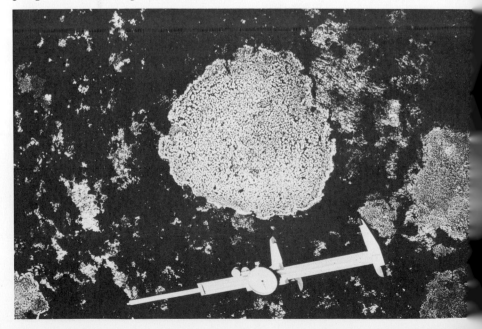

Fig. 5.19: Lichen (a member of the aggregated *Rhizocarpon geographicum* species) and dial calipers used in lichenometrical measurements.

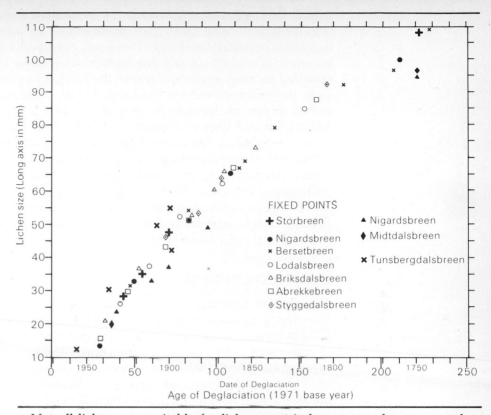

Fig. 5.20: Lichen growth curve based on measurements from sites of known age at localities in Norway (after Matthews 1974).

Not all lichens are suitable for lichenometrical purposes, however; only those that show a gradual and progressive rate of growth can be employed. Moreover, lichen growth is also affected by local environmental conditions. A lichen growth-curve must, therefore, be constructed for specific lichens and will only be applicable to a limited geographical area. Further, it is assumed that in the study of glacier recession, for example, there is no significant delay in lichen colonisation of exposed surfaces following ice retreat. This, however, can never be proved and remains a source of uncertainty. Conversely, lichens have been reported growing on actively-forming medial moraines (Matthews 1973), and therefore in some areas lichen growth must have *preceded* ice wastage. Finally, problems have been encountered in sampling and measurement of lichens in the field and very thorough preliminary investigations are required to establish reproducibility of results within any one region (Jochimsem 1973; Matthews 1975). Further details on lichenometry can be found in Lock *et al.* (1980) and Mottershead (1980).

AGE-EQUIVALENT STRATIGRAPHIC MARKERS

In many Quaternary deposits, distinctive marker horizons are found that are broadly synchronous and form time planes across different sedimentary sequences. The horizons themselves cannot be used, in the first instance, to date Quaternary successions, for other methods are required to establish their age. However, once dated by radiometric or incremental methods at any one locality, they allow age estimates to be extended to other sequences where the marker horizon is present. As such, they form an indirect means of dating. Moreover, in view of their often widespread distribution, they also form a basis for stratigraphic subdivision and time-stratigraphic correlation (see Ch. 6).

Two methods of dating using age-equivalent stratigraphic markers are considered here: **palaeomagnetism**, which is based on the changes in the earth's magnetic field preserved in rocks and sediments, and **tephro-chronology**, the use of volcanic ash layers as a means of dating. Other marker horizons which are more widely employed in stratigraphic subdivision and correlation than in dating, such as palaeosols and shorelines, are discussed further in Chapter 6.

Palaeomagnetism

Geomagnetic field and remanent magnetism

The earth's magnetic field varies constantly both in field strength and in polarity direction. Variations range in periodicity from milliseconds to tens of millions of years (Table 5.4), the shorter-lived phenomena resulting from external influences such as variations in solar radiation, and the longer-lived changes from internal geophysical factors (Watkins 1972). Rocks and unconsolidated sediments containing magnetic minerals are magnetised during formation, and individual crystals or particles therefore become aligned in the ambient magnetic field. Careful analysis can often reveal this **natural remanent magnetism** (NRM) which is a reflection of the geomagnetic field at the time of rock or sediment formation.

Volcanic rocks contain minerals with various ferromagnetic properties. Before cooling, high temperatures lead to thermal fluctuations of ions, and in the absence of an external magnetic field a random orientation of ions would result. The temperatures below which the ambient geomagnetic field would be retained is known as the **Curie Point**, below which thermal agitation is insufficient to destroy the ferromagnetic alignment induced by the geomagnetic field. As long as the crystals are not re-heated above the

Geomagnetic changes	Duration (years)
Change in average frequency of polarity inversions ....5×10^7	
Time between successive polarity inversions ....$\begin{cases} 10^7 \\ 10^6 \\ 10^5 \end{cases}$	
Intensity and direction fluctuations of dipole and non-dipole fields (secular variation) ....$\begin{cases} 10^4 \\ 10^3 \\ 10^2 \end{cases}$	
11-year sunspot cycle ....$\begin{cases} 10^1 \\ 10^0 \end{cases}$	
Annual variation ....10^{-1}	
Diurnal variation .... 10^{-2}	
Magnetic storms .... 10^{-3}	
Micropulsations .... 10^{-4}	

Table 5.4: Time constants of the geomagnetic field (after Thompson 1978b).

Curie Point, then this **thermoremanent magnetisation** (TRM) will be retained by the crystal lattice.

Sedimentary rocks and unconsolidated sediments accumulating on the sea floor or in lakes also contain evidence of former geomagnetic fields, for a record is preserved in the alignment of ferromagnetic sedimentary particles as they settle in water or in water-saturated sediments. The resulting **detrital remanent magnetism** (DRM), though weak, can easily be measured (Thompson 1979). In sediments, NRM can also be acquired by chemical action, where the crystallisation of ferromagnetic oxides results in a **chemical remanent magnetism** (CRM). This process may occur later than, and under a different magnetic field from, that of DRM in the same sediment unit. In this way, a **secondary magnetisation** is introduced into both volcanic and sedimentary rocks which serves to complicate the study of palaeomagnetic variations (Tarling 1971).

At any one point and at any one moment in time the earth's magnetic field can be resolved into three components:

1. Declination.   This is the angle between magnetic north and geographic (true) north.

2. Inclination.   A freely suspended needle at the surface of the earth will align with the prevailing magnetic field (declination). The inclination of the needle is the amount of dip exhibited by the needle relative to the horizontal. The inclination value varies from 0° at the magnetic equator to 90° at the magnetic poles.

3. Intensity.   This refers to the strength of the geomagnetic field. At the present day the field strength at the geomagnetic poles is twice that at the geomagnetic equator. The strength of the field can be estimated in the following way. Suppose a magnetic needle is fixed to a horizontal axle, so that the axle passes through the centre of gravity of the needle and is orientated along magnetic east-west. If the needle is allowed to swing freely it would eventually stabilise at the angle of magnetic dip. Magnetic intensity can be measured by the amount of torque required to prevent the needle returning to the angle of magnetic dip after it has been rotated through 90°.

Magnetostratigraphy

The study of variations in magnetic properties through a sequence of rocks or sediments is termed **magnetostratigraphy**. Geomagnetic field variations can be detected in rocks or sediments that contain even small amounts of magnetic minerals, and the identification of synchronous changes in, for example, declination or inclination in different sequences provides a means of relative dating and correlation. Where sediments have accumulated rapidly, short-term **secular variations** in the earth's field can be employed, but for sediments that have accumulated more slowly (e.g. deep-sea sediments) or for sequences of volcanic rocks, longer-term **field reversals** are used. In addition, however, **mineral magnetic 'potential'**, that is the concentration and magnetic susceptibility of magnetic minerals, also varies within sediments, and under certain circumstances this too can provide a basis for correlation. Each of these aspects will now be considered in more detail.

Secular variations.   Direct measurements of values of field declination, inclination and intensity have been collected in London since 1580, and these are plotted in Fig. 5.21. This diagram shows that magnetic declination has varied through 47° since 1580, with a westerly maximum of about 28° W at 1820 AD. Magnetic inclination has varied through 9° during this period, with a maximum in 1690 AD (Thompson 1978b). Proxy records of geomagnetic variations, obtained from an analysis of archaeological materials (**archaeomagnetic measurements**) reveal a more complicated pattern for the last 1000 years (Fig. 5.21). Geomagnetic fluctuations are thought to have continued throughout geological time, and records for a number of locations allow the plotting of the position of the geomagnetic poles for different times in the past. Variations have also been detected in field intensity. During the past 8500 years, for example, field intensity has varied between 0.5 and 1.5 times that of the present-day value, and even greater variations may have occurred in earlier times (Bucha 1967a, 1967b; 1970).

Lake sediments that have accumulated during the last 10 000 years or so often contain a very detailed record of such geomagnetic variations. Detailed analysis of lake sediments in Britain has established twenty-three key magnetic horizons during the last 10 000 years, consisting of ten secular variations in declination and thirteen in inclination (Thompson and Turner 1979). These key horizons, illustrated in Fig. 5.22, have been calibrated by radiocarbon dating, and should provide a means of dating and correlation of British Flandrian lake sediments, provided that clear palaeomagnetic signals can be deciphered and compared with Thompson and Turner's master curves. It may also be possible that these curves apply to a wider area, for Løvlie and Larsen (1981) have found comparable magnetostratigraphic signals at a site in western Norway.

Field reversals and the palaeomagnetic time-scale.   From time to time the

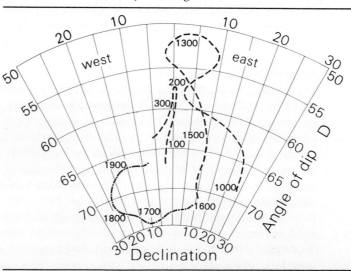

Fig. 5.21: Secular variations in declination and inclination for London based on historical data (dashed /dotted line) and archaeological data (broken line). Ages are shown in years AD (after Thompson 1978b).

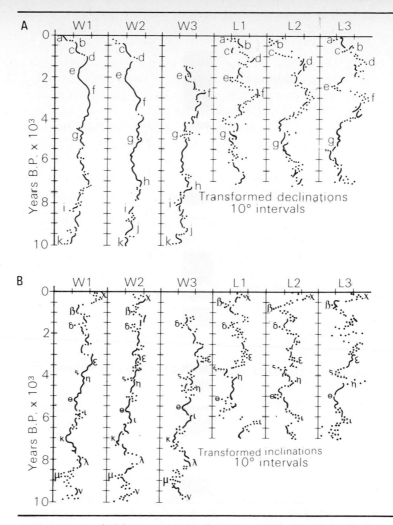

Fig. 5.22: Master curves of British geomagnetic variations during the Flandrian (after Thompson and Turner 1979). A: Relative declination. B: Secular variations in inclination. Curves W1 to W3 are from Lake Windermere in northern England, and L1 to L3 are from Loch Lomond in Scotland. Declination (a to k) and inclination (α to V) have been calibrated by radiocarbon dating.

geomagnetic field reverses so that the geomagnetic poles change relative positions through 180° and these then remain stable for a while. These **polarity reversals** can be detected in the geological record and are of fundamental importance in palaeomagnetic studies. Our present-day magnetic field is regarded as possessing **normal polarity**, and the opposite is referred to as **reversed polarity**. Periods of long-term fixed polarity (10^5 to 10^7 years) are known as **polarity epochs**. These are interrupted by a large number of polarity reversals of shorter duration (10^4 to 10^5 years) which are termed **polarity events**, and also by **polarity excursions**, in which the geomagnetic pole changes direction through 45° or more for a short period only (100 to 100 000 years). Polarity epochs and polarity events are experienced globally and can be used as a basis for world-wide correlations. Polarity excursions are more localised in their effects and therefore are less important in this respect.

   Where polarity epochs and events are found in volcanic rocks they can be dated by the K-Ar method (Cox *et al.* 1965), and a palaeomagnetic time-scale has now been established for the Quaternary and parts of the pre-Quaternary sequence (Fig. 5.23). Dating has concentrated on three

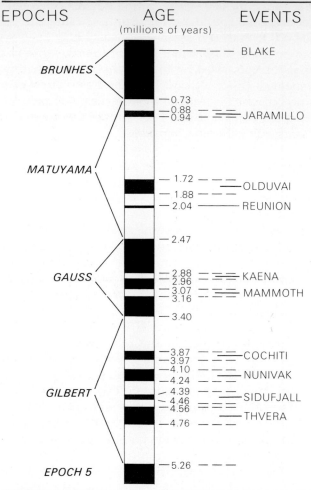

EPOCHS   AGE   EVENTS
(millions of years)

BRUNHES

—————— BLAKE

—0.73
—0.88 — ——— JARAMILLO
—0.94

MATUYAMA

— 1.72 — — ——— OLDUVAI
— 1.88 — — —
— 2.04 ——————— REUNION

— 2.47

GAUSS

—2.88 — —— KAENA
—2.96
—3.07 — —— MAMMOTH
—3.16

— 3.40

—3.87 — —— COCHITI
—3.97
—4.10 — —— NUNIVAK
—4.24 — —
—4.39 — —— SIDUFJALL
GILBERT
—4.46
—4.56 — ——
——————— THVERA
—4.76 — — —

EPOCH 5

— 5.26 — — —

Fig. 5.23: The palaeomagnetic time-scale of the last 5 million years. Shaded areas indicate periods of normal polarity; unshaded areas show intervals of reversed polarity. Only one of several short events during the Brunhes epoch is depicted (modified after Berggren *et al.* 1980).

main boundaries, the Brunhes/Matuyama (*ca.* 0.7 m. yr), the Matuyama/Gauss (*ca.* 2.47 m. yrs) and the Gauss/Gilbert (*ca.* 3.4 m. yrs). In each case the age of the boundary has been established on the basis of averages of a number of K-Ar dates (Mankinen and Dalrymple 1979). The K-Ar method is, however, not sufficiently precise to date some of the polarity events, and their ages on the palaeomagnetic time-scale have been established by extrapolation based on the ages of epoch boundaries. As such, the dating of polarity events is much less secure (Berggren *et al.* 1980).

Within deep-ocean sediments individual particles adopt the direction of the earth's magnetic field as they accumulate, and hence they contain a continuous record of geomagnetic changes. The resolution and length of the polarity time-scale is primarily dependent on rates of sedimentation. In parts of the oceans experiencing comparatively slow sedimentation rates (e.g. 0.1 to 0.2 m per 1000 years), a core may contain a complete record of Pleistocene polarity changes, and even part of the Pliocene sequence (Fig. 6.8). Often, however, the time-scale will be compressed, and it may prove difficult to identify the shorter-lived polarity events. Conversely, a common problem with Atlantic ocean sediments is that because of the more rapid rates of sedimentation (e.g.

0.25 to 0.5 m/1000 years), coring devices often fail to reach the first major geomagnetic boundary, the Brunhes/Matuyama transition (dated to *ca*. 700 000 to 8000 000 years – Johnson 1982). Even where major boundaries can be established, other parts of the sequence can only be dated by extrapolation based on the assumption that sedimentation rates have remained uniform throughout the time period concerned. The use of palaeomagnetic stratigraphy as a means of correlation between individual deep-sea cores and also between the marine and terrestrial records is discussed further in Chapter 6.

Magnetic mineral concentration and susceptibility.   Most recent research on palaeomagnetic variations in Flandrian lake sediments has tended to concentrate not on variations in field properties but on the following magnetic characteristics:

(*a*) Susceptibility. This is the relative 'magnetisability' of a sample, or the relative strength of attraction it exhibits towards a known magnetic force. It is largely a function of the volume of ferromagnetic minerals contained in a sample, such as magnetite and maghemite, although it is also affected by other parameters, such as the size and shape of the magnetic particles (Thompson 1979).

(*b*) Isothermal remanent magnetisation (IRM). IRM is the magnetic moment activated in and retained by a sample placed in a magnetic field at room temperature. With a gradual increase in the strength of the field IRM will increase non-linearly until saturation isothermal remanent magnetisation (SIRM) is reached. This is the level at which, for any particular sample, a further increase in the magnetic field will not result in any increase in IRM, and is dependent upon the sizes and types of magnetic minerals present in a sample (Thompson *et al*. 1975).

(*c*) Coercivity of IRM. This is the reversed field strength required to be applied to a sample in order to reduce SIRM to zero. According to Thompson (1979) low coercivities are characteristic of large-grained magnetite and high coercivities of fine-grained magnetite.

Since measures of magnetic characteristics recorded in lake sediments depend primarily on the volume, rate and type of magnetic minerals in the sediment influx, they cannot be used as a basis for relative chronology or as a means of correlation between sites. However, careful measurement of the properties introduced above allows rapid correlation between cores taken from the same site, a process which is much less time-consuming than conventional methods of core correlation, such as pollen analysis (see Thompson *et al*. 1975; Oldfield 1978; Bloemendal *et al*. 1979). This method would appear to offer considerable potential as a means of correlation in a wide range of sediments and over longer time-scales where variations in mineral magnetic properties reflect synchronous environmental changes within lake basins or catchments.

Tephrochronology

Following a volcanic eruption ash or **tephra** is often spread rapidly over a relatively wide area and forms a thin cover over contemporaneous peat surfaces, lake floor sediments, estuarine sediments, river terraces, etc.

Fig. 5.24: Mazama ash (approximate age 6600 BP) exposed in Holocene fluvial sediments near Lumby, British Columbia, Canada.

Thin ash layers have also been found in deep-sea sediments. Ash beds often stand out as distinctive light-coloured horizons in sedimentary sequences (Fig. 5.24). Individual ash layers can be identified by a variety of methods, including mineralogical composition, the development of hydration layers (see below), and the presence or absence of specific mafic or glass components (Westgate and Gold 1974). The age of an ash bed

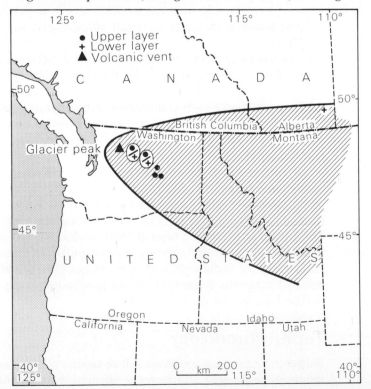

Fig. 5.25: Distribution of Glacier Peak tephra (after Westgate *et al*. 1970).

can be established by radiocarbon dating of associated organic material (wood, bone, shell fragments, etc), or for older deposits by K-Ar or fission track dating of some of the primary mineral constituents (e.g. Izett *et al.* 1972).

Tephrochronology has been most widely employed in North America, where successive phases of vulcanism have resulted in the accumulation of ash layers of different ages in the sedimentary records of the western cordillera and adjacent areas of the high plains. Of particular value have been the Pearlette Ashes which range in age from *ca.* 2 million to 600 000 years (Izett *et al.* 1972) and the Glacier Peak (*ca.* 12 000 BP – Fig. 5.25), Mazama (*ca.* 6600 BP) and St Helens 'Y' (*ca.* 3000 BP) Ashes (Westgate *et al.* 1970). Tephrochronology has also been used in New Zealand (e.g. McGraw 1975) and with more limited success in the Rhineland region of central Europe (Brunnacker 1975). Although a technique of considerable potential, tephrochronology can only form, at best, a basis for regionally-applicable schemes of relative dating, for individual ash layers are restricted spatially by such factors as the magnitude and type of volcanic explosion, the strength and direction of prevailing winds, the particle size of the tephra, etc. The method and its applications are reviewed in Wilcox (1965) and Westgate and Gold (1974).

RELATIVE CHRONOLOGY BASED ON PROCESSES OF CHEMICAL ALTERATION

Fossils, sediments and rocks are affected by a number of chemical reactions that are partly time-dependent. Upon the death of an organism, tissues are broken down by a variety of chemical processes to produce compounds of a more simple chemical structure; the surfaces of fossils or minerals may be altered by the effects of hydration or the accumulation of precipitates of certain chemicals in groundwaters, while weathering and pedogenic processes will gradually effect visible changes on rock and sediment surfaces. In all of these cases, the degree of alteration brought about by different chemical reactions increases with time, and this therefore offers a basis for relative dating.

Several methods of establishing relative chronology based on rates of operation of chemical processes are employed in Quaternary research, and some of the more common ones are briefly discussed below. Particular attention is devoted, however, to a relatively new method which is becoming increasingly important in Quaternary investigation, namely 'amino acid dating'.

Biogeochemistry of amino acids (aminostratigraphy)

Living bone consists of approximately 23 per cent collagen (protein-bearing) fibrils bound within a phosphatic-calcitic matrix. Proteins can survive in bones and shells for extremely long periods,[4] but undergo a number of molecular changes. The discovery of protein residues in fossil bones and shells was first reported by Abelson in 1954, and since then the study of protein transformation in the geological record has grown considerably. Some of the chemical changes in proteins that occur after the death of organisms are believed to be time-dependent, and thus the characteristics of certain protein residues from the Quaternary record provide the basis for a relative chronology.

Chemistry of proteins

Proteins are extremely large and complex molecules and are basic ingredients of all living organisms, being present in the cytoplasm and nucleus of all living cells and in enzyme structures which are vital to metabolic chemical reactions. Proteins are composed of **amino acids**, which have the generalised chemical formula shown in Fig. 5.26A. The 'R' linkage differs for each amino acid, ranging from a simple hydrogen atom in glycine, to a methyl group (CH_3) in alanine, and to extremely complex chemical structures in other amino acids. About twenty amino acids are commonly found in proteins (Table 5.5).

The formation of proteins results from the combination of a number of amino acids, often producing very long 'chains' of acid groups termed **peptide chains. Peptide bonding** refers to the linking of the amino and

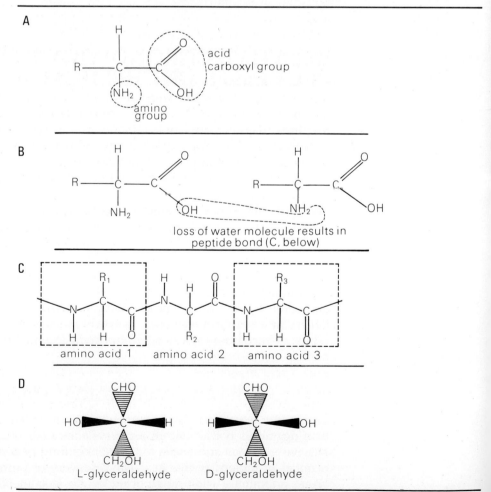

Fig. 5.26: Amino-acid geochemistry. A: Generalised formula for amino acids. B: Schematic representation of chemical combination of amino acids through loss of water molecule to form a peptide bond. C: Combination of carboxyl and amino groups between amino acids to form peptide chains. D: Enantiomers of glyceraldehyde; lined wedges represent bonds extending behind the plane of the page, and bold wedges represent bonds extending in front of the page.

Name	Symbol	Molecular weight
Glycine	Gly	75
Alanine	Ala	89
Serine	Ser	105
Cysteine	CySH	121
Threonine	Thr	119
Valine	Val	117
Leucine	Leu	131
Isoleucine	Ileu or Iao	131
Methionine	Met	149
Aspartic acid	Asp	133
Glutamic acid	Glu	147
Lysine	Lys	146
Arginine	Arg	174
Phenylalanine	Phe	165
Tyrosine	Tyr	181
Tryptophan	Try	204
Histidine	His	154
Proline	Pro	115
Hydroxyproline	Hypro	131

Table 5.5: The twenty common amino acids in proteins.

carboxyl groups of amino acids following the loss of a water molecule (Fig. 5.26, B and C). Any single combination chain of amino acids is referred to as the **primary structure** of a protein, defined by the number and order in which amino acids are arranged in the protein molecule. Large complex chains therefore exhibit polypeptide chains, but the general term **'peptides'** is used when referring to protein structures. If several polypeptides arrange themselves in parallel, some molecular stability is achieved, and the resulting peptide arrangment is referred to as **secondary structure**. Complex folding of polypeptide chains produces three-dimensional **tertiary structures**, and polymeric aggregates of several protein sub-units leads to complex coiled, folded and branched **quaternary structures**. Some of the larger proteins may contain up to 3000 amino acid residues. It is clear, therefore, that the number of ways in which the twenty common amino acids can be joined together is almost infinite, and it is this that leads to the large number of strongly individual proteins present in living organisms.

With the exception of glycine, all amino acids commonly found in proteins can exist in two molecular forms (**isomeric forms**) which are termed the **D-** and **L-enantiomers**. The physical and chemical properties of the enantiomeric forms of each amino acid are similar, the only difference being that they rotate plane polarised light in equal but opposite directions (Fig. 5.26D). The biological significance of this distinction is that only L-isomers occur in living (active) proteins. D-isomers can occur in a free state, as components of non-protein structures, and in fossil collagen as a result of the breakdown of proteins.

Amino acid diagenesis

Chemical alteration occurs in proteinaceous residues, eventually resulting in the breakdown of peptide chains to release free amino acids. If the proteins are exposed to the atmosphere or to biological processes, very rapid degradation will take place. Where they are protected by skeletal hard parts, however, so that a 'closed' system prevails, then much slower chemical alterations result. In these circumstances amino acids in fossil collagens exhibit a wide range of molecular stabilities. Some reaction times for alteration processes are in the range of 50 000 to a few million years whereas other amino acid reactions operate within a time-scale of only a few thousand years.

There are two basic diagenetic processes that are useful for providing indications of relative age of fossils. First, peptide bonds are broken by a process known as **hydrolysis**, which eventually releases free amino acids, so that the ratio of free amino acids to peptide-bound acids increases with time. The second mechanism is called **racemisation** (or **epimerisation**), where L-isomers of individual amino acids change to the D-configuration, a process that is also partly time-dependent. During racemisation the L-enantiomers are converted (interconverted) to the corresponding D-enantiomer, and the reaction continues until there are equal amounts of both enantiomers (*i.e.* D/L ratio = 1.0).

Plotting the reaction pathways of chemical alterations during diagenesis is complicated by a number of factors. First, reaction rates for both hydrolysis and racemisation differ for each amino acid. Secondly, L-to D-enantiomer interconversion can occur both in the free and in the peptide-bound fractions of proteins. Thirdly, all reactions are temperature-dependent. Finally, reaction rates depend on the permeability of the shell or bone matrix and on the original amino-acid composition and type of protein structures (e.g. primary, secondary etc) of samples. Amino-acid composition, relative abundances of amino acids, and rates of various amino-acid reactions are, therefore, for some animals genus-dependent (Lajoie *et al.* 1980), and for some bivalves even species-dependent (Miller and Hare 1980).

A number of chemical measures can be used to determine relative diagenesis in protein residues. These include the ratio between total free amino acids and total acids still peptide-bound, the enantiomeric ratios of a particular amino acid, the relative destruction of particular acids by hydrolysis, or the relative enrichment of a particular amino acid as a result of various decomposition reactions. The specific amino-acid ratios that are the best indicators of the relative passage of time will not be the same in every case, but will depend on the species concerned and on the local site conditions (especially thermal history) that have obtained since the death of the organism.

Although amino-acid determinations can be made on any protein-bearing materials, such as hair, teeth, Foraminifera, bones and tusks, in practice suitable samples are restricted to proteins preserved within 'tight' skeletal carbonate matrices, such as foraminiferal tests and valves of certain molluscs (Andrews and Miller 1980). Bone proteins are complex and highly soluble, and are poorly suited to the amino-acid method. Work so far on Quaternary amino-acid ratios has, therefore, been based largely on mollusc shells, in which the most reliable time-dependent reactions have been found to be racemisation in the total amino-acid fraction (*i.e.* free plus peptide-bound), and to a lesser extent in the free fraction alone (Miller and Hare 1980). Methods of detecting and measuring different amino acids can be found in most modern textbooks on biochemistry, such as Bohinksi (1979), and the methods applied to fossil proteins are summarised in Kvenvolden (1975).

Aminostratigraphy

The use of amino-acid ratios to rank fossils and their associated sediments according to relative age is termed **aminostratigraphy**. Although the technique has only recently been developed, it appears to offer considerable potential as a means of establishing relative chronology, for it is, at least in theory, applicable to the entire span of Quaternary time, and mollusc shells which, next to Foraminifera, have proved to be the most satisfactory dating medium, are found in a variety of Quaternary sediments.

At present, however, the technique has certain limitations. Amino-acid reaction rates are temperature-dependent and, therefore, samples can usually only be compared from areas that have the same climatic characteristics and similar climatic (thermal) histories. Moreover, since intergeneric variation in amino-acid ratios is a significant variable, comparisons between samples have to be made on a monogeneric basis. Finally, diagenesis in an essentially 'closed system' is a prerequisite. Not only must the integrity of the skeletal framework (carbonate matrix) be considered, but also the exact stratigraphic position of the fossils, for samples lying close to the surface of a site may give anomalously high ratios, as they are exposed to greater seasonal and diurnal temperature variations (Miller *et al*. 1979). In spite of these constraints, however, aminostratigraphy has a range of potential applications in Quaternary research:

1. *Development of regional chronologies.* Aminostratigraphy has, so far, been most widely employed as a means of providing a relative-dating framework of local or regional significance. For example, amino-acid ratios in valves of the mollusc *Corbicula fluminalis* taken from interglacial sites

Table 5.6: Isoleucine epimerisation in *Corbicula fluminalis* samples from interglacial sites in south-east England. Group 1: sites assigned to the last (Ipswichian) interglacial, although on stratigraphic grounds, the Clacton site is normally considered to date from an earlier interglacial. Group 2: last-but-one interglacial sites. Group 3: Affinity unknown, but may represent a stage older than 2. Group 4: Early Pleistocene samples. Arkansas modern: modern sample used as a reference standard (after Miller *et al.* 1979).

Interglacial site	Group	Isoleucine epimerisation	
		Lowest	Mean
Aveley	1	0.16	0.19 ± 0.023
Stutton	1	0.17	0.19 ± 0.024
Crayford	1	0.18	0.21 ± 0.031
Clacton	1	0.19	0.19 (1 value only)
Ilford	1	0.19	0.23 ± 0.038
Shoeburyness	1	0.19	0.23 ± 0.038
Selsey	1 or 2	0.21	0.28 ± 0.045
Hackney	1 or 2	0.24	0.27 ± 0.036
Clapton	2	0.27	0.28 ± 0.012
Grays	2	0.28	0.31 ± 0.029
Swanscombe	2	0.29	0.30 ± 0.017
Stoke Newington	2	0.29	0.31 ± 0.026
Purfleet a	3	0.33	0.35 ± 0.032
Purfleet b	3	0.34	0.36 ± 0.015
Wangford	4	0.74	0.76 ± 0.021
Arkansas modern		0.020	0.020 ± 0.001

in south-east England enabled shells of different ages to be separated
(Table 5.6) and a relative chronology of interglacial deposits to be
established (Miller *et al.* 1979). In south-west England and South Wales,
amino-acid ratios detected in shells of *Patella vulgaris* and *Macoma* spp.
from interglacial raised beach deposits formed the basis for a correlation of
interglacial high stands of sea level around the southern part of the Irish
Sea basin, and also allowed a differentiation between interglacial beach
deposits of different ages (Andrews *et al.* 1979). It is anticipated that
eventually sufficient will be learned about racemisation rates to permit
precise age estimates to be derived from specific amino acid ratios for
known thermal conditions.

2. *Resolution of populations of mixed ages.*   In a mixed assemblage of, for
example, mollusc shells, amino-acid ratios provide a means of
differentiating between fossils of different ages (e.g. Miller and Hare
1980).

3. *Aid to radiocarbon dating.*   Radiocarbon dates in excess of about 40 000
BP are often considered unreliable and the dating process is expensive and
time-consuming. By comparison, amino-acid analysis is less costly and
relatively rapid, and may be used to 'screen' fossils prior to radiocarbon
dating. This could be of particular value in the dating of mollusc-bearing
tills, in regions such as Arctic Canada where mollusc valves are the most
widely used materials in radiocarbon dating, but where reworking of
molluscs and mixing of molluscan populations has often occurred (Blake Jr
1980).

4. *Palaeothermometry.*   Since racemisation rates are temperature-dependent,
where the exact age of fossil protein can be established amino-acid ratios
may be used to infer temperature conditions that have affected the
proteins since the commencement of diagenesis. Although this has limited
application where fossils have experienced a number of major temperature
oscillations, aspartic acid D/L ratios in bristlecone pines, for example, have
formed the basis for specific palaeotemperature inferences (Zumberge *et al.*
1980).

5. *Identification of palaeosols.*   Goh (1972) has shown that a study of amino-
acid nitrogen levels in sediments can be used to identify, or confirm the
identification of, buried organic layers of palaeosols.

Fluorine, uranium and nitrogen content of fossil bones.

Hydroxyapatite, the prinicipal mineral constituent of bones and teeth
(p. 206), absorbs fluorine from groundwater progressively over time. The
rate at which the fluorine content increases varies from locality to locality,
but bones that have lain for the same period of time in a particular
deposit will have approximately the same fluorine content. As the fluorin
fixed in the bone is not readily removed, a specimen that is washed out
an older deposit will show a much higher fluorine content than bones th

are contemporaneous with the bed, while bones from a younger stratum will have accumulated substantially less fluorine. The fluorine to phosphate ratio is established by X-ray diffraction techniques. This method is particularly useful when bones have accumulated in sand and gravel, but difficulties have been encountered in limestone cave deposits where it has been found that the passage of fluorine ions is inhibited while the decay of organic matter proceeds at normal rates (Oakley 1969). In more recent years, the analysis of uranium incorporated into fossil bones from groundwater has been employed in a similar way to the fluorine method, the former having the advantage that the counting of uranium emissions does not involve the destruction of the bone material. A final analytical method that establishes the relative ages of fossil bones in an assemblage utilises the amount of nitrogen in bone collagen, as decreasing nitrogen content will indicate increasing age. These techniques and their applications are reviewed in Oakley (1980).

Obsidian hydration

Freshly-exposed surfaces of obsidian absorb water from their surroundings to form a hydration layer known as perlite (Friedman and Long 1976). This external rind should not be confused with the patina that develops on many materials as a result of chemical weathering (see below). Hydration layers require detailed examination by optical microscope, for the thickness of such layers is commonly less than 20 μm. The precise thickness of the hydration layer depends on the length of time since exposure of an obsidian surface and on rate of hydration which, in turn, depends on temperature during the time of exposure and on the chemical composition of the obsidian (mainly the percentage silica content). Hydration thicknesses have been used to determine the relative antiquity of archaeological implements, the age of rhyolite flows (Friedman 1968), and as a means of correlating glacial successions (Pierce *et al.* 1976). It is thought that the dating range of this method may eventually extend to 10^6 years, although uncertainties in age estimates are often as much as 20 to 30 per cent (Andrews and Miller 1980), due, at least in part, to samples having been affected by temperature fluctuations since burial. The practical aspects of the technique are considered in detail in Friedman *et al.* (1969).

Weathering and soil formation

Variations in weathering of rock surfaces and degree of pedogenetic development have been widely employed, particularly in North America, as bases for relative chronology of glacial deposits (e.g. Morrison 1967; Mahaney 1976) and also as means of correlation (Ch. 6). In the Rocky Mountains, Richmond (1965) differentiated older glacial drifts from younger materials on the basis of a range of weathering features including the relative decomposition of exfoliation of boulders (Fig. 5.27), lack of soluble materials (e.g. limestones) on older surfaces, the crumbly nature of sandstone and volcanic clasts on older drifts, the relative concentration of more durable materials (e.g. quartz, chert, dense silicious rocks) at the surface, and the occurrence of partly-peeled thick rinds of desert varnish on

Fig. 5.27: Deeply weathered boulder exposed in late Wisconsinan moraine, Allens Park, Colorado.

some rock types. Weathering rinds in particular appear to offer considerable potential as a means of relative dating (Colman and Pierce 1981). Anderson and Anderson (1981) used variations in weathering rind thickness on calcareous quartzarenite clasts to develop a relative chronology for glacier and rock glacier deposits in part of the Wasatch Range in Utah, while weathering rinds on sandstone boulders were used in the establishment of a Holocene glacial chronology for part of the Southern Alps, New Zealand (Chinn 1981). Studies in which the degree of pedogenetic development have formed the basis for a relative chronology include those by Mahaney and Fahey (1976) who identified five palaeosols of different ages in the Front Range of Colorado, each one differentiated on the basis of distinctive physical and chemical characteristics, and Reider (1975) who was able to distinguish between drifts of Holocene, Late Wisconsinan, Early Wisconsinan and Pre-Wisconsinan ages, also in the Colorado Front Range, by their differing levels of pedogenesis (Table 5.7).

To what extent weathering and soil formation can form the basis for a wholly reliable relative chronology is uncertain, however. Degree of weathering is often arbitrary and recognition highly subjective (Burke and Birkeland (1979); rates of weathering intensity are not linear but decrease with age (Colman 1982); age calibration by radiometric methods is difficult to achieve, and weathering intensity will vary considerably with, for example, altitude, climate and bedrock type. With respect to soil development, time is clearly only one of a number of variables involved in pedogenesis, and other soil forming factors, notably parent material, slope, climate and biotic activity may have been equally if not more important

Soil genesis	Soil characteristics	Inferred age
Weakly-developed	No B horizon	Holocene
Moderately-developed	Presence of cambic B horizons with less Clay than A horizons	Late Wisconsinan (Pinedale)
Strongly-developed	Argillic B horizons (35% clay) present	Early Wisconsinan (Bull Lake)
Intensely-developed	Abundant clay concentration (50%–90%) and complete absence of coarse particles (> 2 mm) in truncated, buried B horizons	Pre-Wisconsinan (Pre-Bull Lake)

Table 5.7: Soil development and the glacial stratigraphy of the Front Range, Colorado (after Reider 1975).

determining agencies. For these reasons, therefore, weathering and soil formation are likely to offer a basis for regional relative chronologies only, and may be of more limited application than has hitherto been considered.

CONCLUSIONS

Although techniques which establish age equivalence on the basis of stratigraphic markers and methods which allow the determination of the relative order of antiquity of rocks or fossils are both widely applied in Quaternary research, clearly the most important methods for establishing the age of Quaternary events are the radiometric and incremental techniques. Because they allow events to be dated in years, their application has often been referred to as 'absolute dating'. However, this term has not been used here as it implies a level of exactitude that is not attainable at present. Nor have we used 'geochronometric', a term that has been employed to describe dating methods from all of the categories described above, including methods which, at best, only provide relative age-relationships of regional application.

No single dating method is applicable to a wide range of materials, and many of the methods are restricted to particular segments of Quaternary time (Fig 5.1). Moreover, each technique has its own individual limitations and error sources which lead to conflicts and uncertainties in interpretation. In this respect it is important to make a distinction between 'precision' and 'validity' in dates. The former applies to the statistical uncertainty that is attached to the particular physical or chemical measurement. This precision in measurement may then be interpreted as a valid indication of age, depending on analytical problems and stratigraphic considerations. For example, precise measurements can be obtained from fossils containing contaminants, but if the contaminants are not recognised as such and corrected for, then the inferred age will be invalid.

The major problem now in Quaternary dating would appear to be how to achieve a greater degree of certainty in estimates based on methods that have rather uncertain foundations. Progress in this respect depends on replication of results from any particular method, but also perhaps more importantly, on the application of more than one method to the dating of a specific horizon or event. Although the methods reviewed in this chapter have gone some way towards providing a realistic dating framework, a major objective in future research must be the development of further dating techniques, thereby providing greater scope for the calibration of existing ones. Bearing these points in mind, we now turn to the wider question of stratigraphic subdivision and correlation of the Quaternary record.

NOTES

1. The principle of superposition states that, in the absence of evidence for disturbance or reworking, the overlying sediments in a succession are younger than those lying beneath them. This applies equally to lithological units, such as tills, solifluction deposits, etc, as to biostratigraphic units such as pollen or molluscan assemblage zones. In this way, pollen analysis, for example, can be employed as a relative dating technique (see Ch. 4).
2. Fractionation is the selective separation of chemical elements or isotopes during natural physical or biochemical reactions, e.g. during evaporation, condensation, osmosis, transpiration, etc.
3. Particle 'accelerators' are designed to generate high voltages enabling the controlled acceleration of nuclear particles. These 'machines', which include the cyclotron, linear accelerator, and van de Graaff generator, can be used to separate and detect atomic particles of different mass or charge with a much higher resolution than conventional mass spectrometers.
4. For example protein residues have been found in Ordovician and Devonian shells, and collagen-like proteins have been recovered from Cretaceous dinosaur bones (Wyckoff 1980).

CHAPTER 6

Approaches to Quaternary stratigraphy and correlation

Introduction

In previous chapters the various methods employed in palaeoenvironmental reconstruction have been examined and the techniques used in the dating of Quaternary events have been discussed. Of particular interest to the Quaternary scientist, however, is the way in which environments have changed through time and how such information can be gained from the stratigraphic record. There are two aspects to the interpretation of that record, namely the ordering of the evidence at any one locality into a time sequence (temporal dimension) and the relation of the evidence at one locality to that at another (spatial dimension). The temporal dimension involves principles of **stratigraphy**, while the spatial dimension involves principles of **correlation**. A proper understanding of the procedures associated with these two aspects of geological investigation is fundamental to a correct interpretation of Quaternary environmental change.

  If the stratigraphic record was complete at all places on the earth's surface, and if every important horizon could be dated accurately, then units of the record could be arranged into stratigraphic order based on increments of time. In this way a **time-stratigraphic** framework for environmental change could be established, and correlation even between widely separated localities would present few problems. In practice, however, this is not possible. The terrestrial record is not complete, and dating control cannot always be established, due either to the absence of suitable dating materials, or to limitations of the actual techniques employed. Only in the deep oceans and in certain exceptional terrestrial situations are long stratigraphic records preserved, and even in these cases only a generalised dating framework has been established. For the most part, therefore, the Quaternary stratigrapher is confronted by a highly fragmented and partial stratigraphic record which only in certain circumstances and over limited time ranges can be dated precisely. As a consequence, very careful palaeoenvironmental interpretation and stratigraphic evaluation are required before sequences can be ordered at one place and correctly related to those at another.

In this chapter the bases of, and the procedures involved in, Quaternary stratigraphy and correlation are examined, and the means whereby time-stratigraphic correlation can be achieved are assessed. This discussion relates essentially to the terrestrial record for it is the terrestrial evidence that provides the greatest obstacles to the application of conventional stratigraphic and correlation procedures. Deep-ocean sediments are discussed in a separate section, and the chapter concludes with an examination of the bases for correlation between the marine and terrestrial records.

Stratigraphic subdivision

Stratigraphy is the study of rock successions and their interpretation as sequences of events in the geological history of the earth. The International Subcommision on Stratigraphic Classification (ISSC), a body set up by the International Union of Geological Sciences (Hedberg 1976), recognises three principal categories of stratigraphic subdivision for use throughout the geological column. These are **lithostratigraphy** (the classification of local sediment units or rock successions according to observed variations in lithology), **biostratigraphy** (the classification of sediment units according to observed variations in fossil content), and **chronostratigraphy** (the classification of stratigraphic units according to age). This last mentioned category is also termed **time-stratigraphy**. In Quaternary stratigraphy, two further categories have been employed. These are **morphostratigraphy** (the classification of landforms according to their relative order of age) and **geologic-climatic units** (the allotment of stratigraphic units to positions within a sequence of inferred climatic episodes). The latter classification is sometimes referred to as **climatostratigraphy**. Each of these will now be discussed in more detail.

1. Lithostratigraphy

This is the fundamental building block of stratigraphy, the organisation of the geological record into stratigraphic units on the basis of observed changes in the character of the sediments. Lithostratigraphic units are traditionally recognised and defined by readily observable features of petrological, mineralogical, geochemical or even palaeontological character, and not by inferred mode of genesis. In Quaternary stratigraphy, however, a more interpretative approach is generally adopted and lithostratigraphic units are usually defined on the basis of both physical attributes and inferred mode of origin (e.g. tills, aeolian sands, glaciofluvial gravels). Boundaries between lithological units are placed at positions of lithological change and commonly cut across the limits of fossil ranges and the boundaries of other types of stratigraphic unit. Moreover, they frequently cut across time-horizons (**time-transgression**), a feature which poses particular problems in Quaternary correlation (see 'Correlation' section below). These relationships are illustrated in Fig. 6.1.

Major lithostratigraphic units in conventional geological codes are

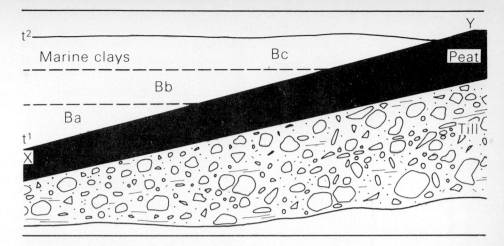

Fig. 6.1: Time-transgression in lithostratigraphic and biostratigraphic boundaries. A till unit is overlain by a peat layer which has, in turn, been buried by marine clays. The marine clays accumulated during a gradual marine transgression (t^1 to t^2) and therefore the lithostratigraphic boundary between the peat and the overlying marine sediments (X–Y) will be time-transgressive. During the deposition of the clays, changes in fossil content have occurred, represented by biozones Ba, Bb and Bc. The lithostratigraphic boundary X–Y therefore cuts across the biozone boundaries.

termed **formations**, and these constitute the primary formal unit of lithostratigraphic classification (Hedberg 1976). Formations are defined on the basis of well-marked upper, lower and lateral boundaries that can be traced and mapped over considerable distances. They can be subdivided into **members**, entities within the formation possessing lithological characteristics that distinguish them from adjacent parts of the formation, and **beds**, units or layers within the member or formation. Distinctive beds (termed **marker beds**) that are particularly useful in stratigraphic subdivision are usually given proper names. In formal geological procedure, the term 'bed' is usually applied to layers of a centimetre to a few metres in thickness. Very thin beds of one centimetre or less are known as **laminae**. In practice, however, the application of all of these terms depends on the local complexity of the stratigraphy and on often subjective assessments of the record. In early work on Quaternary sediments, stratigraphic accounts were rather descriptive and often lacked precision. This engendered an uncritical, almost 'impressionistic' approach that led to misinterpretation of complex sequences (see e.g. discussion of till units in Ch. 3). More recently, however, there has been a move towards the application of standard geological procedures in Quaternary stratigraphy, particularly in the recognition and definition of beds, members and formations (e.g. Bowen 1978). The way in which these terms might be applied to a glacigenic sequence is shown in Fig. 6.2. Individual beds are those homogeneous layers of sediment interpreted as having originated during the same depositional event. For example, careful examination of a sand unit may reveal a series of current-bedded or ripple-bedded layers, reflecting changes in the depositional environment. Similarly, within a flow till alternations between sand and gravel beds and layers of finer material will indicate variations in supraglacial deposition. Members, on the other hand, are units encompassing beds that are regarded as being genetically related in that they have formed during the same overall depositional phase and by the same agent of deposition. Hence the stratified sediments in Fig. 6.2 are grouped together to form a member which can be differentiated lithologically from the unstratified till members. The whole sequence is classified as a formation (scheme 1) on the basis that all of the sediments

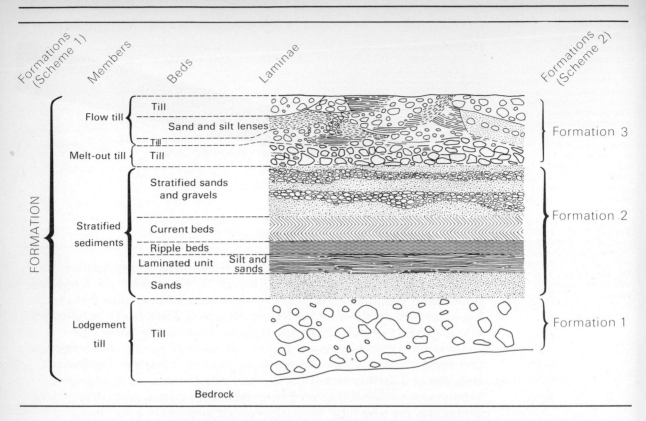

Fig. 6.2: Lithostratigraphic subdivision of a glacigenic sequence.

are interpreted as having been derived from episodes of glacial and glaciofluvial activity, associated with the advance and retreat of one major ice mass.

This approach, however, is not without its problems. The interpretation of the Quaternary depositional record is often complicated by the fact that similar sediments and sedimentary sequences can result from different processes. Fluvial, glaciofluvial and aeolian sediments, for example, can appear lithologically similar despite the fact that they have accumulated in widely-contrasting depositional environments. Many Quaternary sedimentary sequences show considerable lateral and vertical variation, and these local **facies**[1] changes frequently pose problems of interpretation and classification. Even where the origin of the sediments can be established, the subdivision of very complex sequences into lithostratigraphic units is not always straightforward. In Fig. 6.2, for example, it could be argued that the upper and lower till units represent two separate glacial intervals and should be classified as units of formation rank (scheme 2). Consistency of interpretation between individual workers may, therefore, be difficult to achieve. Nevertheless, in so far as it emphasises the need for careful analysis and interpretation, it is considered that this approach offers one of the most satisfactory means of subdividing the Quaternary rock-stratigraphic record.

2. Biostratigraphy

Biostratigraphic classification organises rock strata into units based on the variety and abundance of fossils. Biostratigraphic units are usually termed

biozones and the following four types are employed in stratigraphic classification:
1. **Assemblage biozone** or **Cenozone**: a group of strata characterised by a distinctive, natural fossil assemblage.
2. **Range biozone**: a group of strata containing the stratigraphic range of a particular fossil.
3. **Acme biozone**: a group of strata based on the acme or maximum development of a particular taxon.
4. **Interval biozone**: the stratigraphic interval between two biohorizons.

In pre-Quaternary geology, the biostratigraphic record is subdivided largely on the basis of evolutionary changes in organisms. Hence, acme biozones and range biozones reflect those parts of the geological record where a species appeared, thrived for a while, and then died away. Hence, acme and range biozones constitute valuable stratigraphic markers, and often form the basis for correlation. Although there is some evidence of evolutionary development in the Quaternary record, this is either too restricted in terms of the number of species affected, or too slow within the limited time-scale of the Quaternary to form a meaningful basis for stratigraphic subdivision. As a consequence only assemblage biozones are widely employed in Quaternary stratigraphy, and these reflect the ecological response of organisms to environmental change rather than evolutionary changes in flora and fauna. As such, they are essentially a reflection of local environmental conditions.

Because of the time-transgressive nature of climatic and environmental change, the boundaries of biozones (e.g. pollen assemblage zones, molluscan assemblage zones or diatom assemblage zones), cut across time horizons and frequently transgress the boundaries of other stratigraphic units. In the deep oceans, however, where sedimentation rates are often extremely slow, assemblage biozone boundaries are to all intents and purposes time-parallel. Difficulties are encountered in the recognition and classification of Quaternary biozones as a result of the reworking and selective destruction of fossils, and the problems of identification and ecological interpretation of fossil assemblages, all of which were discussed in Chapter 4.

3. Chronostratigraphy

The purpose of chronostratigraphy is to divide sequences of strata into units (**chronostratigraphic units**) that correspond to intervals of geological time. Such units are bounded by isochronous surfaces or **chronohorizons**. Chronostratigraphic units can be defined on the basis of precise geological age, where this can be established, or in terms of time intervals between isochronous horizons. In other words, where chronostratigraphic comparisons are being effected between sites, reference can be made to the time-period encompassed by, for example, particular biozone boundaries, where actual age remains unspecified. If there are grounds for believing that these boundaries are time-parallel, then the biozone boundaries can be employed as chronostratigraphic boundaries, and the biozone considered as a chronostratigraphic unit. Chronostratigraphy therefore differs from both lithostratigraphy and biostratigraphy in so far as the characteristics of the rock- and biostratigraphic record are essentially observable, whereas the basis for chronostratigraphy is wholly inferential (Vita-Finzi 1973).

The stratigraphic column can be divided into units of **chronostratigraphy** and **geochronology** according to the hierarchy of terms shown in Table 6.1. The chronostratigraphic unit refers to the rock sequence laid down during a particular time interval. Hence the term 'Quaternary system' is used to signify all the rocks or sediments that have accumulated over a time range that is called the Quaternary. The geochronological equivalent, that is the time interval itself, is referred to as the 'Quaternary period'.

The basic working units in Quaternary stratigraphy are **stages**, for they are the smallest subdivisions of the standard stratigraphic hierarchy that can be recognised on both a regional and inter-regional scale. Conventional stages typically range from 3–10 million years, but in the Quaternary, stages are measured in thousands of years, a level of precision that often causes difficulties. Stages can be divided into smaller units or substages. The basic chronostratigraphic unit is the **chronozone**, the time-span of which is usually defined in terms of the time-span of a previously designated stratigraphic unit such as a formation, or a member, or a biozone (Hedberg 1976). Chronozones have been most widely employed in Quaternary research where pollen assemblage zones have been dated by radiocarbon.

Division of Quaternary strata into chronostratigraphic units is seldom straightforward, however, and problems arise that are not encountered in the earlier geological record. In the main, these result from the relatively short time-span of the Quaternary, and from the fact that the Quaternary scientist has to deal with increasingly finer divisions of the stratigraphic record. No geological dating method produces consistently reliable results, but aberrations are more likely to pose problems in Quaternary research because of the more limited time range involved. Moreover, if the statistical uncertainty associated with the dates (*i.e.* the quoted ± value) is as great or greater than the stratigraphic intervals that are under examination, then clearly these age determinations are insufficiently precise to define the boundaries to such units, or to distinguish subdivisions within those units (see e.g. Lowe and Gray 1980). A further difficulty arises over the recognition of isochronous horizons in the stratigraphic record. In pre-Quaternary strata, lithostratigraphic and biostratigraphic boundaries, although inherently time-transgressive, *appear* to be synchronous when set against the vast span of geological time. As such, they are frequently used as time-stratigraphic markers. In the Quaternary stratigraphic record, most boundaries are time-transgressive. The only time-parallel marker horizons are palaeosols, volcanic ash layers, palaeomagnetic horizons and, in certain cases, shorelines. In Quaternary

Table 6.1: Conventional hierarchy of chronostratigraphic and geochronological terms (after Hedberg 1976).

Chronostratigraphic	Geochronologies	Examples
Eonothem	Eon	Phanerozoic
Erathem	Era	Cenozoic
System	Period	Quaternary
Series	Epoch	Pleistocene
Stage	Age	Devensian
Chronozone	Chron	Younger Dryas

marine sequences, the products of catastrophic sedimentary events, such as turbidites, may also be time-parallel. Chronostratigraphy is therefore not without its complications, although it clearly offers the most secure basis for time-stratigraphic subdivision of the Quaternary terrestrial record. Some of the chronostratigraphic methods employed in Quaternary correlations are discussed more fully below.

4. Morphostratigraphy

Morphostratigraphy is not recognised as a major category of stratigraphic subdivision by the ISSC since landforms are not a part of the general sedimentary record. Sediments accumulate through time, so that an interval of time is represented in a tangible way by a body of sediment. Landforms on the other hand are abstract surfaces, and these may be difficult to isolate within the lithostratigraphic column. A morphostratigraphic unit has been defined as 'a body of rock that is identified primarily from the surface form it displays' (Willman and Frye 1970). Commonly, such units hold lateral relationships with each other in, for example, the case of recessional moraines, shoreline sequences related to falling sea-levels, or sequences of terraces resulting from river incision. Morphostratigraphic units should not be confused with lithostratigraphic units, and although they have found general acceptance amongst Quaternary workers, their status remains informal, and their application is restricted to certain rather exceptional geomorphological situations.

5. Climatostratigraphy

A widely-employed basis for the subdivision of the Quaternary is climatic change for in many instances the characteristics of the stratigraphic record can be related directly to former climatic conditions. In recognition of this fact, the American Commission on Stratigraphic Nomenclature established a new stratigraphic subdivision termed a **geologic-climatic unit**. This was 'an inferred widespread climatic episode defined from a subdivision of Quaternary rocks' (American Commission 1961). In areas affected by Quaternary glaciation, glacials and interglacials constitute the principal geologic-climatic units, while stadials and interstadials form units of lesser rank. In areas not affected by glacier ice, it was anticipated that other geologic-climatic units, such as pluvials and interpluvials, would be established.

   Geologic-climatic units are undoubtedly useful concepts and, in so far as the Quaternary at mid- and high-latitudes tends to be subdivided into glacials and interglacials, they form the basis for stratigraphic subdivision at the regional and continental scales. However, because climatic change is time-transgressive, the boundaries of geologic-climatic units are diachronous. It is not, therefore, appropriate to use geologic-climatic terms (glacial, interglacial etc) and chronostratigraphic terms (stage, sub-stage) interchangeably (Morrison 1968b). For example, in most areas formerly occupied by glacier ice, deposits may be laid down only at a late stage in glaciation. Therefore, the actual time interval represented in the stratigraphic record by a till (geologic-climatic unit) reflects only a part of the period of time of the glacial stage (see Fig. 6.5). Moreover, although geologic-climatic units are in some ways similar to chronostratigraphic units, in that they are both inferential (*i.e.* they are not based on observable characteristics of the stratigraphic record), a more sophisticated level of interpretation is required in order to derive meaningful geologic-climatic units, and the precision that can often be attained in chronostratigraphy can seldom be achieved in climatostratigraphy.

The fundamental problem is that it is not climate that is recorded in the stratigraphic record, but manifestations of climate, namely the results of climatic influences on, for example, biota, soils, sediments and glaciers. Climatic reconstructions are, therefore, two steps removed from the observable data, and at each stage in the analysis, interpretation is required. If, for example, pollen assemblage zones are being used as the basis for geologic-climatic units, the first step is to infer vegetational communities and patterns of vegetational change from the pollen record, and the second is to use these reconstructions to infer climatic changes. Interpretations are therefore being made from interpretations. Errors can enter into both stages of the analysis, but those resulting from the first will be compounded in the second. These complications will arise irrespective of what form of lithostratigraphic or biostratigraphic evidence is being employed.

The American Commission intended the boundaries of geologic-climatic units to follow the boundaries of the rock- or biostratigraphic unit that formed the basis for their definition. In practice, however, boundaries are much more difficult to locate. Consider, for example, Fig. 6.3 which depicts temperature change through a glacial-interglacial cycle. At what point on the curve does the geologic-climatic unit of the interglacial begin? It could be argued that the boundary should be placed at that point on the curve where temperature increases following a thermal minimum (point X). Alternatively, the boundary could be located where a temperature similar to that of today is achieved (point Y). A third view would be to place the boundary at a point where a particular temperature threshold is crossed (point T) as indicated, for example, by the first occurrence of a certain indicator species in the fossil record. The problem, of course, is that an essentially 'static' boundary is being placed on what is a continuum of climatic change. Any of the above could be used as a basis for defining the onset of an interglacial, but there is no guarantee that an interglacial defined in one region will be an equivalent geologic-climatic unit to that in another. Similar problems are encountered in

Fig. 6.3: Different ways of defining the onset of an interglacial. The curve is schematic and represents temperature oscillations between glacials and interglacials. The three hypothetical points of onset (X, T, Y) are explained in the text.

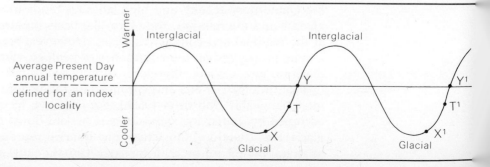

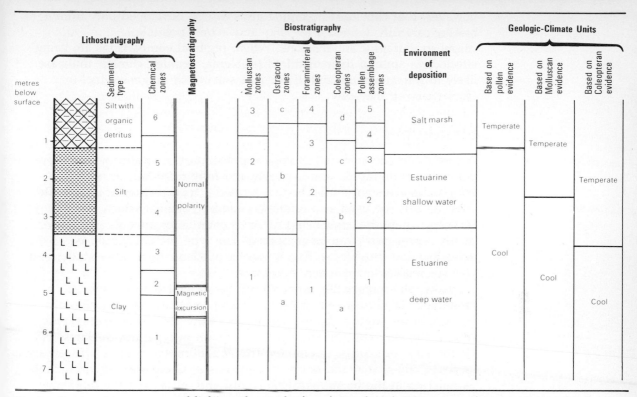

Fig. 6.4: Stratigraphic subdivision of a sedimentary sequence based on different criteria.

establishing the end of an interglacial. Moreover, the extent to which climatic episodes will be represented in the stratigraphic record will depend both on the amplitude and duration of climatic shifts. These will vary from locality to locality, thereby making consistency of interpretation and classification very difficult to achieve.

Further complications arise when different types of evidence are being employed in environmental reconstruction. In Fig. 6.4. a transgressive sequence of marine sediments has been divided on the basis of a number of lithostratigraphic and biostratigraphic criteria. Different lines of evidence may indicate different geologic-climatic units, however. On the basis of the molluscan records, for example, the boundary between warm and cold climatic conditions may be placed between assemblage zones 1 and 2; the pollen data may indicate a climatic change between pollen zones 4 and 5, while the boundary between cold and temperate coleoptera may occur between assemblage zones 1 and 2. These contrasts reflect different response rates of biota to climatic change and also, in so far as the taxa are derived from both marine and terrestrial environments, time-delay between atmospheric and oceanic temperature changes. Within a single sequence, therefore, the boundary of a geologic-climatic unit can be placed at any one of several levels. Once again, the problems that this can pose for inter-regional comparisons are clear.

The designation of geological-climatic units is, therefore, frequently intuitive, often arbitrary, and much less precise than other forms of stratigraphic subdivision. Yet climatic change is the dominant characteristic of the Quaternary, and it is difficult to envisage a stratigraphic scheme that does not explicitly acknowledge this fact. It must be emphasised, however, that climatostratigraphic units can be developed only from a basis of carefully defined litho-, bio- and, where possible, chronostratigraphic subdivisions. Where regional comparisons are being effected, in spite of their manifold problems, geologic-climatic units are likely to remain an integral, if perhaps somewhat enigmatic tool in Quaternary stratigraphy.

Stratotypes

According to conventional geological procedures, a locality where a particular stratigraphic unit is clearly and fully recorded, or where its boundaries are soundly defined, is termed a **type section** or a **type site**. This can then be used as a reference standard against which other sections, where the equivalent unit or its boundaries are only partially or poorly represented, can be compared. The type section should be accessible and durable, so that it may be available for further study, and perhaps reassessment where necessary.

   Ideally, all stratigraphic units should be defined with reference to a **stratotype**. This, however, is often difficult in Quaternary stratigraphy because of the highly fragmented nature of the terrestrial stratigraphic record, the very considerable spatial variation in type and thickness of Quaternary sediments, the limited lateral continuity of many Quaternary deposits, the spatial and temporal contrasts in Quaternary environments as reflected in the biostratigraphic record and, above all, the markedly time-transgressive nature of terrestrial stratigraphic boundaries. In many situations, therefore, a stratotype will have little more than local application and, as a consequence, the concept of the stratotype has found less favour with Quaternary workers than with geologists dealing with older segments of the stratigraphic record. However, in the interests of clarity and precision, a case could be made for the more widespread adoption of Quaternary stratotypes than has been the practice in the past, particularly in chronostratigraphy and climatostratigraphy where the boundaries are located on an inferential rather than an observable basis. Certainly, if conventional geological procedures are to be applied to the subdivision of the Quaternary stratigraphic record, the establishment of regional stratotypes as reference standards is an integral component of that approach.

Correlation

The stratigraphic methods outlined above all provide a basis for correlation, that is the relationship of stratigraphic sequences or events at one locality to those at another. Throughout most of the geological column, lithostratigraphic and biostratigraphic boundaries are, to all intents and purposes, time parallel, and are therefore regarded as being almost of equivalent status to chronostratigraphic units in time-stratigraphic subdivision and subsequent correlation. This assumption cannot, however, be made in the correlation of Quaternary successions except perhaps at the local scale. Not only are lithostratigraphic and biostratigraphic boundaries time-transgressive, but the repetitive nature of Quaternary climatic change has meant that at any given locality similar depositional sequences may be preserved that are of markedly different age. In view of the fragmented and highly diverse nature of the Quaternary stratigraphic record, the problems of effecting meaningful correlations between often widely separated sites are therefore considerable.

Some examples of the complications that can arise in the correlation of Quaternary successions are illustrated in Fig. 6.5. This shows, in a diagrammatic way, the extent of glaciation over time, the shaded area representing the time-period during which ice covered the ground at increasing distance from the ice dispersal centre. Site 1 was affected by glacier ice for almost the whole of the 'glacial' time interval; sites 2 and 3 were occupied by ice for a shorter period, while site 4, and especially site 5, were ice-covered for only a small proportion of the glacial phase. Sites

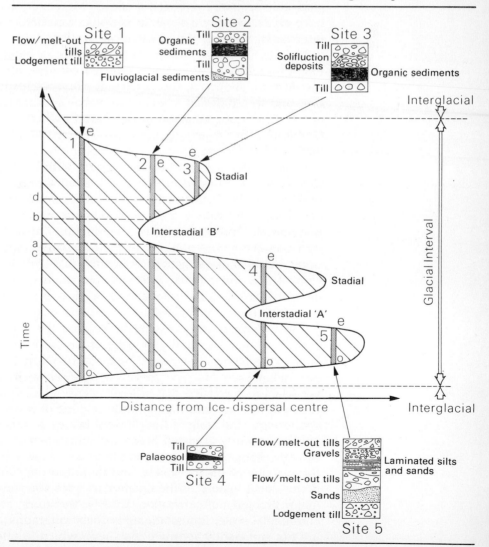

Fig. 6.5: Diagram showing the onset (o) and end (e) of glaciation at sites at increasing distance from the ice-dispersal centre (modified after Andrews 1979). Possible glacigenic sequences and relative sediment thicknesses at each site are also depicted. For further explanation see text.

2, 3 and 4 also experienced interstadial conditions between successive glacial advances. At each site, the environmental history is recorded in sequences of glacigenic sediments, intercalated at sites 2, 3 and 4 with organic deposits (peats or soils).

At site 1, which for much of the glacial interval lay some distance up-glacier, erosion probably dominated and the only record of glacial deposition may be a thin layer of lodgement till that accumulated at a very late stage in glacier wastage. By contrast, a complex sequence of glacigenic sediments may have accumulated at site 5 near the ice margin. There are two points to note here: first, the thickest and most complex sequence of sediments is preserved at the site that was glaciated for the shortest length of time, and secondly, although site 1 was affected by glacier ice before site 5, on a time-stratigraphic basis, the deposits *preserved* at site 1 may, in fact, be *younger* than those at site 5. At sites 2, 3 and 4, two till units are preserved, but from the stratigraphic evidence alone, it is far from obvious which till unit is the lithostratigraphic correlative of the glacigenic sequence recorded at site 5. Moreover, because sites 2, 3 and 4 have experienced two separate periods of glaciation separated by an interstadial interval, almost identical stratigraphic successions (till/organic sediments/till) have evolved. In the absence of other evidence, correlation might therefore be unwittingly effected between these sequences. Yet, as the diagram shows, the succession at site 4 relates to an earlier period of time and to a different stadial/interstadial oscillation from that at sites 2 and 3. Errors of correlation of this nature, where lithostratigraphy is confused with time-stratigraphy, are referred to as **homotaxial errors**. Such errors can also arise in biostratigraphy. It is entirely possible that a very similar environment existed during both interstadials A and B and this would be reflected in the fossil evidence preserved in the organic horizons at sites 2, 3 and 4. Again, the inference of a single interstadial based, for example, on the pollen evidence at the three sites would be homotaxially incorrect. Finally, the diagram illustrates the time-transgressive nature not only of the lithostratigraphic and biostratigraphic units, but also of the geologic-climatic units. At site 2, the geologic-climatic unit (interstadial) inferred from the fossil content of the organic deposit spans the time interval a–b, while at site 3, the same interstadial covers the time interval c–d.

These complications arise partly because the deposits of varying ages are not arranged vertically in order of superposition, and partly because lateral correlation often has to be made between numerous short-lived depositional records. The above example relates to a single glacial interval, yet when it is recalled that at least twenty glacial/interglacial cycles affected the mid- and high-latitude regions of the world during the Quaternary, the scale of the problem begins to emerge. Moreover, it is not only in formerly glaciated areas that difficulties are encountered. Repeated climatic changes had a profound effect on those regions that lay beyond the margins of the ice sheets, and there too the complicated erosional and depositional history of the Quaternary presents the stratigrapher with major problems of correlation. Clearly, therefore, at the regional and continental scale, lithostratigraphy, biostratigraphy and climatostratigraphy are not sufficiently sensitive tools with which to effect meaningful time-stratigraphic correlations. A chronostratigraphic basis is required, although as yet no single radiometric dating method is universally applicable to the whole of the Quaternary time-scale. The following, however, are believed to offer the most likely means whereby time-stratigraphic correlation can be achieved in the terrestrial depositional record:

Palaeomagnetic stratigraphy

The palaeomagnetic record can be divided into a series of magnetozones (units of rock with a specific magnetic character), whose boundaries are clearly-defined due to the relatively abrupt changes in the earth's magnetic field (Ch. 5). Because these changes are experienced globally, palaeomagnetic stratigraphy offers one of the most promising methods for establishing world-wide correlation of Quaternary events. It is, moreover, the most widely used means of correlating the marine and terrestrial records (see final section below). On the other hand, the method is restricted to certain rock types and specific depositional environments, and the precise nature and timing of some palaeomagnetic events remain to be established. Further, it is a relatively 'coarse' record at present, although a much finer resolution (e.g. the determination of global or regional palaeomagnetic changes within the Holocene) may ultimately be achieved.

Tephrochronology

In so far as distinctive tephra layers constitute marker horizons in the stratigraphic record that are essentially isochronous, they can serve as the basis for time-stratigraphic correlation between often diverse depositional sequences (Wilcox 1965). Correlation on the basis of tephra deposits is not as universally applicable as palaeomagnetic correlation, however, for individual ash beds have relatively limited geographical ranges. Certainly, no ash bed is sufficiently extensive to be useful for inter-continental correlation (Morrison 1968b). Moreover, although specific ash layers can usually be identified, subsequent eruptions from the same volcanic centre (and sometimes even from different centres) can produce a series of ash layers that have very similar compositions. Very careful identification by an experienced petrologist is therefore necessary if homotaxial error is to be avoided. Finally, even in areas where volcanic ash layers are relatively widespread, preservation is often variable, and tephra horizons may only be found in isolated sections. Overall, therefore, although tephra deposits are potentially excellent time-datum surfaces, in practice tephrochronology can only form a basis for local or regional correlation.

Palaeosols

Well-developed palaeosols that evolved during a specific soil-forming interval and which possess sufficiently distinctive pedologic characteristics to enable them to be traced over a wide area can be considered as **soil stratigraphic units**. A soil stratigraphic unit is defined by the American Commission on Stratigraphic Nomenclature (1961) as 'a soil with physical features and stratigraphic relationships that permit its consistent recognition and mapping'. This concept has been developed by Morrison (1967) who proposed the term **geosol** to replace the term soil in stratigraphic usage. Morrison (1978) has also argued that as soil stratigraphic units or geosols are more nearly time-parallel than almost any other common sedimentary and biostratigraphic units in the Quaternary terrestrial record, they can be used for the time-stratigraphic subdivision of sequences as well as for time correlation, providing that they can be identified reliably. It is assumed that episodes of rapid soil profile development during interglacials and certain interstadials alternated with periods of negligible pedogenesis in the intervening cold phases. Well developed buried soils are therefore regarded as interglacial in age and these marker horizons form the basis for time-stratigraphic correlation.

Palaeosols have been most widely used in time-stratigraphic correlation in North America. For example, the well-developed Sangamon Soil which is believed to be of last interglacial age (Follmer 1978 and refs therein) has formed the basis for a scheme of correlation encompassing broad areas of the western and central United States (e.g. Morrison and Frye 1965; Richmond 1970; Birkeland *et al.* 1971). However, the use of palaeosols in Quaternary correlation is not always straightforward. The assumption that degree of soil development is a direct function of time can only really be made where other soil-forming factors (parent material, climate, slope and biological factors) can be shown to have been constant within a given area (Valentine and Dalrymple 1976). Moreover, it has already been pointed out (Ch. 3) that most soils are polygenetic and therefore any buried profile is likely to be the product of more than one phase of pedogenesis. Finally, in any one region where a Quaternary succession is patchy and includes several interglacial soil horizons, homotaxial errors may once more be difficult to avoid (Vita Finzi 1973).

Shorelines

If world sea-levels are stable for long enough to allow the development of pronounced shoreline features, then essentially isochronous reference surfaces will form which, in theory, should provide a basis for inter-regional schemes of correlation. At the local scale isochronous shorelines can often be traced across an area, even in regions affected by glacial isostasy. If the shoreline cuts certain sedimentary units, but is overlain by others, then the shoreline can be used as a time-stratigraphic reference plane for separating units of different ages. At the broader scale, evidence of former sea-levels in coastal areas, such as raised beaches and corals, submerged corals, submerged clifflines, etc, offer a means of correlation between terrestrial and deep-sea records. Problems can arise in correlation, however, from complexities in the coastal records induced by variations in amounts and rates of local tectonic activity, and from the more general phenomenon of geoidal eustasy (Mörner 1976 – see Ch. 2).

Radiometric dating

These methods, which have been fully discussed in the previous chapter, are an extremely important independent means of long distance correlation, for they form time planes across the stratigraphic record against which the time-transgressive litho-, bio- and morphostratigraphic boundaries can be measured. Whether radiometric dates should form the basis for the definition of Quaternary stage boundaries, however, has been strongly debated. Morrison (1968b), for example, has argued that the subdivision of the Quaternary should rest on the stratigraphic record, and radiometric dates are merely a means whereby that record can be underpinned. Certainly, any radiometric age determination relates only to the locality and to the horizon from which it was taken and it can only be related to other successions on the basis of the observed stratigraphic sequence at the different localities. As has already been shown, no radiometric date is free from analytical errors and in some cases the error ($\pm$) associated with the date may be so great that the date cannot be used in time-stratigraphic correlation. Equally, all dated samples are prone to errors arising from, for example, isotopic exchange and contamination. Every date should not only be carefully checked against other age determinations, but must be thoroughly evaluated in the light of its stratigraphic context before it is used as an aid in correlation. Overall, radiometric dating is perhaps best regarded as a means of corroborating and validating other stratigraphic and correlative procedures, rather than as the primary basis for time-stratigraphic correlation.

The 'complete' Quaternary succession: ocean sediment records

It has already been shown that, in the deep oceans of the world, long sequences of relatively undisturbed sediments are preserved that frequently extend back beyond the beginning of the Quaternary. Within these sediments the microfauna and flora contain a record of changing oxygen isotope ratios that not only provide evidence for former glacial and interglacial oscillations (Ch. 3), but also form the basis for stratigraphic subdivision and long distance correlation.

For stratigraphic purposes, Emiliani (1955) divided the isotopic curves from Atlantic and Pacific cores into sixteen stages (Fig. 6.6). The parts of the curve interpreted as representing warm stages (lower $\delta^{18}O$ PDB values- Fig. 6.8) were given odd numbers, and the cold stages (higher $\delta^{18}O$ PDB, values) were assigned even numbers. Stage 1 designated the present Holocene or Flandrian period, and higher numbers indicated successively

Fig. 6.6: 'Palaeotemperat ure' curves based on the pioneering work of Emiliani (1955), showing his subdivision into Oxygen Isotope Stages 1–13. Curve (a) is based on *Globigerinoides rubra*; curve (b) on *Globigerinoides sacculifera*; and curve (c) on *Globorotalia menardii*.

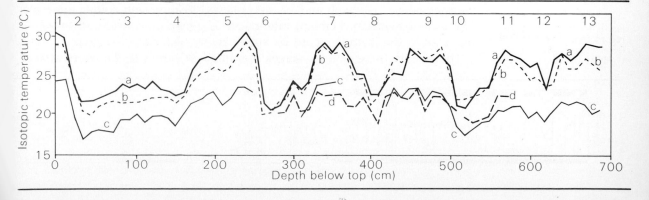

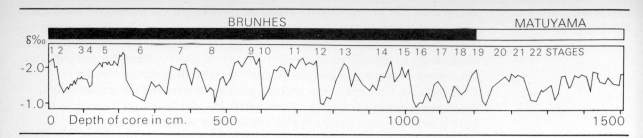

Fig. 6.7: Oxygen isotopic composition of *Globigerinoides sacculifera* in core V28–238 (after Shackleton and Opdyke 1973).

older cold and warm stages. The method of designating stage numbers is thus essentially a 'count from the top method', where reliance is placed on subdivision of a sinusoidal curve, and the underlying assumptions are that successive fluctuations are directly correlated with temperature changes or ice sheet growth and decay, and that no major interruption in sedimentation has occurred.

The stratigraphic subdivision of the oxygen isotope record was subsequently extended by Shackleton and Opdyke (1973) following analysis of sediments in core V28–238[2] obtained from a depth of 3120 m at 01°01′N, 160°29′E on the Solomon Plateau in the western Pacific (Fig. 6.7). Within this core, twenty-three isotopic stages younger than the Jaramillo Geomagnetic Event were recognised, and the record was interpreted as reflecting more or less continuous sedimentation since *ca.* 870 000 BP. A core from the Equatorial Pacific was chosen because it was considered that the complicating factor of temperature changes between glacials and interglacials would be less marked in ocean waters in those latitudes, and that a clearer oxygen isotope stratigraphy dominated by the 'glacial effect' would therefore be obtained.

A characteristic observed in this, and in many other isotopic records in deep-ocean cores, is that the transition from glacial to interglacial extremes occurs abruptly. These seemingly rapid changes from inferred glacial to interglacial conditions have been referred to as **terminations** (Broecker and van Donk 1970). Termination 1, for example, refers to the sudden relative decrease in ¹⁸O content of the oceans at the end of the last glacial stage (Stage 2). A complicating factor in the use of this terminology, however, is that Isotopic Stage 3 is not interpreted as a full interglacial, and thus Termination 2 marks the transition between Oxygen Isotope Stages 6 and 5 and not between 2 and 3 (Fig. 6.7). It should also be noted that Isotope Stage 5 has been subdivided, with substages 5a, 5c and 5e interpreted as warmer phases while 5b and 5d are believed to represent colder intervals. The last (Ipswichian, Eemian) interglacial is generally considered to be reflected in the isotope record by substage 5e (Shackleton 1969b).

The twenty-three isotopic stages identified in core V28–238 were also

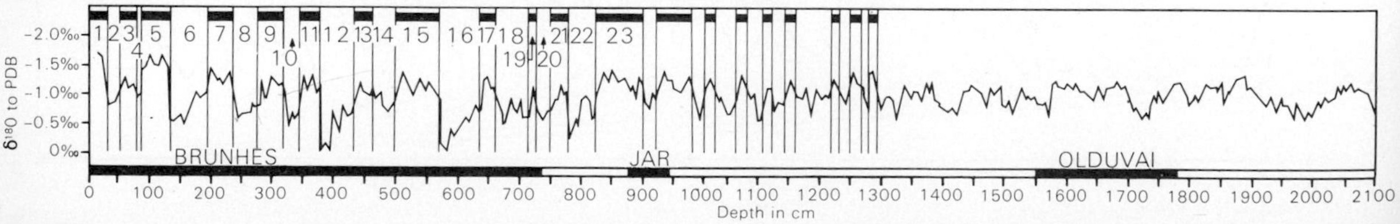

Fig. 6.8: Oxygen isotope record from core V28–239 (after Shackleton and Opdyke 1976).

recognised in the upper 9 m of a nearby longer core, V28–239 (Fig. 6.8) which contained over 20 m of sediment (Shackleton and Opdyke 1976). Table 6.2 indicates the depths of isotopic boundaries and terminations identified in V28–239 and also their inferred ages. These are based on interpolations from the known age of the Brunhes/Matuyama geomagnetic boundary (*ca*. 700 000 BP) assuming a constant rate of sedimentation. The older sediments in V28–239 contained a record of isotopic changes believed to extend back over the last 2.1 million years. However, the observed isotopic fluctuations older than Stage 23 are considered to be of too high a frequency and too low an amplitude to be of use at present in stratigraphic correlation.

In the Equatorial Atlantic, van Donk (1976) has obtained an isotopic record from over 12 m of sediment extending back over *ca*. 2.3 million years. The core, V16–205, was raised from a depth of 4045 m at 15°24′N, 43°24′W, and oxygen isotope analyses were carried out on a single foraminiferal species, *Globigerinoides sacculifera*. A good geomagnetic control is available on the core and van Donk was able to identify 42 isotopic stages interpreted as 21 isotopically determined 'interglacials' and an equal number of 'glacial' or 'near glacial' stages. Although this work demonstrates the very considerable potential of the oxygen isotope method for the investigation of climatic stages through the whole of Quaternary time, Berggren *et al*. (1980) have advocated caution before the number of globally-applicable isotopic stages are extended further. 'Comparison between the low-accumulation rate Atlantic core V16–205 . . . and the rather higher accumulation rate core V28–239 . . . supports the notion that the use of the oxygen isotope record as a high resolution stratigraphic tool should be restricted at present to the 23 stages extending to the Jaramillo magnetic event or to about 0.9 million years. Future work on cores with high rates of sedimentation may enable the system to be extended to the base of the Pleistocene.' (p. 281).

Table 6.2: Estimated ages of stage boundaries and terminations in core V28–239. Ages interpolated assuming constant sedimentation rates and based on data for core V28–238. Terminations according to criteria of Broecker and van Donk (1970) (after Shackleton and Opdyke 1976).

Boundary	Age (years BP)	Termination
1–2	13 000	I
2–3	32 000	
3–4	64 000	
4–5	75 000	
5–6	128 000	II
6–7	195 000	
7–8	251 000	III
8–9	297 000	
9–10	347 000	IV
10–11	367 000	
11–12	440 000	V
12–13	472 000	
13–14	502 000	
14–15	542 000	
15–16	592 000	VI
16–17	627 000	
17–18	647 000	
18–19	688 000	
19–20		
20–21		
21–22		
22–23		

Deep-ocean sediments, therefore, hold a number of advantages over terrestrial sequences from the point of view of stratigraphic subdivision and correlation. First, the records are more commonly continuous and appear relatively undisturbed. Secondly, a common technique (oxygen isotope analysis) can be used to compare profiles from widely scattered localities on the deep-ocean floors. Thirdly, the pronounced terminations can be used as reference points in inter-core correlation. Fourthly, although the isotopic changes are a consequence of climatic changes, and are therefore time-transgressive, this to a very large extent is masked by the slow rate of sediment accumulation. As a consequence, isotopic stage boundaries and terminations can be interpreted as essentially time-parallel horizons. Fifthly, the sedimentary records can be dated and correlated by the independent method of palaeomagnetic stratigraphy.

The oxygen isotope evidence provides an impressive record of Quaternary climatic changes, both in terms of the level of detail and its remarkable replication in sediments on widely-distributed parts of the ocean floors. Interpretation of the isotopic evidence, however, is not always straightforward. Continuity of sedimentation can never be proved, and it is questionable whether gaps in the sedimentary record could be reliably detected. Moreover, although the profiles can be dated and correlated by reference to the palaeomagnetic time-scale, there are considerable time-spans during which no detectable magnetic fluctuations occur. In view of the fact that correlation between individual isotopic profiles is based largely on a 'count from the top' principle, the possibilities of homotaxial error are always present. A further possible source of error arises from the differentiation between interglacials and interstadials. It has already been shown that Isotope Stage 3 is regarded as being of interstadial rather than interglacial status, hence the possibility cannot be excluded that some previous interglacials and interstadials have been confused, particularly in the earlier part of the Quaternary record. If this has occurred, then it clearly has fundamental implications for the number of glacial/interglacial cycles so far inferred from the isotopic record. By the same token, the complexities of Isotope Stage 5 (and possibly also of Isotope Stage 7 – Bowen 1978) are surely not unique to the upper parts of the isotopic sequence. Consistency of interpretation between different isotopic profiles may not, therefore, be easy to achieve, particularly in those cores where stratigraphic resolution is poor.

Although the above difficulties have yet to be satisfactorily resolved, there is no doubt that the isotopic trace in the ocean sediments provides a unique record of Quaternary climatic change, and it is now widely accepted that this, and not the terrestrial sequence, provides the basic framework for a global scheme of Quaternary correlation. Because the isotopic stages are essentially a reflection of climatic change, they are, in effect, geologic-climatic units and, as such, should have correlatives in the terrestrial record. Clearly, if meaningful global correlations are to be effected, a basis for correlation between the deep-ocean and terrestrial successions is required, and it is to the ways whereby this may be achieved that attention is finally directed.

Correlation between the marine and terrestrial records

If correlations are to be established with the oceanic successions, terrestrial sequences must possess certain characteristic features. First, a lengthy stratigraphic record must be available. Secondly, within that sequence evidence of climatic change must be clear and unequivocal so that geologic-climatic units can be inferred for comparison with those based on oxygen isotope profiles. Thirdly, depositional sites should be sought where the record of sedimentation is believed to be continuous. Every hiatus or interval of non-deposition poses a potential problem, particularly where no radiometric dates are available to provide an estimate of the duration of the break in the stratigraphic sequence. Fourthly, it is preferable, although not essential, if the succession can be dated by palaeomagnetic methods, for this offers a direct means of comparison between the terrestrial and oceanic records.

Depositional sequences that possess these particular characteristics are least likely to be found in areas formerly occupied by glacier ice, yet ironically, most of the schemes of stratigraphic subdivision and global correlation have been based upon the evidence from such regions. In recent years, however, the focus of attention has shifted to those parts of the world that lay beyond the margins of the Quaternary ice sheets, and it is the stratigraphic records in these areas that now appear to offer the greatest potential for establishing the links between the marine and terrestrial successions. Of particular importance in this respect are the deposits in major tectonic continental basins, lake sediments, aeolian sediment accumulations, and shoreline sequences.

Major continental tectonic basins

In some of the great tectonic basins of the world, sediments have been accumulating for several millions of years. A striking example is the Carpathian Basin, for beneath the present Hungarian Plain lie several thousand metres of Quaternary, Pliocene and Miocene sediments (Cooke 1981). Two recent boreholes through the deposits have provided a high resolution palaeomagnetic record. Pollen, molluscan and ostracod analyses suggest that warmer conditions prevailed until approximately 1.5 million years ago, and the onset of consistently cooler conditions began approximately 900 000 years ago. This type of long sedimentary sequence constitutes a basis for correlation between terrestrial and marine successions.

Lake sediments

Perhaps the most impressive example of the potential of lake sediment sequences as a basis for world-wide correlation of terrestrial deposits is the record in Lake Biwa, the largest lake in Japan. There, over a thousand metres of sediment have been recovered from the lake floor, and these are believed to span some 1.5 million years (Fig. 6.9). The lake sediments contain a wide range of fossils and also a number of distinctive tephra layers. These can be dated by the fission track method, and further time-

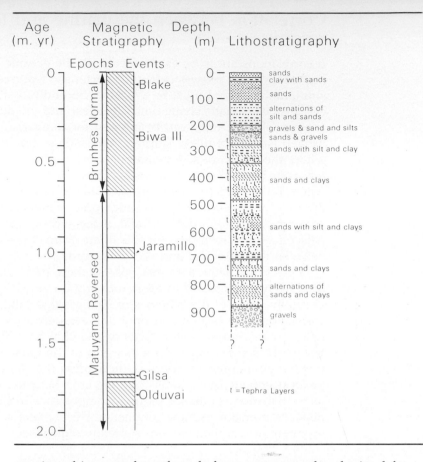

Fig. 6.9: Lithostratigraphy of a 1000 m core from Lake Biwa, Japan, in relation to the palaeomagnetic time-scale. The tephra layers (t) range in thickness from <1.0 to approximately 10.0 cm, and some have been dated by the fission track method (based on data in Horie 1976).

stratigraphic control on the whole sequence can be obtained from palaeomagnetism (Horie 1976). Other major lake basins appear to contain long Cenozoic histories of sediment accumulation, including Lake George (Australia), Lake Titicaca (Bolivia-Peru), and a number of lakes in New Zealand, Africa and the south-west United States (Horie 1979). Not only do these long lake sequences form a basis for correlation between the marine and terrestrial records, but as they often accumulated more rapidly than most ocean sediments, they provide a much greater resolution of the stratigraphic changes that occurred over successive glacial/interglacial cycles.

Aeolian deposits

In certain areas of the world, lengthy aeolian sedimentary sequences provide a basis for Quaternary correlation, perhaps the best published examples being the loess deposits around Prague, Brno and Nitra in Czechoslovakia, and near Krems in Austria. Within the loess sequences are numerous soil horizons (Fig. 6.10), and the complete succession appears to contain a record of glacial and interglacial conditions stretching back to the early part of the Quaternary (Kukla 1970, 1975; Fink and Kukla 1977). The loess units are interpreted as representing full glacial conditions, while the interbedded palaeosols (mainly braunerdes, parabraunerdes and chernozems) are considered to be indicative of interglacial or interstadial episodes. The close correspondence between

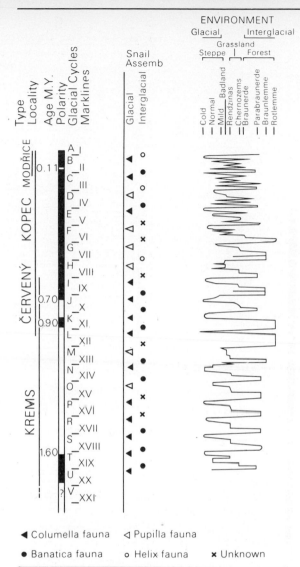

Fig. 6.10: Environmental changes around Brno and Krems, Czechoslovakia, reconstructed from the loess, palaeosol and gastropod faunal records (modified after Kukla 1975).

◀ Columella fauna   ◁ Pupilla fauna

● Banatica fauna   ○ Helix fauna   ✕ Unknown

sedimentary sequences of different ages enabled first order (glacial) and second order (stadial) sedimentary cycles to be established. Each cycle follows a similar pattern:

(a) deposit of hillwash loam;

(b) forest soil of braunerde type (early/mid interglacial);

(c) steppe soil (chernozem: late interglacial/early glacial);

(d) marker horizon – thin layer of calcareous silt separating humus-rich steppe soils from pellet sands (probably derived from dust storms);

(e) pellet sands – sand-size pellets of reworked silt and clay formed during episodes of heavy rainfall after a dry season;

(f) loess (glacial)

The distinct boundaries between the thick layers of 'cold' loess and the overlying braunerdes are known as **marklines** and these form the basis for major cyclic subdivisions. Submarklines are the boundaries between loess and any kind of braunerdes or chernozems and these delimit the stadial

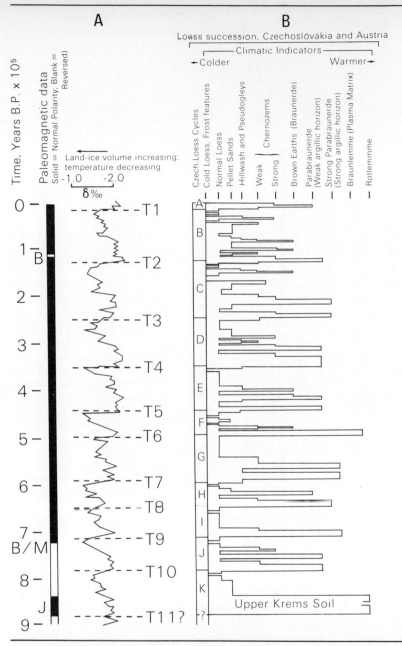

Fig. 6.11: Comparison of the deep-sea (A) and central European loess records (B) for the past 900 000 years. Isotope curve (which also shows Terminations – T) from Shackleton and Opdyke (1973); loess data from Kukla (1970, 1975). On the palaeomagnetic scale; B = Blake, B/M = Brunhes/Matuyama; J = Jaramillo (modified after Morrison 1978).

cycles (Kukla 1975). Additional palaeoenvironmental information has been derived from the rich gastropod faunas (Ch. 3) and the dating framework is provided by radiocarbon in the more recent sediments and magnetostratigraphy in the older deposits. The records obtained from a number of different sections are remarkably consistent, and indicate that within the last 1.6 million years, at least seventeen major glacial/interglacial climatic shifts have affected central Europe. Most remarkably, these data bear a very close resemblance to the results obtained from deep-sea records (Fig. 6.11), and would appear to confirm the palaeoclimatic conclusions based on the oxygen isotope variations. A further

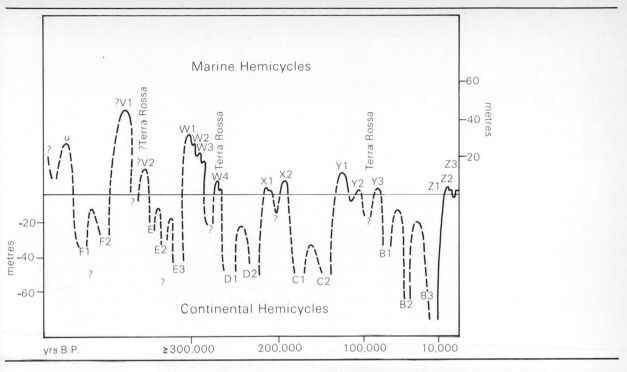

Fig. 6.12: Mediterranean sedimentary cycles and relative sea-level changes, as recorded on Mallorca. There is no time-scale prior to 300 000 BP and the earlier record is both incomplete and chronologically distorted (after Butzer 1975).

Marine cycle	Apparent sea-level (metres)	Faunal characteristics	Radiometric age
Z3	2	Banal	Post-Roman
Z2	2	Banal	
Z1	4	Banal	
Three aeolianite generations HEMICYCLE B			
Y3	0.5–3	Probably banal	80 000 ± 5000 BP
Y2	1.5–2	Partial *Strombus* fauna	110 000 ± 5000 BP
Y1	9–15	Partial *Strombus* fauna	125 000 ± 10 000 BP
Two aeolianite generations HEMICYCLE C			
X2	6.5–8.5	Impoverished Senegalese fauna	190 000 ± 10 000 BP
X1	2–4.5	Full *Strombus* fauna	210 000 ± 10 000 BP
Two aeolianite generations HEMICYCLE D			
W4	4–8	Banal	>250 000 BP
W3	15–18	*Patella ferruginea*	?
W2	22–24	*Patella ferruginea*	
W1	30–35	?	
Three aeolianite generations HEMICYCLE E			
V(?2)	*ca.* 15	Banal	
V(?1)	45–50	Banal	
Two aeolianite generations HEMICYCLE F			
U	30 (other levels?)	*Patella ferruginea*	
?	60–65	?	
?	75–80	(*?Purpura plessisi, Ostrea cucullata*)	
?	100–105		

Table 6.3: High Pleistocene sea-levels and marine cycles of the western mediterranean based on the Mallorcan evidence (after Butzer 1975).

Table 6.4: External
correlations of the
Mallorcan sequence. The
deep-sea core stages are
after Shackleton (1975)
and the European loess
cycles are after Kukla
(1975). The prefix 'L'
denotes the loess cycles;
the intervening
palaeosols are prefixed
'B' and are numbered
according to increasing
age (after Butzer 1975).

Radiometric dates	Mallorcan cycle	Deep-sea stage	Loess cycle
since 10 000 BP	Z	1	B-1
10 000–70 000	B	2–4	L-B
75 000–125 000	Y	5	B-3
	C	6	L-C
180 000–220 000	X	7	B-5
	D	8	L-D
>250 000 BP	W	9(?)	B-7
	E		L-E
	V	?	B-9
	F		L-F
	U		B-11

example of the importance of aeolian sediments, in a different
geomorphological context, is given below.

Shoreline sequences

Since eustatic sea-levels are partly related to climatic conditions, there
should be some broad correlation between dated sea-level variations in
regions not affected by glacio-isostasy and the main glacial/interglacial
cycles recognised in the oceanic record. Of particular importance in this
respect are the complex coral reef sequences recording sea-level
oscillations over the mid- and late Pleistocene (see Ch. 5), and curves of
sea-level variations based principally on ^{230}Th/^{234}U dating are being
constructed from a number of areas (e.g. Chappell 1974a; Bloom et al.
1974). There should be some correspondence between times of
significantly lowered sea-level and climatic cooling, although climatic
change and sea-level variations may be slightly out of phase since time is
required for the melting of the ice sheets following climatic improvement.
Nevertheless, detailed coastal sequences that contain long histories of sea-
level changes clearly offer a basis for correlation between oceanic and
terrestrial records.

An example of the types of comparisons that can be made using
shoreline evidence has emerged from work on the Pleistocene sedimentary
record of coastal Mallorca. In common with the shorelines of many
tropical and subtropical lands with arid or semi-arid climates, parts of the
Mallorcan littoral are characterised by well-developed calcareous dunes.
Successive generations of windblown sands are widespread, and in many
areas these have been cemented by carbonates to form a friable calcareous
rock known as aeolianite (Butzer 1962; Butzer and Cuerda 1962). These
aeolian sediments accumulated through deflation of freshly-exposed
marine deposits during glacio-marine regressions (Butzer 1963), and each
generation can therefore be correlated with a glacial period in higher
latitudes. Interglacial high sea-levels are recorded by marine abrasion
platforms and beach deposits. In a detailed synthesis of the Mallorcan
evidence, Butzer (1975) divided the shoreline stratigraphy into six
terrestrial and six marine 'hemicycles' (Fig. 6.12 and Table 3), and these
have been dated partly on the basis of marine biostratigraphy and partly
by ^{230}Th/^{234}U dating of marine shells (Ch. 5). The sequence can be
correlated with the loess sequence in central Europe and also with the
oxygen isotope stratigraphy of the deep-ocean floors (Table 6.4). Although
there may be problems with the interpretation of some of the dates used

in the Mallorcan reconstruction, it is this type of record combining evidence of both terrestrial and oceanic changes that offers one of the most attractive means of establishing correlations between the terrestrial and marine successions.

CONCLUSIONS

Although the Quaternary possesses rather special features that make it unique in the geological record, there is a growing consensus that, as far as possible, the subdivisions of the most recent part of the stratigraphic record should follow conventional geological procedures. In terrestrial successions, therefore, lithostratigraphy and biostratigraphy constitute the basic units of classification at the local scale, and these can be clearly defined on the basis of observable criteria. Stratotypes form reference standards, although these may have more limited temporal and spatial application than in the earlier geological column. Stratigraphic boundaries, in the main, are time-transgressive, and for the purposes of correlation should ideally be underpinned by radiometric dates. Ultimately, geologic-climatic units can be established and, despite the problems associated with this form of stratigraphic subdivision, these are likely to continue as the basis for correlation at the regional and continental scales. They are, moreover, an essential basis for correlation between the marine and terrestrial successions. The oxygen isotope stratigraphy of the deep-ocean sediments constitutes a reference standard for global correlation, and it is against this record that most future stratigraphic schemes will be measured.

NOTES

1. The term facies refers to the sum total of features such as rock type, mineral content, sedimentary structure, bedding characteristics, fossil content, etc that characterises a sediment as having been deposited in a given environment.
2. In deep-sea core investigations, the core numbers code the research vessel, journey number and core number raised on a particular voyage. For example, V28–238 indicates that the core was the 238th obtained during the 28th cruise of the research vessel *Vema*.

CHAPTER 7

Environmental changes in Britain during the last (Devensian) cold stage

INTRODUCTION

Thus far we have considered the various forms of evidence that can be used in the analysis of Quaternary environments, the means by which a time-scale for environmental change can be established, and the stratigraphic procedures that allow sequences to be built up and meaningful correlations to be effected between often widely-scattered sites. In this final chapter, the aim is to demonstrate how these different methods can be employed to produce a synthesis of environmental change over a particular segment of Quaternary time. For this purpose we have selected the last cold stage in the British Isles (the **Devensian**) and have done so for several reasons. First, a considerable body of data has now accumulated for environmental conditions in the British Isles during that period, and much of it has emerged within the past fifteen or twenty years (e.g. Mitchell and West 1977). The data are extremely diverse and yet often there is remarkable agreement between the different forms of evidence, thereby strengthening considerably subsequent palaeoenvironmental reconstruction. Secondly, although the Devensian was predominantly a period of arctic climatic conditions, short-lived warm episodes (interstadials) occurred during which summer temperatures rose to be as high, or perhaps even slightly higher, than those of the present day. Hence, this particular time period is ideal for exemplifying both the range of techniques that can be used in the analysis of Quaternary environments, and also the types of results that they can reasonably be expected to produce. Thirdly, a considerable part of the stage lies within the range of radiocarbon dating, arguably the most versatile and certainly the most widely used radiometric dating technique in late Quaternary studies. Hence it is often possible to discuss environmental change within the framework of age estimates (Ch. 5). Fourthly, although fragmentary, morphological and stratigraphical evidence is often better preserved by contrast with that for earlier Quaternary stages, and thus there is greater scope for detailed environmental reconstruction. Fifthly, the period coincides with the emergence of *Homo sapiens,* and the analysis of late Quaternary environments therefore becomes increasingly the study of the landscape as the home of man. As a consequence, environmental reconstructions during this period are of interest to a wide range of disciplines, and provide a framework for archaeological and anthropological investigations.

It must be appreciated at the outset that both the distribution and quality of the evidence are highly variable. For much of the period under consideration, particularly the earlier part of the record, little is known about events in highland Britain, as the passage of the last ice sheet across northern and western parts of the country has removed most of the evidence from the terrestrial record. Hence palaeoenvironmental reconstructions are biased towards the lowland regions of central and southern England. Even in these areas, however, the spatial distribution of sites containing meaningful information is far from perfect, and consequently regional variations in, for example, climatic régime and vegetation composition, may be difficult to detect. Indeed, in many lowland areas, there are long time periods for which we have virtually no evidence, while in others, the evidence that is available is partial and therefore only allows the most generalised of inferences to be made about former environmental conditions. Moreover, because of the highly fragmented nature of the stratigraphic record, correlations have to be made between widely-separated sites on grounds which are frequently less than secure, and homotaxial errors are a recurrent problem. In many places, the evidence is equivocal or conflicting, and therefore controversial. Any reconstruction, therefore, contains a strong element of subjectivity, and in this respect the following account is no exception. It nevertheless represents an attempt to make sense of the evidence that we currently have at our disposal, and provides a broad outline of the sequence of environmental changes that have occurred in Britain over the last 100 000 years or so. It illustrates the types of problems that are encountered in stratigraphic investigations within a relatively restricted time-scale, and it highlights the need for a multidisciplinary approach in the reconstruction of Quaternary environments. The palaeogeographic picture that emerges is hazy in places but it is offered as a basis for discussion and as a framework for future research.

The Devensian Stage in Britain

As yet no site has been discovered in Britain in which a continuous sequence of deposits is preserved that spans the whole of the Devensian cold stage. The generally-accepted stratigraphic framework is based on a number of key sites, each of which contains a palaeoenvironmental record for only a part of the Devensian. The most important elements in the stratigraphic subdivision of the Devensian stage are as follows:

Chelford Interstadial. Sections exposed in sand pits between Chelford and Congleton in Cheshire (Fig. 7.1) show a thick suite of alluvial sands within which is a stratum of organic muds (Fig. 7.2). The sands are overlain, and intermittently underlain, by till (Fig. 7.3) and appear to have

Fig. 7.1: Location of Devensian sites referred to in the text.

accumulated under periglacial conditions. The organic muds contain macrofossil remains of trees and reflect a thermal improvement known as the Chelford Interstadial which has been dated to 65 000 to 60 000 BP (Worsley 1977). The Chelford evidence therefore suggests that a periglacial environment existed in lowland Britain prior to *ca.* 65 000 BP, a significant climatic amelioration then occurred, and there subsequently followed a return to cold conditions. The tills indicate that the periglacial phases were preceded and succeeded by full glacial conditions. The upper till, which resembles a flow till, is considered to have been deposited by Late Devensian ice (see below), but the age of the basal till is unknown.

Fig. 7.2: Section near Chelford, Cheshire, showing dark peat layer of the Chelford Interstadial exposed within alluvial sands.

Upton Warren Interstadial. At several sites in the English Midlands and south-east England, there is evidence for a warm interval which occurred approximately 42 000 years ago. This milder phase is known as the Upton Warren Interstadial after the site in Worcestershire at which evidence for thermal improvement was first discovered (Fig. 7.1). However, the stratigraphic context of this episode is best displayed at Four Ashes in Staffordshire (A. Morgan 1973). At that site, several organic lenses are interstratified with gravels showing signs of periglacial disturbance (Fig. 7.4). The lowermost organic horizon (1 in Fig. 7.4) contains

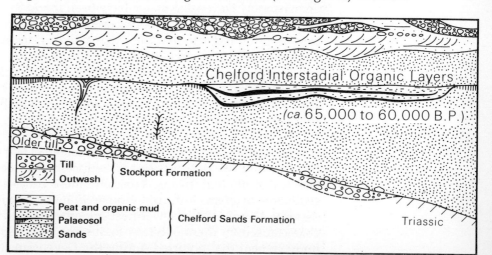

Fig. 7.3: Schematic cross-section through the succession exposed in the Oakwood Quarry, Chelford during the period 1974–76 (after Worsley 1977).

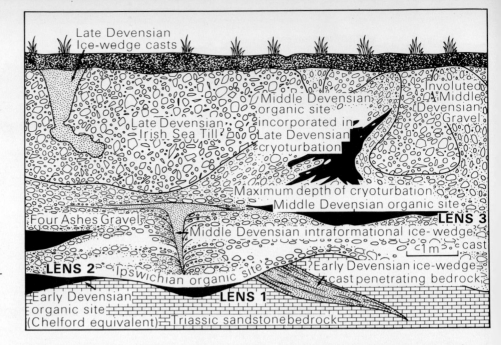

Late Devensian
Ice-wedge casts

Middle Devensian
organic site
incorporated in
Late Devensian
cryoturbation

Late Devensian
Irish Sea Till

Involuted
Middle
Devensian
Gravel

Maximum depth of cryoturbation
Middle Devensian organic site

LENS 3

Four Ashes Gravel

Middle Devensian intraformational ice-wedge cast
<1 m

LENS 2   Ipswichian organic site

?Early Devensian ice-wedge
cast penetrating bedrock

LENS 1

Early Devensian
organic site
(Chelford equivalent)   Triassic sandstone bedrock

Fig. 7.4: Schematic cross-section of the deposits exposed in the Four Ashes pit during the period 1967–70 (after A. V. Morgan 1973).

palaeobotanical evidence which suggests a last (Ipswichian) interglacial age, while a second organic lens (2) which overlies it has been correlated on the basis of biostratigraphy with the Chelford Interstadial. Some 2 m further up the sequence a third organic lens (3) rich in coleopteran remains indicative of relatively warm conditions represents the Upton Warren Interstadial. The Four Ashes site contains the most detailed record of environmental changes in Britain during the Devensian, and is now accepted as the type locality for the Devensian Stage (Mitchell *et al.* 1973).

Late Devensian Glaciation.   Although there are now indications that glacier ice began to accumulate in northern Britain early in the Devensian, the conventional view is that widespread glaciation during the Devensian occurred only in the later part of the period. Certainly, no tills that are demonstrably Devensian in age have been found that pre-date the Upton Warren Interstadial, and radiocarbon dates on organic materials beneath Devensian till indicate that the expansion of glaciers into lowland Britain occurred some time after 30 000 BP. At Four Ashes, for example, a till unit which is part of a widespread glacigenic formation found throughout the English Midlands and adjacent parts of north-east Wales (the 'Irish Sea Till') overlies gravels containing organic materials that have been dated to 30 000 BP or older (A. Morgan 1973).

Late Devensian Lateglacial.   By approximately 13 000 BP glaciers had receded from most of the British Isles and sediments had begun to accumulate in lake and kettle hole basins. Typically, these consist of basal minerogenic sediments overlain, in turn, by organic, minerogenic and further organic deposits (Fig. 7.5). The lowermost minerogenic and organic sequences are characterised by thermophilous fossils and record the most recent climatic improvement that occurred during the Devensian, the Lateglacial

Fig. 7.5: Core of Lateglacial lake sediments showing two minerogenic layers (lighter bands) separated by organic lake muds of Lateglacial Interstadial age (see also Fig. 3.28).

Interstadial, which spans the time period from *ca*. 13 500 to 11 000 BP (Fig. 3.28). The overlying minerogenic deposits contain evidence of a short-lived climatic deterioration (the Loch Lomond Stadial) during which periglacial conditions became established once more and glaciers reformed in many highland areas. Together, the Lateglacial Interstadial and Loch Lomond Stadial constitute the Late Devensian Lateglacial. The uppermost organic sediments began to accumulate during the warm interval of the Flandrian Interglacial, the lower boundary of which has been dated to around 10 000 BP.

It has been suggested that the sequence in Lake Windermere in the English Lake District (Fig. 7.1) be used as the reference section for the Lateglacial Interstadial (Coope and Pennington 1977), and the site at Drymen near Glasgow has been suggested as a possible stratotype for the Loch Lomond Stadial (Jardine 1981). However, in view of the known spatial variations in environmental developments in the British Isles during the Lateglacial (see 'The Late Devensian Lateglacial' below) and the transitional nature of climate during that short time period, it is questionable whether any single site can serve as a stratotype for the whole or a part of the Lateglacial which has application at the national scale. A number of regional stratotypes are really required, but these have yet to be designated. Moreover, the choice of the Windermere sequence as a stratotype has been criticised on the grounds of ambiguous climatostratigraphic interpretation and an uncertain dating framework (Lowe and Gray 1980). Although the term 'Windermere Interstadial' has been introduced into the literature therefore (Coope and Pennington 1977), the more widely known 'Lateglacial Interstadial' is retained here.[1]

The sequence outlined above is summarised in Figure 7.6. The Devensian is conventionally divided into three principal substages, following Mitchell *et al*. (1973), and the stratigraphic framework shown in Fig. 7.6 is generally used as a basis for the correlation of Devensian

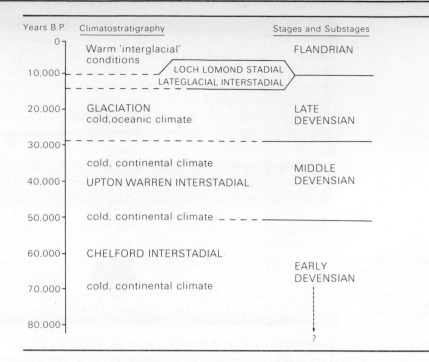

Fig. 7.6: Climatostratigra phic subdivision of the last 80 000 years or so in Britain.

deposits throughout the British Isles. It should be noted however, that the dates of the substage boundaries are somewhat arbitrary and will probably be modified in the light of future research. It should also be noted that no boundary stratotypes currently exist for the Early and Middle Devensian, and that correlations throughout the Devensian are made very largely without reference to type sections. In view of the fragmentary nature of much of the stratigraphic record therefore, the possibility cannot be excluded that the presently-accepted Devensian sequence is incomplete, and the evidence for hitherto unrecognised climatic oscillations may eventually be discovered.

The Early Devensian

Onset of the Devensian

The base of the Devensian in Britain is not easy to define, for very few sites have so far been discovered where deposits of the last cold stage can be shown to overlie conformably sediments of the last (Ipswichian) interglacial. At Wretton in Norfolk (Fig. 7.1), deposits attributed to the last interglacial are overlain by Early Devensian sediments, but there is a crucial gap in the stratigraphic record, and the later part of the interglacial pollen sequence is missing (West *et al*. 1974). At Four Ashes the organic lens (1, Fig. 7.4) of inferred last interglacial age may be truncated by the overlying gravels, and although the deposit contains coleopteran and plant remains, only limited palaeoclimatic data have been obtained (A. Morgan 1973), and the status of the contact between the organic silts and the overlying gravels is unclear.

At Histon Road, Cambridge (Sparks and West 1959) and at Wing in south Leicestershire (Hall 1980), deposits of Ipswichian age appear to be overlain conformably by Early Devensian sediments. Of the two sites, the profile at Wing is the more complete and has been investigated in greater detail. A deep depression cut into the Jurassic bedrock is floored by till, and contains about 5.5 m of silts and clays underlain by over 2 m of peats. Detailed pollen and plant macrofossil analysis suggest that the organic sediments are of interglacial age, while the upper minerogenic deposits accumulated during the initial part of the succeeding cold stage. The sequence spans the interval from an early part of the Ipswichian to the Early Devensian, and it is anticipated that the site at Wing may serve as a suitable stratotype for the Ipswichian/Devensian boundary in lowland England. Dating remains a problem, however, for the boundary lies beyond the range of conventional radiocarbon methods, and there are, as yet, no other dating techniques which are sufficiently reliable as a means of establishing the precise age of the beginning of the Devensian cold phase (Shotton 1977b).

In the absence of a closely-dated boundary stratotype, the timing of the onset of the last cold phase in Britain must be inferred from other evidence. Palaeoclimatic data from a range of sources spanning the transition from the last interglacial to the middle of the last cold phase are shown in Fig. 7.7. These reveal a broad measure of agreement which can be summarised as follows:

1. Interglacial conditions (Eemian/Ipswichian ?) at around 125 000–120 000 BP. Oxygen Isotope Stage 5e.
2. Cooling at around 110 000 BP (Glacial ?). Oxygen Isotope Stage 5d.
3. Warmer conditions at around 105 000–100 000 BP (Interglacial/Interstadial ?). Oxygen Isotope Stage 5c.
4. Cooling at around 90 000 BP (Glacial ?). Oxygen Isotope Stage 5b.

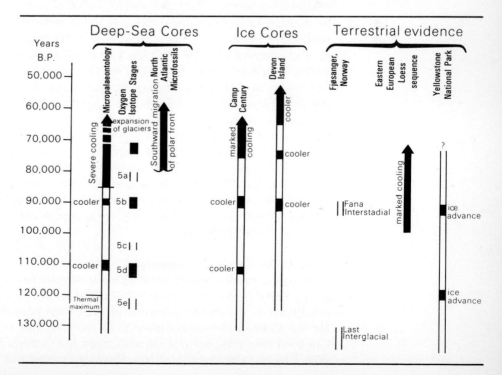

Fig. 7.7: Principal lines of evidence for marked climatic cooling during the period 130 000 to 50 000 BP. For further explanation see text.

5. Warmer conditions around 85 000–80 000 BP. (Interglacial/Interstadial ?). Oxygen Isotope Stage 5a.
6. Pronounced cooling between 80 000 and 65 000 BP. Oxygen Isotope Stage 4.

Before these data can be used to provide an indication of the age of the onset of the Devensian cold stage, however, a major stratigraphic problem must be resolved concerning the status of the Ipswichian Interglacial in Britain. At the time of writing (1982), no unitary pollen diagram for the Ipswichian Interglacial has been published. The most detailed and extensive pollen record is that from Wing, but even at that site the transition from the previous cold stage and the initial part of the interglacial sequence are missing. In the absence of a reference diagram for the whole of the Ipswichian, the palaeoenvironmental record for the last interglacial has been pieced together from fragmentary evidence at a number of sites scattered across southern and eastern England. None of the terrestrial 'Ipswichian' sequences from which pollen and plant macrofossil evidence have been obtained have been radiometrically-dated, and correlation between sites has been entirely on the basis of biostratigraphy. In view of the known similarities in vegetational developments between successive warm episodes, and in the absence of a time-stratigraphic framework for correlation, there is a possibility that homotaxial errors have occurred and that, as has been suggested by Sutcliffe (1975, 1976), sequences from different interglacials (or even interstadials, as these may be indistinguishable from interglacials in the proxy record) have been erroneously correlated.

The question therefore arises as to which of the warm intervals in the scheme outlined above is represented by the 'Ipswichian' in the British stratigraphic record. Two alternatives seem possible. One is that the Ipswichian Interglacial *sensu stricto* is the correlative of the Eemian of north-west Europe (e.g. Mangerud *et al* 1979) and Oxygen Isotope Stage 5e, the warmest episode in Stage 5 of the oceanic record (Shackleton 1969b), dated to 130 000 to 120 000 BP. This would imply that the Devensian as a climatostratigraphic unit began at *ca.* 115 000 BP and that the warm intervals 3 and 5 (above) would be designated as interstadials within the last cold stage. The alternative is that the warm intervals between 115 000 and 70 000 BP can be considered as interglacials (e.g. Richmond 1977; Woillard 1978). According to this interpretation, the Ipswichian could equate with a later episode and the Devensian may be inferred to have begun approximately 70 000 years ago. Although a majority of workers perhaps incline to the former interpretation (see e.g. Grüger 1979[2]), there is, at present, no unequivocal evidence in support of either of these alternatives.

Early Devensian environments

Although a predominantly cold climatic régime prevailed throughout most of the Early Devensian, the most detailed record of environmental conditions in the British Isles during that time period has been obtained for the short-lived warmer interval of the Chelford Interstadial. At the Chelford site, arboreal macrofossils from the interstadial organic layer (Fig. 7.8) include *Pinus sylvestris*, *Betula pubescens* and *Picea abies*[3] (Simpson

Fig. 7.8: Fossil tree stump exposed in Chelford Interstadial organic sediments at Chelford, Cheshire.

and West, 1958), the last mentioned being most useful as a stratigraphic marker since the last natural occurrence of *Picea* in Britain appears to have been during the Chelford Interstadial. A number of radiocarbon dates have been obtained from peat and wood samples from the organic layer and these have yielded a range of finite and infinite ages (Worsley 1980). A date of 60 800 ± 1500 BP is the most widely quoted however, and although this is the oldest finite radiocarbon age determination from a site in the British Isles, it may yet prove to be a minimum age for the deposit (Grootes 1978). Sediments at three other sites, Four Ashes, Beetley and Wretton (Fig. 7.1), have been correlated with the Chelford organic deposits on the basis of their stratigraphic context and similarities in fossil assemblages. There are, however, no dates from any of these sites to aid in correlation. At Four Ashes, an organic layer (2, Fig. 7.4) has yielded pollen and coleopteran assemblages that suggest affinities with those from the temperate horizons at Chelford (Andrew and West 1977). At the Beetley site, pollen and macrofossil evidence indicate regional coniferous woodland dominated by *Pinus*, but with a strong representation of *Betula* and *Picea*, and these sediments are also inferred to be of Chelford Interstadial age (Phillips 1976).

The sequence at Wretton is somewhat more complex than those described above, and its interpretation illustrates the types of problem that can arise when correlations between sites are based on biostratigraphy alone. In the Wretton area of North Norfolk, organic sediments rich in plant remains, mollusc shells, and bones are found in fluviatile terrace sands and gravels, and also in small depressions that are considered to be features formed by the melting of ground ice. These deposits are believed to be of Early Devensian age. Pollen evidence from the site was initially interpreted as indicating two phases of woodland development, and three periods of open habitat conditions (Table 7.1). The upper woodland zone, dominated by *Betula*, *Pinus* and *Picea*, was correlated with the Chelford Interstadial and the lower woodland zone therefore appeared to represent an earlier, hitherto unrecognised warm phase which was termed the 'Wretton Interstadial' (West *et al.* 1974). Coleopteran analyses however suggest a different interpretation of the upper 'woodland biozone' (Table 7.1), for the insect faunas indicate a harsh climate of arctic severity and barren tundra conditions (Coope 1974b). Vertebrate remains (see below) support the coleopteran evidence, for they are suggestive of an open herbaceous vegetation, and include no woodland animals (Stuart 1977). No arboreal plant macrofossils have been recovered from the site, and it is therefore possible that the arboreal pollen in the upper 'woodland biozone' reflects transport from long distance. Correlation between the upper pollen-defined 'interstadial' at Wretton with the Chelford Interstadial may therefore be incorrect (Coope 1975a). Hence, the status of the 'Wretton Interstadial' remains questionable.

In a number of caves, deposits have been found which are believed to be of Early Devensian age. Cave earths in Long Hole, Glamorgan (Fig. 7.1) contain pollen assemblages which indicate that a period of arctic tundra conditions was interrupted by the development of boreal coniferous forest with *Pinus*, *Betula* and, significantly, *Picea*. A Chelford Interstadial age has been suggested for this forest episode (Campbell 1977). Two cave sites in Somerset also contain evidence suggestive of woodland development during the Early Devensian. In Picken's Hole and Tornewton Cave, cave earths sandwiched between cold climate thermoclastic screes contain a number of large vertebrate remains including red fox (*Vulpes vulpes*), brown bear (*Ursus arctos*), red deer (*Cervus elaphus*), spotted hyaena (*Crocuta crocuta*), reindeer (*Rangifer tarandus*), woolly rhinoceros (*Coelodonta*

Pollen assemblage biozone (principal components)	Environmental reconstruction	
	(a) pollen evidence	(b) beetle evidence
H. Grameae-Cyperaceae-Compositae	open, herbaceous vegetation: cold climate	damp grassland: cool, moderately continental climate
G. *Pinus-Betula, Picea-Calluna-Sphagnum* F. *Pinus-Betula-Picea*	woodland: thermal improvement of (?) 'Chelford Interstadial'	barren tundra: more extreme continental climate
E. Gramineae-Compositae	open, herbaceous vegetation: cold climate	no evidence
D. *Betula-Pinus-Alnus-Gramineae-Calluna* C. *Betula-Pinus*	woodland: thermal improvement of (?) 'Wretton Interstadial'	no evidence
B. Gramineae-Cyperaceae-Compositae A. Gramineae-Cyperaceae-*Artemisia-Selaginella*	open, herbaceous vegetation: cold climate	damp grassland: cool, moderately continental climate

Table 7.1: Pollen assemblage zones and environmental reconstructions at Wretton, Norfolk, based on data in West *et al.* (1974) and Coope (1974b).

antiquitatis), horse (*Equus caballus*), and elk or moose (*Alces alces*). Of these, red fox, elk and perhaps also red deer are forest creatures. The cave earths may, therefore, be of Chelford Interstadial age (Stuart 1974). In caves in Yorkshire and in the Mendip Hills, a phase of speleothem deposition which reflects milder and wetter conditions has been dated by uranium series methods to *ca.* 60 000 BP, and may also be the correlative of the Chelford Interstadial (Atkinson *et al.* 1978).

If these correlations are correct, a range of evidence from a number of sites suggests the following generalised environmental reconstruction for the Chelford Interstadial. Around or perhaps even before 60 000 BP, a birch, pine and spruce woodland developed over large areas of central and southern Britain. This type of forest cover appears to have been very similar to the modern boreal forest of southern Finland (Simpson and West 1958), where mean July temperatures are in the range 12° to 16 °C, January temperatures are between −10° and −15 °C, and annual precipitation is between 400–700 mm. This climatic reconstruction has been fully supported by evidence from coleopteran assemblages at Chelford which indicate mean July temperature of around 15 °C (1° to 2 °C lower than at the present day) and a more continental climatic régime than currently prevails on the Cheshire Plain (Coope 1959). Moreover, coleopteran evidence from the organic lens at Four Ashes which is correlated with the Chelford Interstadial is in good agreement with this general climatic assessment (A. Morgan 1973).

Prior to and following the Chelford Interstadial, a more severe climatic régime prevailed. The presence of ground ice mounds and involutions at some of the sites suggests winter temperatures of well below 0 °C, and the occurrence of ice wedges indicates mean annual air temperatures at times of below −8 °C. Continuous permafrost may have been widely developed and it is possible that discontinuous permafrost was present throughout the birch-pine-spruce phase of the Chelford Interstadial (Watson 1977). Pollen evidence from Swanton Morley (Phillips 1976), Beetley, Wretton and Wing[4] indicates an almost complete absence of trees from the landscape of south-eastern England, and also very few shrubs, save those of dwarf varieties, such as *Salix herbacea* and *Betula nana* (dwarf willow and dwarf birch). A continental climatic régime is indicated by the plant macrofossil and pollen record, by the molluscan assemblages and by coleopteran remains. At Wretton, coleopteran evidence suggests phases of cool but not arctic climatic conditions, perhaps similar to those currently experienced in north-east Europe, interspersed with more severe climatic episodes in which conditions were analogous to those in the alpine zone of the northern mountains or on the tundras of the present day. Mean July temperatures of less than 10 °C are implied for these latter periods. Vertebrate remains from Wretton support this general reconstruction, for they include such typical steppe and tundra elements as arctic fox, mammoth (*Mammuthus primigenius*), horse, bison (*Bison priscus*) and reindeer (Stuart 1977).

Whether glaciers existed in Britain during the Early Devensian is a matter for conjecture. As yet, there is no unequivocal evidence in the terrestrial record which indicates glaciation at that time, although an Early Devensian ice advance across parts of eastern England has been suggested by Straw (1979, 1980) on the basis of both morphological and stratigraphic evidence, while Sutherland (1981a) has inferred glacier activity in Scotland during that time period on the basis of the distribution of shell-bearing marine clays buried by till. These are believed to reflect a marine transgression immediately prior to and consequent upon loading of the earth's crust by the build-up of the last Scottish ice sheet, and the 'infinite' radiocarbon dates on the shells suggest that this may have occurred during the Early Devensian. Ruddiman *et al.* (1980) have obtained detailed faunal evidence from ocean cores for a major glacial build-up around 75 000 BP, while Oxygen Isotope Stage 4, which spans the later part of the Early Devensian also indicates expansion of the world's glaciers after *ca.* 75 000 BP. In the North Atlantic evidence from coccolith assemblages suggests a major southward pulse of polar water shortly after 75 000 BP which would have taken the oceanic Polar Front well to the south of the British Isles (McIntyre *et al.* 1972). Assuming that the atmospheric Polar Front also migrated southwards, a combination of increased precipitation from the associated cyclonic activity, along with the generally low temperatures discussed above, may well have been sufficient to generate and sustain glacier ice in Highland Britain. The microfaunal and microfloral evidence from the ocean sediments suggests, however, that the southward penetration of polar waters into the mid-latitude regions of the North Atlantic was relatively short-lived, particularly when compared with the expansion of polar water masses in the Late Devensian (Fig. 4.23). Moreover, the increase in $\delta^{18}O$ values appears less pronounced in many ocean cores in Isotope Stage 4 by comparison with Isotope Stage 2 (the generally-accepted Devensian glacial maximum) implying greater ice volumes during the latter stage. While there is, therefore, a certain amount of circumstantial evidence for glacier development in highland Britain during the Early Devensian, there are indications that glaciation may have been less extensive at that time than during the Late Devensian. Whether the basal till at Chelford was emplaced by Early Devensian ice or during a previous glacial episode cannot, at present, be established.

No direct evidence has yet been found for sea-levels around the coasts of Britain during the Early Devensian. At the height of the Ipswichian Interglacial the marine limit around southern England may have been as much as 8 m above that of the present day (West and Sparks 1960), but sea-levels were falling by the end of the interglacial and this trend continued into the Devensian. Pollen assemblages indicative of salt marsh and estuarine conditions have been found in a borehole in the English Channel at a depth of more than -89 m NGF. This sequence has been tentatively correlated with the Brørup Interstadial of the Netherlands (Morzadec-Kerfourn 1975), which is generally regarded as being of an equivalent age to the Chelford Interstadial (Shotton 1977b). Certainly, the oxygen isotope evidence suggests eustatic sea levels at least as low, and probably lower, for much of the Early Devensian (Shackleton 1977).

The Middle Devensian

Following the climatic amelioration of the Chelford Interstadial, a relatively long period of predominantly cold conditions prevailed. During

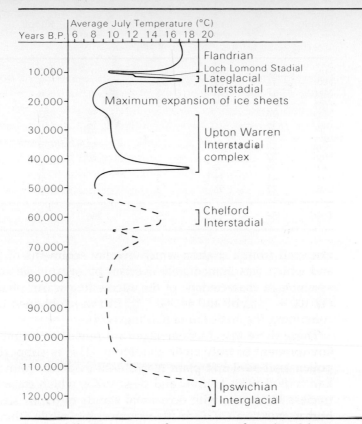

Fig. 7.9: Fluctuations in average July temperature in lowland Britain since the last interglacial based on evidence of coleopteran assemblages (after Coope 1975a).

the Middle Devensian, however, a short-lived but very pronounced climatic oscillation occurred (Upton Warren Interstadial) during which summer temperatures in lowland Britain rose to be as warm, or perhaps even warmer, than those of the present day (Fig. 7.9). Subsequently, a cold continental climatic régime became re-established prior to the build-up of ice sheets in northern Britain in the Late Devensian. This climatic fluctuation has been recognised at a number of fossiliferous sites in central and southern England, principally on the basis of assemblages of fossil Coleoptera, for the palaeobotanical evidence suggests a landscape almost totally devoid of trees, and therefore the palynologist's traditional index of climatic change, the ratio of arboreal to non-arboreal pollen, is not applicable in this case. The coleopteran evidence indicates that the climatic changes occurred extremely rapidly, and at several sites stratigraphic resolution of the warm and cold episodes has proved difficult, even with the aid of radiocarbon dating.

The initial cold phase

Detailed evidence for the cold conditions that preceded the Upton Warren Interstadial has been found at relatively few sites. At Four Ashes, an infinite radiocarbon date (>43 500 BP) was obtained from an organic silt (organic lens 3, Fig. 7.4) which contained an insect fauna of a distinctly arctic aspect (A. Morgan 1973). A Middle Devensian age was also inferred for an arctic insect fauna found in organic silts in a gravel quarry near Earith in Huntingdonshire (Fig. 7.1) dated to >45 000 BP (Coope *et al.* 1971). An arctic fauna from a third site, Tattershall in Lincolnshire, was

Month	Prec.	Snow storage	Dir. evap. of snow	Pot. evap.	Runoff	Soil moisture	Temp. (°C)
Jan.	20	108	3		0	Frozen	−17
Feb.	14	119	3		0	frozen	−17
Mar.	17	133	3		0	Frozen	−14
Apr.	21	149	5		0	Frozen	−7
May.	24			50	103	100	−1
Jun.	38			65	0	73	7
Jul.	41			70	0	44	12
Aug.	59			55	0	48	10
Sep.	52			20	0	80	6
Oct.	39	34	5		0	Frozen	−2
Nov.	41	72	3		0	Frozen	−9
Dec.	22	91	3		0	Frozen	−15
Total	388		25	260	103	(average)-4	

Table 7.2: Water balance of a site in lowland Britain during the cold period in the early Middle Devensian about 50 000 years ago (after Lockwood 1979). Except for temperatures all values are in rainfall equivalents in mm.

extracted from a stratum which overlay sediments of last interglacial age, and which was immediately overlain by organic silts containing an insect assemblage characteristic of the succeeding warm phase. Two finite dates (42 100 + $^{+1400}_{-1100}$ BP and 44 300 $^{+1600}_{-1300}$ BP) were obtained from the sediments containing the arctic fauna (Girling 1974).

These three insect assemblages are remarkably similar and suggest an environment of truly arctic conditions. This is supported by the limited pollen analytical and plant macrofossil evidence from Four Ashes and Earith (Bell 1970; Andrew and West 1977), which show an essentially treeless landscape with occasional stands of dwarf shrubs. Many of the herbaceous taxa are distinctly xeromorphic, while others, for example sea milkwort (*Glaux maritima*), are markedly halophytic (Bell 1969). Their presence may be indicative of saline soils induced by lower summer precipitation and by high levels of evaporation. Such soils are typically associated with permafrost in northern Eurasia today (West 1977b), and certainly ice wedge casts have been found associated with the arctic plant beds.

The coleopteran evidence, in asssociation with the limited palaeobotanical data, indicate a climatic régime of arctic severity, similar to that of the north-east coast of European Russia today (Coope 1975a). This would imply a mean July temperature of around, or perhaps even below, 10 °C, winter temperatures in the range −13° to −24 °C, mean annual temperatures ranging between −4° and −8 °C, an annual precipitation of 400–600 mm with more than 50 cm of winter snowfall and, if these inferences are correct, a water balance in lowland Britain approximating that shown in Table 7.2.

The warm interval (Upton Warren Interstadial)
Radiocarbon-dated fossiliferous sediments from a number of British sites are in remarkably close agreement in suggesting a warm interval during the Middle Devensian with a thermal maximum between approximately 42 000 and 43 000 years ago (Table 7.3). However, it must be emphasised that these finite dates are close to the limits of resolution of the radiocarbon method, and it may well be that additional data and the application of new techniques (Ch. 5) will result in this date being

Site (see Fig. 7.1)	Date (yrs BP)	
Isleworth, Middlesex	43 140	+ 1520 −1280
Tattershall, Lincolnshire	43 000	+ 1400 −1100
	42 000	± 1000
Four Ashes, Straffordshire	42 530	+ 1345 −1115
	40 000	+ 1400 −1200
	38 500	+ 1200 −1050
Earith, Huntingdonshire	42 140	+ 1890 −1530
Upton Warren, Worcestershire	41 900	± 800
	41 500	± 1200
	>40 000	

Table 7.3: Radiocarbon-dated insect faunas from deposits in Britain assigned to the Upton Warren Interstadial (after Coope 1975a, 1975b).

adjusted, most probably towards the older end of the time-scale (Coope 1975a).

At some sites, such as Isleworth (Coope and Angus 1975) and Tattershall (Girling 1974), the beetle assemblages are dominated by species of a distinctly southern European aspect. Many have present-day distributions in which the northern limits cross southern England and the southern tip of Fennoscandia. The fauna as a whole is typical of north-central Europe, akin to that of the North German Plain at the present day. A July average temperature of *ca*. 18 °C is suggested (1° to 2 °C higher than in central and southern England at present) and a degree of continentality no greater than that of central Europe is clearly implied.

At other sites, such as Upton Warren (Coope *et al* 1961), Earith (Coope *et al* 1971), Four Ashes (A. Morgan 1973) and Kempton Park (Gibbard *et al*. 1982), the insect assemblages are more complex and include elements of both the warm and the succeeding cold episodes (see below). They may therefore represent slightly younger assemblages, transitional between the Upton Warren Interstadial and the succeeding cold phase, although the two groups cannot be differentiated by the radiocarbon method because of the large standard deviations normally associated with radiocarbon dates of this age (Table 7.3). The increased importance at these sites of species with a south-east European and central Asiatic distribution, many with geographical overlaps in eastern Asia, indicates July temperatures of around 15 °C, and there is clear evidence of a trend towards increasing continentality.

A curious feature of all of these assemblages is the total lack of any species that are dependent on trees, either as a direct source of food, or indirectly as a source of specialised habitats (Coope 1975b). Open habitats are also implied by molluscan and ostracod evidence. In this respect the faunal and floral evidence are in complete agreement in suggesting a grassland devoid of trees (see e.g. Gibbard *et al*. 1982). At Earith, for example, Bell (1970) has described a flora which is closely related to both our present-day lowland chalk and limestone grassland flora, and to the basiphilous flora of the mountains of the north and west of Britain. The herbaceous floras are often a mixture of southern, steppe and halophyte elements, with the southern species, such as the slender naiad (*Najas flexilis*) and gipsywort (*Lycopus europaeus*), suggesting July temperatures perhaps as high as 16 °C (Bell 1969). Continentality is also implied by the large number of taxa which are characteristic of the steppe region of the south-west USSR and by the fact that many of the southern elements have a distinctly continental distribution in Europe as a whole. The lack of macroscopic remains of trees at all of the sites investigated suggests that the low frequencies of *Betula* and *Pinus* pollen at sites such as Four Ashes (Andrew and West 1977), are likely to be the product of long-distance transport. Evidence of dwarf-shrub communities is often found with, for example, the arctic willows (*Salix herbacea* and *S. polaris*) and the dwarf birch (*Betula nana*). Apart from a few taller willows, however, other shrubs appear to have been scarce.

The absence of trees from lowland England during such a warm climatic episode has been widely discussed. Three main hypotheses have been advanced to account for this apparent anomaly:

1. *Effects of climatic régime.*   The botanical evidence suggests that although summer temperatures may have been high, winters were much cooler, with mean January temperatures perhaps as low as –15 °C (Bell 1969), and the mean annual temperature may not have risen above 0 °C. Ice wedges have been observed at both Four Ashes and Earith, and it is possible that at least discontinuous permafrost existed in lowland Britain throughout the Middle Devensian (Watson 1977). The absence of calcite deposition in British caves during the Upton Warren Interstadial has been taken as further evidence for extensive permafrost at that time (Lockwood 1979). Hence, a combination of low winter temperatures, soil instability, high evaporation and waterlogged soils in the summer could have created an environment unfavourable to tree growth. Severe wind exposure may also have been an adverse factor. Indeed, analogies have been drawn with present-day conditions in parts of Siberia, Alaska and Labrador (Bell 1970; West 1977b). However, the coleopteran evidence, particularly from the thermal maximum of the Interstadial, does not suggest such extreme continentality nor the development of a steppe-type of vegetation cover.

2. *Influence of migration rates:*   During the cold phase prior to the Upton Warren Interstadial, the forest limit would have lain well to the south of Britain, possibly no nearer than southern Europe. The thermal maximum of the Interstadial apparently followed this period of arctic climate with dramatic suddenness and involved an increase in summer temperatures of about 8 °C or more (Fig. 7.9). This is implied by the Tattershall sequence, where an insect assemblage indicating temperate conditions (dated at 43 000 $^{+1400}_{-1100}$ BP and 42 000 ± 1000 BP) replaces a cold, impoverished fauna (dated at 44 300 $^{+1600}_{-1300}$ and 42 100 $^{+1400}_{-1100}$ BP) within a few centimetres of organic silt (Girling 1974). The overlapping radiocarbon dates underline the closeness in age of the two horizons despite their marked faunal contrasts. The dates on the thermal maximum at Tattershall are supported by an age determination of 43 140 $^{+1520}_{-1280}$ BP from Isleworth, and the intensity of the climatic amelioration is demonstrated by the fact that at the latter site, of 248 named coleopteran species only one has a marked northern geographical distribution. However, the radiocarbon-dated

coleopteran assemblages from Earith, Four Ashes and Upton Warren imply that the thermal maximum was short-lived and that within the space of 2000 years or less climatic deterioration had already set in. If the foregoing is correct, then the seeming discrepancy between the coleopteran and palaeobotanical evidence (in that the former indicate relatively high temperatures, while the latter suggests an absence of trees) may largely reflect differing migration rates of plants and insects (e.g. Coope 1975b, 1977a). The mobile thermophilous Coleoptera may have reacted almost instantaneously to climatic amelioration and spread rapidly northwards from their southern refugia, while the trees in particular, with their long regeneration time and relatively poor seed dispersal rates, would have spread northwards much more slowly. Only some of the more thermophilous herbs would have reached southern and central Britain before climatic deterioration once again pushed the treeline southwards. Under this interpretation, the absence of trees in lowland Britain during the Upton Warren Interstadial resulted from (a) the shortness of the temperate interval, (b) the severity of the climate that came before and immediately after it, and (c) the very great distances that the trees would have had to cover in their migration across continental Europe.

3. *Browsing animals.*   Vertebrate remains have been recovered from a number of Middle Devensian sites, particularly Upton Warren, Tattershall (Rackham 1978) and Isleworth. Typical large vertebrates include mammoth, bison, reindeer, woolly rhinoceros and horse, all of which are characteristic of open, treeless habitats. Bison, which are not normally arctic animals, may well have moved south during the winter months towards the forest margins where, as with grazing animals in woodlands today, the regeneration of trees would have been diminished by their activities. In this way, large herbivores could have played a contributory, if subsidiary, role in slowing down the northern spread of woodland (Coope and Angus 1975).

The above explanations should not, however, be seen as mutually exclusive, and a combination of factors may well have conspired to prevent the European treeline reaching central and southern England during the Upton Warren Interstadial.

The landscape of lowland Britain at the height of the Upton Warren Interstadial was therefore one of open grassland with a climatic régime similar to that of north-central Europe at the present day. This suggests July temperatures in the region of 17 ° to 18 °C, January temperatures of around 0 °C to −2 °C, with a mean annual temperature of about 7 ° or 8 °C (1 ° or 2 °C below that of the present day). A precipitation of 450–650 mm, with an average of *ca.* 550 mm per year, is also implied (Lockwood 1979). A suggested water balance for lowland Britain during this time period is shown in Table 7.4. During the later part of the Interstadial, summer temperatures declined to *ca.* 15 °C, while winter temperatures must have fallen well below freezing. A more detailed elaboration of climatic conditions during this phase, however, awaits the results of further research.

Table 7.4: Water balance of a site in lowland Britain during the Upton Warren Interstadial (after Lockwood 1979). Except for temperatures all values are in rainfall equivalents in mm.

Month	Prec.	Snow storage	Dir. evap. of snow	Pot. evap.	Runoff	Soil moisture	Temp. (°C)
Jan.	41	38	3		0	Frozen	−1
Feb.	34	69	3		0	Frozen	−1
Mar.	28			30	60	100	3
Apr.	42			52	0	90	5
May.	52			85	0	57	12
Jun.	62			95	0	24	16
Jul.	65			97	0	0	17
Aug.	63			80	0	0	17
Sep.	48			45	0	3	14
Oct.	37			20	0	20	9
Nov.	39			5	0	54	4
Dec.	42			3	0	93	1
Total	553		6	512	60	(average)	8

The succeeding cold phase

By shortly after 40 000 BP conditions of arctic severity had once more become established in lowland Britain. A large number of sites have been investigated that contain insect faunas dating from this cold period (Table 7.5). All of these coleopteran assemblages are strikingly similar and they contain some of the most exotic elements so far discovered in British Quaternary deposits (Coope *et al.* 1971). Thermophilous Coleoptera are almost totally absent and the faunas are dominated by exclusively boreal or boreo-montane species, and contain species that are today restricted to eastern Asia. The faunas as a whole are characteristic of those found in the alpine zone of the mountains of Scandinavia, or its equivalent near the forest limit in the Siberian arctic (Coope 1977a). This suggests that the climate of the times in lowland England was markedly continental, with average July temperatures at or just below 10 °C, while winters were extremely cold with temperatures between −20° and −30 °C. Mean annual temperatures may have ranged from −8° to −12 °C, with the average annual precipitation probably being between 250 and 350 mm (Lockwood 1979). The coleopteran evidence suggests that this cold period was both

Table 7.5: Radiocarbon-dated insect faunas from the cold phase immediately following the Upton Warren Interstadial (after Coope 1975b; *after Gibbard *et al*. 1982).

Site (see Fig. 7.1)	Date (yrs BP)	
Queensford, Oxfordshire	39 330	+1350 −1150
Oxbow, Yorkshire	38 600	+1720 −1420
Fladbury, Worcestershire	38 000	±700
Syston, Leicestershire	37 420	+1670 −1390
Four Ashes, Staffordshire (3)	36 340	+770 −700
·Kempton Park, Surrey	35 230	±185
Kirby-on-Bain, Lincolnshire	34 800	±1000
Sutton Courtney, Berkshire	33 190	±3450
Brandon, Warwickshire	32 270	+1029 −971
Coleshill, Warwickshire	32 160	+1780 −1450
Four Ashes, Staffordshire (2)	30 655	±700
Standlake, Oxfordshire	29 500	±300
Great Billing, Northamptonshire	28 225	±330
Beckford, Gloucestershire	26 000	±300
Thrapston, Northamptonshire	25 780	±870

drier and more continental than that which preceded the Upton Warren Interstadial, and that the severity of the climate increased throughout the later stages of the Middle Devensian.

This palaeoclimatic reconstruction is supported by sedimentological evidence. Many of the fossiliferous horizons are associated with periglacial indicators of frozen ground, including ice-wedge pseudomorphs, sand wedges and involutions: freeze-thaw activity and repeated cycles of deposition and frost heaving are indicated by vertically-orientated pebbles. Although strict contemporaneity between the periglacial structures and the fossil-bearing deposits cannot be proved, the stratigraphic evidence suggests that many date from the cold period following the Upton Warren Interstadial thermal maximum. Following Williams (1975), the periglacial evidence implies mean annual temperatures of below −8 °C, an average temperature of the warmest month of not more than 10 °C, a high degree of continentality, and a difference between the January and July averages of around, or in excess of, 30 °C (Fig. 3.16); a mean annual precipitation of 250 mm or less is also indicated. All of these climatic estimates are in remarkably close agreement with those described above.

Both palaeobotanical and palaeozoological evidence suggest a barren, tundra landscape gradually approaching that of a polar desert. Pollen and plant macrofossil remains from sites such as Syston in Leicestershire (Bell *et al.* 1972), Oxbow near Leeds (Gaunt *et al.* 1970) and Great Billing in Huntingdonshire (Morgan 1969) are dominated by arctic-alpine, northern montane and present-day tundra elements. These include grasses and sedges and species of a wide range of ruderal habitats, including small herbs typical of bare ground and moving soils. Plants characteristic of saline soils provide further evidence of low levels of precipitation. Trees are absent, and of the shrubs only those of the dwarf variety are present, including some which are typical of snow-patch vegetation. Characteristic faunal remains from sites such as Fladbury in Worcestershire (Coope 1962), Oxbow and Tattershall (Rackham 1978), and from numerous caves (Campbell 1977; Stuart 1977), include mammoth, reindeer, woolly rhinoceros, horse, bison, deer, arctic lemming (*Dicrostonyx torquatus*) and northern vole (*Microtus oeconomus*), all species characteristic of cold climates and open-tundra landscapes. A number of sites contain molluscan assemblages and these frequently include species that are now absent from Britain but which are found today in arctic-alpine areas of Europe (e.g. Briggs and Gilbertson 1980).

A similar type of environment appears to have existed in Ireland. At two sites near Maguiresbridge in County Fermanagh, organic silts dated at greater than 41 500 BP (Hollymount Silts) and 30 500 $^{+1170}_{-1030}$ BP (Derryvree Silts) underlie till of Late Midlandian (Late Devensian) age. The pollen content of the silts indicates open tundra conditions (Colhoun *et al.* 1972). At Castlepook Cave, County Cork, a date of about 33 500 years BP was obtained from collagen of a mammoth femur (Stuart 1977). A rich faunal assemblage within the cave contained elements which may also be of Middle Devensian age including arctic lemming, Norway lemming (*Lemmus lemmus*), wolf, arctic fox, reindeer and giant deer (*Megaceros giganteus*). In Scotland, lenses of peat incorporated into the basal layers of a till at Burn of Benholm between Montrose and Aberdeen (Fig. 7.1) yielded pollen indicative of a tundra environment. The radiocarbon date of >42 000 BP suggests an Early or Middle Devensian age for the deposit (Donner 1979).

Throughout the entire Middle Devensian therefore, the British landscape was essentially treeless and the climate was markedly continental. Such climatic conditions would only obtain if the depressions which move eastwards towards the British Isles from the north Atlantic were blocked by a stable anticyclone situated perhaps slightly to the north producing an easterly or south-easterly airstream across the country. Maximum ocean level during the Middle Devensian was probably never higher than 10 to 20 m below present sea-level (Dreimanis and Raukas 1975) and for much of the period was probably very considerably lower. Hence, a large area of what is now the North Sea was dry land and thus little moisture would be collected along the track of the easterly winds blowing towards central and southern Britain. Low precipitation and very low winter temperatures would therefore be expected. Undoubtably temperatures were low enough to permit glacier development in highland areas, but it is questionable whether precipitation was high enough over the uplands to generate and sustain glaciers. The decrease in $\delta^{18}O$ values in north Atlantic ocean cores during Isotope Stage 3 (approximately equivalent with the Middle Devensian) suggests a widespread melting of land ice rather than glacier build-up, particularly in the early stages. Certainly, there is no unequivocal evidence from the terrestrial record for glacier expansion in Britain during the Middle Devensian.

The question arises, however, as to why two phases of more oceanic conditions occurred within the otherwise markedly continental climatic régime of the Early and Middle Devensian. The answer may lie in the changes that occurred in the circulation of the North Atlantic Ocean. For much of the Devensian, sub-polar water masses probably occupied the north-east Atlantic as far south as latitude 42 °N, and there were frequent southward pulses of truly polar water (Fig. 4.23). McIntyre and Ruddiman (1972) have estimated that temperatures of north-east Atlantic surface waters were at least 5 °C below those of the present day, which would clearly have had a marked cooling effect on the adjacent landmasses of north-west Europe. Periodic northward movements of this cold water occurred, however, which may have allowed more moderate climatic influences to impinge on western Britain. During such phases a south-westerly airstream and consequently more maritime conditions would have been established (Lamb 1977a). Fig. 4.23 shows that a major northward retreat of polar water occurred during the period equivalent to the Chelford Interstadial, and a further, short-lived migration is suggested shortly after 40 000 BP. The mild and relatively oceanic climatic régimes indicated by the terrestrial evidence for the Chelford Interstadial and also, somewhat less emphatically, for the Upton Warren Interstadial, may therefore be related to changes in the positions of the principal water masses in the north Atlantic, and the contrasting expressions of the Chelford and Upton Warren Interstadials in the terrestrial fossil record may be at least partly explicable in terms of the timing, rapidity and duration of each of these major changes in ocean waters. However, the driving force behind these oceanic changes is not at present fully understood.

The Late Devensian

The major feature of the Late Devensian was the expansion of the last ice sheet to affect the British Isles which, at its maximum extent, covered over two-thirds of the present land area of the country (Fig. 7.10). The development of such a large ice sheet clearly reflects a significant climatic shift from the predominantly continental conditions that had prevailed throughout much of the Early and Middle Devensian to a more oceanic climatic régime, particularly in the more northern and western areas of the British Isles. This in turn suggests at least a partial breakdown of the major anticyclonic system that had formed the dominant element of the north-west European meteorological situation hitherto.

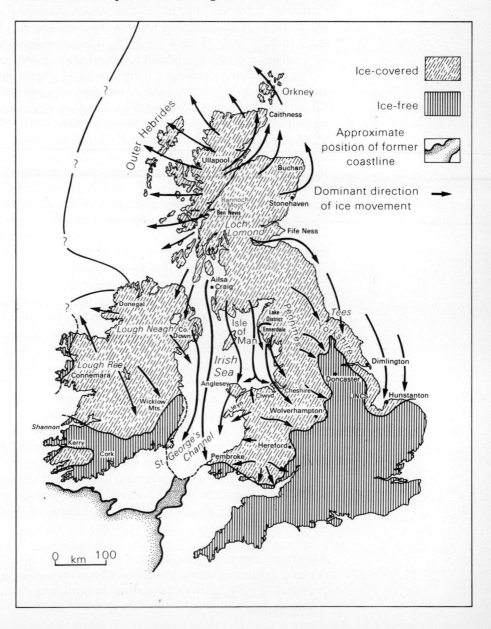

Fig. 7.10: Maximum limit of the Late Devensian ice sheet, principal paths of ice movement, and localities mentioned in the text. The approximate coastline (assuming a 100 m fall in sea-level) is also shown. It is likely, however, that at the height of the glacial phase, sea-level was lower than this.

Onset of glaciation

Precisely when glacier ice began to accumulate in Highland Britain is not known, but, as was discussed above, the possibility cannot be excluded that glaciers were present in upland regions before the beginning of the Late Devensian. In their discussion of the steady-state model developed for the British ice sheet Boulton *et al*. (1977) suggest that about 15 000 years would have been required for its build-up (Fig. 2.11). If a glacial maximum at around 18 000 to 20 000 BP is accepted, this would imply that glaciers began to form in the British Isles well before 30 000 BP. It is worth noting, however, that this is a very much longer period than the 10 000 years which has been estimated for the build-up of the considerably larger Laurentide ice sheet (Andrews and Mahaffey 1976). Certainly, if the concept of 'instantaneous glacierisation' is applied, in which rapid glacier growth follows the coalescence of snowbanks over large upland areas (Ives *et al*. 1975), a much shorter time interval for the accumulation of the last British ice sheet may be envisaged.

The only direct evidence of the timing of glacier build-up are radiocarbon dates on organic sediments buried beneath Late Devensian till, and three sites in Scotland (Fig. 7.1) have been considered particularly significant in this respect (e.g. Sissons 1974c, 1976). The dates are 27 333 ±240 BP from a peat layer beneath till on the Island of Lewis (von Weymarn and Edwards 1973); 28 140 $^{+480}_{-450}$ BP from a palaeosol buried beneath till (?) at Teindland in Morayshire (FitzPatrick 1965); and 27 500 $^{+1370}_{-1680}$ BP from the bone of a woolly rhinoceros buried in glaciofluvial gravels beneath till at Bishopbriggs near Glasgow (Rolfe 1966). If correct, these dates suggest that ice expansion did not occur until after *ca*. 27 000 BP and, moreover, in view of the proximity of the Bishopbriggs site to the major ice accumulation and dispersal centres in the Grampian Highlands, that around 27 000 years ago most of Scotland was ice-free. It is clear, however, that such dates can only provide maximum ages for the spread of ice across the localities from which they were obtained and also, in the absence of a boundary stratotype, it is not known to what extent there was a time interval between organic sediment accumulation and till deposition. It should also be pointed out that a recent critical evaluation of radiocarbon dates from Scotland suggests that the Teindland date, and possibly also the age determination from Bishopbriggs, may be too young (Sissons 1981).[5] On the other hand, it is interesting to note that micropalaeontological analyses from north Atlantic sediments suggest that the major southward pulse of polar waters (and hence the oceanic Polar Front) in the Devensian did not occur until around 30 000 BP (McIntyre *et al*. 1972). If the last British ice sheet grew as a result of increased precipitation induced by enhanced cyclonic activity associated with the southward migration of the oceanic Polar Front (as is suggested below), a date for the onset of glacier build-up of between 30 000 and 25 000 BP would be implied.

Maximum extent of glaciation

At its maximum extent, the Late Devensian ice sheet covered almost all of Scotland, a very substantial area of Ireland and Wales, and large parts of northern England (Fig. 7.10). The greatest areas of accumulation and dispersal were in the highlands of western Scotland where individual ice domes, including the largest over the Rannoch Moor area, coalesced to form a major ice sheet from Loch Lomond to the area near Ullapool (Sissons 1976). The thickness of the ice sheet is, however, difficult to determine accurately. Erratics and striations at elevations of over 1000 m are found throughout western Scotland, but there is no certainty that this evidence relates to the last ice sheet. The ice sheet model (Fig. 2.11) of Boulton *et al.* (1977) indicates a maximum ice surface altitude of *ca.* 1800 m across western Scotland (over 450 m above the summit of Ben Nevis) but this figure must be regarded as speculative as their reconstruction of the ice sheet depends on the accurate determination of glacier limits, and in many areas this is not possible on present evidence. In view of the obvious extent of the ice cap, however, it is likely that the surface of the Scottish ice sheet stood far above the highest mountain summits of the present day.

Directions of flow within the ice sheet (Fig. 7.10) were complex and have been inferred from a variety of sources including glacial erratic suites, striations, till fabrics, and evidence from such landforms as drumlins and roches moutonnées. Essentially the flow was radial, with the greatest ice volumes discharging down the troughs to the western seaboard where the ice sheet reached perhaps to the Outer Hebrides and beyond. On present evidence, however, it is not possible to say how far out into the north Atlantic the Late Devensian ice sheet extended. Ice flow appears to have been less strong to the east, and parts of Orkney, north-east Caithness, Buchan and even northern areas of the Hebrides may have remained unglaciated (Synge 1977). In many places, however, the evidence is equivocal and it is equally possible that some, if not all, of these areas were affected by Late Devensian glacier ice (see e.g. Clapperton and Sugden 1977). An almost universally-held belief is that the presence of Scandinavian ice to the west of Norway deflected ice from the Scottish mainland northward and southward so that it flowed approximately parallel with the British coast. Yet no conclusive evidence has yet been proposed for a zone of confluence between the two ice sheets and, moreover, the limit of Late Devensian ice to the east of Scotland has never been established. Recent work beneath the North Sea has revealed the presence of a complex morainic ridge system just off the east coast of Scotland which can be traced over a distance of some 80 km between Stonehaven in the north and Fife Ness in the south (Thomson and Eden 1977). It has been suggested that this morainic ridge may mark the outer limit of the last Scottish ice sheet and, furthermore, that an ice-free area lay between the Scottish and Scandinavian ice sheets at the last glacier maximum (Sissons 1981). A zone of confluence between the two ice sheets, therefore, remains to be proven.

A major feature of the glaciation of western Britain was the great mass of ice which moved down the Firth of Clyde where it merged with glaciers from the Southern Uplands and Northern Ireland to form a vast ice lobe flowing southwards into the Irish Sea basin (Fig. 7.10). This Irish Sea ice was further augmented by ice from the Lake District, from North Wales and from the Wicklow ice cap in south-east Ireland, and spread as far south as Pembrokeshire and the entrance to St George's Channel

(Bowen 1973b). Evidence for the passage of ice down the Irish Sea basin and, in some areas, a considerable distance inland, is convincing. In Anglesey, striations, erratic trains and the alignment of ice-moulded forms such as drumlins demonstrate that the island was glaciated from north-east to south-west (Bowen 1974). Similar evidence for a north-south component of ice movement can be found in north-east Ireland (Davies and Stephens 1978). Erratics of northern origin, including the highly distinctive Ailsa Craig microgranite, have been found at numerous localities around the coasts of Cardigan Bay and in eastern Ireland (e.g. Hill and Prior 1968; John 1970), while erratics from the Southern Uplands and the Lake District have been found in Irish Sea tills as far inland as the Wolverhampton area (A. V. Morgan 1973). Glacially-transported marine clays, often containing shell fragments, are widespread and occur, for example, along the coasts of north Pembrokeshire (John 1970) and inland in County Down (Stephens and McCabe 1977), while large spreads of glacial sediments including both lodgement and flow tills have been found on the sea bed throughout the southern Irish Sea basin (Garrard and Dobson 1974; Garrard 1977).

The fact that these glacial sediments overlie deposits believed to be of Ipswichian Interglacial age in a number of localities suggests that the drifts date from the Devensian. More secure evidence of the age of the last Irish Sea ice advance comes from the radiocarbon dates that have been obtained on organic material either incorporated within or underlying till. Marine shell fragments in till beneath a drumlin in the Ards Peninsula of County Down were dated at 24 050 ± 650 BP (Hill and Prior 1968), while glacially-transported marine shells on the Cheshire Plain yielded a radiocarbon age of 28 000 $^{+1800}_{-1500}$ BP (Boulton and Worsley 1965). The latter date is in good agreement with dates of 29 000 ±1200 BP and 31 800 $^{+1800}_{-1200}$ BP from shells in Irish Sea till and outwash in the Lleyn Peninsula (Saunders 1868a) and with the age determination of 30 500 ±400 BP from organic silts underlying till of Irish Sea provenance at Four Ashes in Staffordshire (A. Morgan 1973). If correct, these dates provide a maximum age for the passage of Irish Sea ice across these areas of eastern Ireland and the Cheshire-Shropshire plain. A younger date of 18 000 $^{+1400}_{-1200}$ BP has been obtained however from a mammoth bone in a cave sealed by Irish Sea till at Tremerchion in the Vale of Clwyd (Rowlands 1971). This age determination is very similar to two radiocarbon assays (18 500 ± 400 BP and 18 240 ± 250 BP) on moss fragments included within silts underlying glacial till at Dimlington in Holderness (Penny et al. 1969). The Dimlington dates have been taken to indicate that the Late Devensian ice maximum in eastern England was reached around, or shortly after, 18 000 BP.

One part of the Irish Sea basin where problems are encountered with this general reconstruction is the Isle of Man. As the island lies over 220 km to the north of the postulated limit of Irish Sea ice in the St George's Channel, it would seem probable, on glaciological grounds, that the Isle of Man was submerged beneath the Late Devensian ice sheet. Indeed, an ice thickness over the island of ca. 1600 m has been suggested for the last glacial maximum (Bowen 1977a). Yet, Thomas (1976, 1977) has argued that the morphological and stratigraphic evidence shows that the Isle of Man was not overridden by Irish Sea ice during the Devensian.

Some support for this contention comes from five radiocarbon dates ranging from *ca.* 18 400 to 18 900 BP obtained from the base of a kettle hole developed in the youngest glacial sediments (Shotton and Williams 1971, 1973). If correct, they imply that the island has been ice free for the past 18–19 000 years. However, the validity of the dates has been questioned, for they were derived from a moss (*Drepanocladus*) which is capable of photosynthesising subaquatically, and hence a hard water error (Ch. 5) must be considered (Shotton and Williams 1973). This could mean that the dates are too old, although the magnitude of any error cannot be calculated; nor, indeed, can an error be proven. The Isle of Man evidence is therefore enigmatic and requires further detailed analysis. A partial solution to the problem may, however, lie in the effects of a readvance (or readvances) of the Irish Sea glacier, discussed below.

On its western flank, Irish Sea ice was confluent with a large mass of land ice which covered over three-quarters of Ireland. The principal ice shed lay across an area stretching from Lough Ree in central Ireland north-eastwards towards Lough Neagh in Ulster, although there was a radial flow from the centres in Connemara, Donegal and the Wicklow Mountains. A separate ice cap also developed over the hills of the extreme south-west in Kerry and West Cork (Synge 1979). The limits of this glaciation (the **Midlandian**) have been mapped using both morphological and stratigraphic evidence, aided by relationships which exist between the parent drifts and the soils that developed upon them (Davies and Stephens 1978). The principal morphological evidence is a large terminal moraine which swings around the northern end of the Wicklow Mountains and across Ireland to the mouth of the Shannon. First described by Charlesworth (1928) it is generally known as the 'Southern Ireland End Moraine' and often separates fresh and relatively unweathered drift to the north from more deeply-weathered and cryoturbated drift, derived from an earlier glacial phase, to the south. There is evidence to suggest that the maximum limits of Late Devensian ice in Ireland were not everywhere contemporaneous, but absolute dating of the various stages has yet to be achieved (Davies and Stephens 1978).

Off the coasts of west Wales, till of Irish Sea provenance overlies Welsh till, suggesting that ice flowed out from the hills of central and northern Wales across the old sea floor of Cardigan Bay prior to the arrival of Irish Sea ice (Garrard 1977). These local Welsh glaciers took the form of small piedmont lobes, the margins of each frequently being defined by lateral moraines or 'sarns', a number of which are still preserved on the sea bed of present-day Cardigan Bay (Foster 1970). In south Wales, the limits of Irish Sea and local Welsh ice have been most securely fixed in Pembrokeshire and the Gower on the basis of lithostratigraphy. The presence of till overlying Ipswichian raised beach deposits in coastal exposures in north Pembrokeshire and the eastern Gower contrasts with the southern coasts of Pembrokeshire, Carmarthenshire and the west Gower where the raised beaches are overlain only by periglacial head (Bowen 1973b, 1977a). Inland, the limits are less clearly defined but appear to lie generally close to the south coast. Suggestions that the limit of Late Devensian ice lay in North Wales and that much of central and south Wales was ice-free during the Devensian (see e.g. Watson 1972) are not supported by stratigraphic evidence (Bowen 1974). The extent of Late Devensian ice in the Welsh borderland and the Hereford area is not firmly established, and proposed limits often rest on morphological evidence, such as the contrasting degrees of dissection of drift terrain (e.g. Luckman 1970). There are, however, extensive deposits of sand and gravel which may be related to the Devensian ice limit (Fig. 7.11). Further north, in south Staffordshire and eastern Shropshire the so-called 'Wolverhampton Line', which is held to mark the maximum extent of Late Devensian ice, is supported by stratigraphic evidence, and is marked by a pronounced thickening of Irish Sea till to the north and by a concentration of large erratic blocks of northern provenance (A. V. Morgan 1973).

Around the southern margins of the Pennines, the Late Devensian ice limit is not clearly defined, but there are indications from the presence of loess deposits of possible Devensian age, which are intermixed with limestone residues, that the uplands of Derbyshire may not have been ice covered (Pigott 1962). In the Vale of York, the prominent Escrick Moraines have long been regarded as marking the outer limits of Late Devensian ice, but it now seems that these indicate a retreat stage and that the ice maximum lay some 30 to 40 km to the south. Sands and gravels interpreted as ice-marginal sediments suggest that an ice lobe extended as far south as the Isle of Axholme to the east of Doncaster and that a glacial lake (Lake Humber) developed between the ice margin and the uplands to the south and west (Gaunt 1976). A radiocarbon date of 21 835 ±1660 BP on a bone fragment in one of the lake shorelines appears to confirm the Late Devensian age of this glacial phase (Gaunt 1974).[6] Further east, the lobe of ice which extended down the coast of Holderness, Lincolnshire and into the Hunstanton area of north Norfolk (Fig. 7.10) has been substantiated by careful analysis of the till stratigraphy (e.g. Catt and Penny 1966; Madgett and Catt 1978), and its Late Devensian age has been confirmed by the Dimlington dates of around 18 000 BP from beneath

Fig. 7.11: Section in outwash sands and gravels near the Late Devensian ice limit at Aymestry, Herefordshire.

the upper till of Holderness (Penny *et al.* 1969). The true nature of this ice lobe has been the subject of much discussion. One suggestion is that it represents a surge resulting from some instability in the ice sheet margin whose steady-state position lay some distance to the north, perhaps in the vicinity of the River Tees (Boulton *et al.* 1977). Confirmatory evidence in support of this idea has, however, yet to be produced. Moreover, it has yet to be established whether this ice advance in eastern England was contemporaneous with those in the Irish Sea area to the west (Shotton 1977b).

The Late Devensian environment

Relatively few fossiliferous sites have so far been found from which inferences can be drawn about the nature of the Late Devensian environment. Pollen analyses have been carried out on the organic sediment at Teindland (Edwards *et al.* 1976) and Tolsta Head, the results of which suggest that around 27 000 BP (if the radiocarbon dates are valid) the landscape of both eastern Scotland and northern Lewis was open, essentially treeless and perhaps resembled that of the later part of the Middle Devensian in central England. Both pollen records imply cold, periglacial conditions, but at Tolsta Head where local shrub development of *Juniperus* and *Salix* is recorded, a more oceanic climatic régime is indicated.

Slightly less continental conditions than had prevailed hitherto are also inferred from the coleopteran assemblage in the Dimlington silts, for although fossil remains are sparse, there is a lack of the exclusively Asiatic species that were so common in the later part of the Middle Devensian records (Coope 1977a). A very harsh climate is none the less implied with summer temperatures probably several degrees below 10 °C. Arctic or sub-arctic coleopteran assemblages have also been found in the Barnwell Station 'Arctic Bed' in Cambridge and in the Lea Valley 'Arctic Bed' in North London (Coope 1968b, 1975b). The former have been dated to 19 500 ±600 BP (Bell and Dickson 1971), although there are reasons for believing that this age determination may be too young, while dates of 21 530 ±480 BP (Coope 1975b) and 28 000 ±1500 BP (Godwin 1964) have been obtained from the Lea Valley beds. These arctic plant beds contain a range of northern and montane plant remains, including arctic willows and dwarf birch, along with a high proportion of steppe and halophytic elements. Macrofossil remains of tree species are absent, however, and the representation of shrub taxa is small. Moreover, the flora lack true thermophiles, and there are few southern species. The Lea Valley 'Arctic Beds' have also yielded a vertebrate fauna which includes arctic lemming, mammoth, woolly rhinoceros and horse (Stuart 1977). Similar faunal remains have been found in cave sites in south-west England and in Wales, but a Late Devensian age for the assemblages cannot always be proved. Fossil mollusca which include elements of a distinctly arctic aspect, and which are believed to be of Late Devensian age, have been found at a number of sites in central and southern England. These include assemblages from the chalklands of Kent (Kerney 1965) and also mollusca from river terrace deposits in the Thames Valley and adjacent drainage systems (Briggs and Gilbertson 1980; Brown *et al.* 1980).

A polar desert-like landscape and a climatic régime of arctic severity have also been inferred on the basis of the widespread distribution of periglacial features in central and southern England. Although precise dating of features, particularly ice wedges, is clearly difficult, much of the periglacial evidence is believed to be of Late Devensian age. Thick deposits of thermoclastic scree in many limestone caves are indicative of severe frost weathering (Campbell 1977), while the widespread occurrence of ice wedges and involutions suggests extensive continuous permafrost. According to Watson (1977) this implies a mean annual air temperature of at least 16°–17 °C lower than that of the present day (*i.e.* -6° to -7 °C), and the fall may have been up to 20 °C. The presence of involutions suggests a mean annual temperature range of around 30 °C, with a July average of around 10 °C and a January mean perhaps as low as -20 °C (Fig. 3.16) – a temperature curve very similar, in fact, to that derived from coleopteran evidence (Coope *et al.* 1971).

In an attempt to establish climatic gradients across the British Isles, Williams (1975) has drawn attention to the significantly greater number of ice wedges in East Anglia by comparison with the south-west of England. If the growth of ice wedges in the south-west was inhibited by higher temperatures (*i.e.* above -6 °C), mean annual temperatures may have been 3°–4 °C or even higher in the west than in the east. Alternatively, the disparity could reflect differences in precipitation, if the relationship between snowfall thickness and ice wedge development outlined in Chapter 3 is correct. Williams has suggested that the considerable number of ice-wedge pseudomorphs in eastern and central England could imply an annual precipitation of less than 250 mm in East Anglia and the Midlands (less than one half and one third, respectively, of that at present), while in excess of 250 mm of precipitation fell in the south-west. Under either interpretation (and the two are not necessarily mutually exclusive) a marked increase in continentality across the country is implied. Continentality and associated easterly winds are also indicated by the spreads of Late Devensian loess deposits in eastern England (Fig. 3.18) whose source lay largely to the east and which thin out rapidly westwards across the country (Catt 1977). The presence of ventifacts of possible Late Devensian age in the English Midlands (A. V. Morgan 1973) and in parts of south Yorkshire (Williams 1975) is testimony both to wind strength and sparse vegetation at the height of the last glaciation. Overall, therefore, there is a broad measure of agreement between evidence from a range of sources about environmental conditions in those areas of Britain that lay beyond the margins of the Late Devensian ice sheet.

At first it might seem incongruous that the prevailing climatic conditions in central, southern and eastern England during the Late Devensian were predominantly continental, when clearly oceanic conditions would be required to generate and sustain the massive ice sheet that developed over northern Britain at that time. However, this conclusion is not out of keeping with palaeoclimatic reconstructions based on climatic modelling. Work by Lamb and Woodroffe (1970), for example, suggests that the climax of the last glacial the British Isles lay under the influence of a high pressure cell centred over north-west Europe during the winter months, which led to the development of a generally easterly surface wind, but during the summer months, mainly low pressure conditions produced lighter and more variable winds over much of the country. At both seasons, however, much northerly wind flow over the sea areas between Greenland and north-west Europe is indicated (Lamb 1977b). Sophisticated simulation models of atmospheric circulation in the vicinity of the North American and European ice sheets during the last glacial suggest strong cyclonic activity in the north Atlantic, just off the coasts of the British Isles (Williams *et al.* 1974), and a precipitation maximum just to the west of Britain with a very marked decrease in precipitation over the continental areas to the east (Williams and Barry 1974). Such increased cyclonic development was probably related to the southward movement of the oceanic Polar Front which had lain largely to the north of the British Isles in the Early and Middle Devensian, but during the Late Devensian migrated southwards (Fig. 7.12) to reach a position off the coast of Portugal at around 18 000 BP (Ruddiman *et al.* 1977; Ruddiman and McIntyre 1981). The high levels of precipitation in western Britain suggested by the models may therefore be envisaged as a belt of rain which moved slowly south across the country. Because of the prevailing low temperatures, much of the precipitation would probably have fallen in the form of snow.

Precipitation estimates for Britain during the build-up of the ice sheet vary, but values of up to 150 per cent of those of the present day have been suggested for the major accumulation and dispersal areas (Manley 1975). Using a climatic simulation model, Williams and Barry (1974) derived annual precipitation estimates ranging from 1800 mm in south-west Ireland (present-day values are in the region of 1500–2000 mm) to 700 mm in north-east Scotland (present annual rainfall 800 mm), while Lockwood (1979) has suggested figures for upland Britain of about 550 mm per year at an altitude of 500 m and 740 mm per year at an

Fig. 7.12: Position of the Polar Front and limit of winter sea ice during the period *ca*. 20 000 to 10 000 BP (based on Ruddiman and McIntyre 1981). 1: 20 000–16 000 years ago 2: 16 000–13 000 years ago 3: 13 000–11 000 years ago 4: 11 000–10 000 years ago. Thin lines represent the pronounced thermal gradient to the south of the Polar Front; PIL: the approximate southern limit of pack ice at the present day.

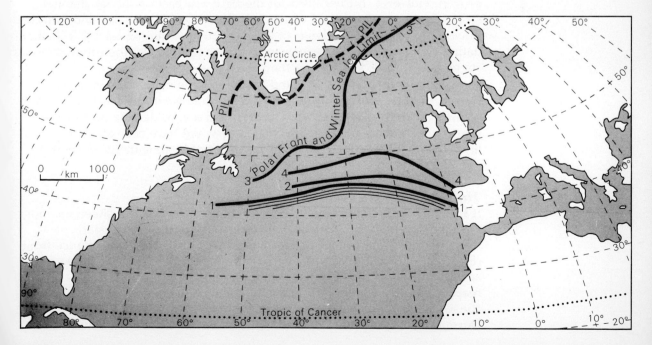

altitude of 1000 m, much of which may have fallen in the form of snow during the summer months. The general consensus, however, is that the Late Devensian ice sheet grew as a result of significantly increased precipitation in the northern and western regions of Britain, but that there were strong precipitation gradients between the maritime margins of the ice sheet and the unglaciated lowland areas to the south and east. Thus the evidence both from the climatic models and from the terrestrial record suggests that over the central and eastern parts of mainland Britain, a predominantly continental climatic régime prevailed throughout the Late Devensian.

Many attempts have been made to establish the position of sea-level during the last glacial maximum, and estimates typically range between −100 m and −130 m, with many authorities favouring the latter figure (e.g. Milliman and Emery 1968; Bloom 1971). Some workers are of the opinion, however, that sea-level lowering would have been greater than this (Guilcher 1969), and indeed the values for $\delta^{18}O$ in many ocean cores during Oxygen Isotope Stage 2 have been interpreted as reflecting a eustatic lowering of 165 m during the Late Devensian (Shackleton 1977). Around the coasts of southern and western Britain there is ample evidence of low sea-levels in the form of submerged platforms, clifflines and beach deposits, and the continuation of valley systems and river channels for some considerable distance offshore (Kidson 1977; Mottershead 1977). It is apparent, however, that much of this evidence relates to pre-Devensian low sea-levels, and indeed many of the features may be polycyclic or polyphase in origin. Dating is particularly difficult. In the area between the Scilly Isles and south-west Ireland submarine features, including ridges and 'littoral deposits', have been found beneath the sea at depths of 110–120 m, and although a Late Devensian age has been inferred (Stride 1962), this cannot be satisfactorily proven. However, further north in the Irish Sea basin a fluvial channel system has been discovered which is incised into Irish Sea till and into the underlying bedrock to a depth of at least 130 m (Whittington 1977). If the dating of the till as Late Devensian is correct (see above) then this suggests a eustatic lowering at the height of the last glacial of a comparable order to that indicated by the isotope evidence. Sea-levels as low as −130 to −160 m would have meant that the entire North Sea south of the ice margins, the English Channel, and a considerable part of the western approaches to the south of Ireland would have been above sea-level (Fig. 7.10), further enhancing the effect of continentality in central and eastern England described above.

The retreat of the ice sheet

It has long been considered that the wastage of the last ice sheet was not a uniform process, but that deglaciation from the Late Devensian maximum was interrupted by readvances of different parts of the ice front (Fig. 2.3). In recent years, however, the evidence for renewed glacier activity during ice wastage has been the subject of much critical reappraisal, and a great deal has been found to be equivocal. As a consequence, a number of the postulated readvances of the Late Devensian ice are now considered to lack foundation. In Scotland, for example, the morphological evidence relating to the 'Aberdeen-Lammermuir Readvance' has been shown to be capable of an alternative, more satisfactory explanation, and a readvance of ice is no longer envisaged. Similarly, both the morphological and stratigraphic bases for the 'Perth Readvance', believed at one time to be securely dated to between 13 000 and 13 500 BP,[7] have been reinterpreted (e.g. Paterson 1974), and this readvance too cannot be supported by existing evidence. In the Anglesey and Lleyn Peninsula areas of north-west Wales some of the tripartite sequences of till/sand and gravel/till, previously held to be unequivocal proof of a readvance (e.g. Saunders 1968b; Whittow and Ball 1970) may now be explicable in terms of Boulton's (1972a, 1977) models of glacier wastage, and hence may represent only a single melt-out phase. A similar explanation may be advanced for the tripartite drift sequence and chaotically-assembled glacial deposits in parts of the Cheshire-Shropshire lowland (Bowen 1973c), and possibly also for the deposits hitherto assigned to the 'Scottish Readvance' (e.g. Huddart 1971) in the Cumbrian lowland and adjacent areas of north-west England. Finally, in Ireland there is uncertainty in many areas about the age of the various till sheets, and it is possible that some tills formerly believed to represent readvance deposits may in fact relate to the main advance of Late Devensian (Midlandian) ice and that underlying glacial sediments date not from the last ice sheet, but from an earlier glacial episode.

Conversely, there are other localities where the lithological and morphological evidence are difficult to explain other than by a readvance of the Late Devensian ice sheet. In the Tremadoc area of Cardigan Bay, the presence of local Welsh till *overlying* Irish Sea till suggests a readvance of Welsh ice while to the west of the Lleyn Peninsula a second succession of Irish Sea till has been interpreted as reflecting renewed activity of the Irish Sea glacier (Garrard 1977). In the Lleyn Peninsula and on Anglesey, the regional distribution of two quite distinct till units is difficult to account for by the Boulton hypothesis alone and is strongly suggestive of a readvance, as indeed is the presence of Irish Sea till in the Vale of Clwyd (Fig. 7.10) at a topographic level considerably lower than the previous occupation by Welsh ice (Bowen 1977a). Further east in the Shrewsbury district, Welsh till overlies Irish Sea glaciofluvial deposits, while the massive nature of the Ellesmere-Whitchurch morainic complex (Fig. 2.3) and the marked contrast in morphological expression on either side of the moraine (Boulton and Worsley 1965) suggest a readvance of ice following retreat from the Late Devensian maximum at the 'Wolverhampton Line'.

Further north, investigations of multiple till units in the Cumberland lowland have revealed the presence of a widely distributed till, which appears to post-date that deposited by the initial pulse of the Late Devensian ice sheet. This upper till unit displays properties characteristic of a basal rather than a supraglacial till and again is therefore suggestive of a readvance (Huddart *et al.* 1977). Finally, in Scotland, an ice sheet readvance along the coasts of Wester Ross has been proposed on the basis of an extensively developed end moraine (Robinson and Ballantyne 1979). Although corroborative stratigraphic evidence is not available, a readvance, or at least a standstill of the ice sheet along the west Highland seaboard, has been suggested on the basis of the abrupt inland termination of high level raised beaches (Peacock 1970). A marked fall in the marine limit has also been interpreted as indicating a stillstand or ice-front readvance in the Loch Fyne area (where a terminal moraine is present) and near Stirling (Sissons 1981; Sutherland 1981b). Collectively this array of evidence strongly suggests glacier readvances during wastage from the Late Devensian maximum, although the timing and glaciological significance of these ice-marginal oscillations has yet to be established.

Radiocarbon dates from a large number of sites in northern Britain imply that the last ice sheet had largely disappeared by *ca*. 13 000 BP, and may have wasted away completely shortly thereafter (Sissons and Walker 1974). However, polar waters did not recede from the British coasts until around 13 500 BP or later (Ruddiman and McIntyre 1973, 1981), a date which is confirmed by the terrestrial biostratigraphic evidence for climatic amelioration in northern Britain at the close of the Late Devensian (e.g. Gray and Lowe 1977; Pennington 1977b). It is clear therefore that the greater part of the Late Devensian ice sheet wasted away while the British Isles were surrounded by polar water, and that glaciers had abandoned most or all of lowland Britain before the rise in temperature at or slightly before 13 000 BP (Coope 1975a). The implication must be that the last ice sheet wasted away not because of increased temperatures, but largely as a consequence of decreased snowfall. This in turn can be related to the oceanic and atmospheric Polar Front being located well to the south of the British Isles (Fig. 7.12). In this situation, cloud cover would have been thinner, and thus an increase in direct solar radiation would have assisted glacier decay (Sissons 1981). Confirmatory evidence for predominantly dry but very cold conditions can be found in the ice-wedge pseudomorphs penetrating Late Devensian till that have been reported from a number of localities. Many other periglacial features in, for example, the uplands of Wales, may also date from this phase (Bowen 1977b).

The position of the Polar Front is therefore of paramount importance in determining the amount, rate and overall distribution of glacier development in Britain during cold periods. Centres of maximum ice growth during the Late Devensian probably shifted southwards with the migration of the Polar Front, and it is conceivable that the ice sheet in northern Britain became starved of moisture and was therefore contracting at the same time as increased cyclonic activity in England and Wales was producing glacier expansion. If this was so, then it is equally likely that the northward movement of the Polar Front brought increased snowfall along its path, which would have the effect of reinvigorating those glaciers that had previously been wasting through lack of nourishment. This may be one explanation for readvances of ice in different parts of Britain during the general wastage of the Late Devensian ice sheet, although this hypothesis has yet to be tested. If correct, however, then the readvance (or surge) of the ice in eastern England which buried the Dimlington silts, the reactivated Welsh ice, the evidence from the Loch Fyne area for a readvance at a time when glaciers throughout the rest of Scotland were in retreat, and the Wester Ross Readvance may all reflect localised and short-lived influences of the Polar Front. It is also possible that time-transgressive glacier retreat and readvance associated with the passage of the Polar Front may explain some of the apparent anomalies in radiocarbon-dated stratigraphies, for example between the sequences at Dimlington, Tremerchion and the Isle of Man (see above).

Although much remains to be learned about the growth and decay of the last ice sheet and the chronology of associated events, it is apparent from present evidence that throughout the Late Devensian, dramatic landscape changes were occurring within a relatively short (in geological

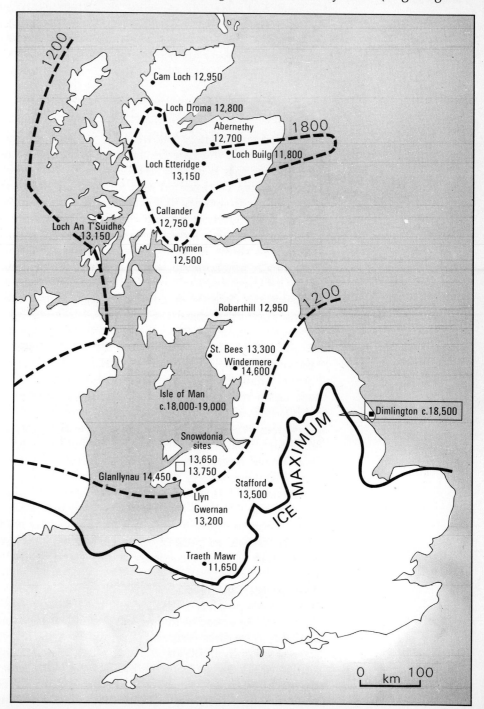

Fig. 7.13: Radiocarbon dates (rounded to nearest 50 years) from the earliest organic sediments that accumulated in lakes following the wastage of the Late Devensian ice sheet (based on various sources). The maximum limit of Late Devensian ice, the 1200 and 1800 m ice-surface contours (after Boulton *et al.* 1977) and the Dimlington site are also shown.

terms) time period. The data shown in Fig. 7.13 relating to ice wastage exemplify this point. If the radiocarbon dates from Dimlington and Tremerchion are valid, then a very large area of the British Isles was covered by glacier ice some time after 18 000 BP. By *ca.* 14 000 BP, however, that ice sheet, which had achieved thicknesses in excess of 1200 m in the southern Irish Sea basin and parts of the North Sea adjacent to eastern England, had almost disappeared. By 13 000 BP, according to the radiocarbon evidence, most of the Scottish Highlands was ice-free. In other words, the last British ice sheet, whose modelled volume at glacier maximum was some 346 000 km^3 (Boulton *et al.* 1977), wasted away almost completely in less than 5000 years!

The Late Devensian Lateglacial

The period immediately following the wastage of the last ice sheet is properly termed the Late Devensian Lateglacial, although for the sake of convenience this is usually abbreviated to 'Lateglacial'. In one sense the Lateglacial can be regarded as a transitional period from the climatic régime of arctic severity that prevailed for much of the Late Devensian to the markedly warmer conditions of the Flandrian ('Postglacial') that followed. There is a wealth of geomorphological, lithological and biological evidence to show, however, that this transition was not straightforward. The climatic amelioration which resulted in the final, rapid disintegration of the ice sheet reached a peak at about 13 000 BP, after which temperatures declined steadily (although perhaps with some slight fluctuations) throughout the Lateglacial Interstadial. Around, or perhaps somewhat before 11 000 BP cold conditions returned to the British Isles, glaciers reformed in many upland areas, and beyond the ice margins a tundra landscape developed once more. This cold phase (the Loch Lomond Stadial) was, however, short-lived, and by 10 000 BP temperatures were again rising, this time in a relatively uninterrupted manner, culminating in the 'climatic optimum' during the Flandrian some 2–3000 years later. It is almost as though the present interglacial experienced a false start, with the Loch Lomond Stadial forming a short-lived but pronounced interruption in an overall trend towards climatic amelioration (Fig. 7.9). Conversely, it is clear that the thermal oscillation between about 13 500 and 11 000 BP represented an interstadial phase, comparable with those of the Chelford and Upton Warren Interstadials, in other words a brief, warm interlude within the predominantly cold Devensian. As such, the Lateglacial can properly be regarded as a part of the Devensian Stage, rather than as the initial phase of the Flandrian Interglacial.

The Lateglacial Interstadial

It is not yet possible to assign a precise date to the beginning of the Lateglacial Interstadial in Britain. The oldest radiocarbon dates that are available from sites formerly occupied by Late Devensian ice are 14 468 ± 300 BP at Glanllynnau (Fig. 7.13) in north Wales (Coope and Brophy 1972) and 14 623 ± 360 BP from the base of the Windermere profile in the Lake District (Coope and Pennington 1977), but at neither of these dated levels does the fossil evidence indicate that climatic amelioration had begun.

Indeed, it seems likely that at around 14 500 BP severe periglacial conditions still prevailed over much of the British Isles. Between 14 000 and 13 000 BP, however, the climate seems to have changed dramatically. Polar waters withdrew progressively from the shores of the British Isles, and by 13 000 BP the oceanic Polar Front lay far to the north-west of Scotland (Fig. 7.12). At around 13 000 years ago, the indications are that summer temperatures over much of the country were little different from those of the present day (see below). The 1000-year period prior to 13 000 BP was therefore one of transition, with climatic amelioration occurring in a time-transgressive manner across the country. There is, however, no consistent pattern in the timing of thermal improvement. For example, a radiocarbon date from the base of an organic sequence at Stafford has given an age of 13 490 ± 375 BP and the climate indicated by the faunal evidence is one of arctic severity (A. V. Morgan 1973). Cold climatic conditions are also indicated by plant macrofossil evidence from an organic lens in river terrace gravels at Colnbrook in Buckinghamshire which was radiocarbon-dated to 13 405 ± 170 BP (Gibbard and Hall 1982). A similar date (13 560 ± 210 BP) has been obtained from a peat deposit at Colney Heath, Hertfordshire, but the coleopteran fauna at that site suggest a climate not much cooler than that of the present day (Pearson 1962; Godwin 1964). Further north in Windermere, however, horizons characterised by thermophilous insect assemblages and the presence of tree birch have yielded dates of 13 938 ± 210 BP and 13 863 ± 270 BP (Coope and Pennington 1977). Some of the apparent discrepancies may well be attributable to errors in the radiocarbon method, particularly the 'hard-water' factor in limnic sediments (Ch. 5), but it also appears likely that the effects of climatic change, *i.e.* its ultimate influence on local sedimentation, were manifest at different times in different places. All that it is possible to conclude at present, therefore, is that climatic amelioration at the beginning of the Lateglacial Interstadial began before 13 000 BP and after 14 000 BP (Coope 1977a).

Details of the levels of climatic improvement are provided principally by evidence of coleopteran assemblages. These show that the thermal maximum had occurred by 13 000 BP and, moreover, that the transition from the arctic climate that had prevailed prior to that date was remarkably rapid. At many sites, the biostratigraphic evidence reveals an often abrupt change from arctic coleopteran assemblages to those characterised by species with a distinctly more southern aspect; indeed, in many profiles, species are found whose present-day range does not reach as far northwards as the coastline of southern Britain (Coope 1975a). At Glanllynnau in north Wales, an attempt to calculate the rate of climatic improvement based on sedimentation rates between radiocarbon-dated horizons produced the remarkable estimate of *1 °C per decade* (Coope and Brophy 1972). Although this figure must be regarded with caution, there can be little doubt that the change from arctic to temperate conditions over much of the British Isles took place very rapidly indeed. The coleopteran evidence suggests that in southern England July temperatures reached 18 °C, while in northern England and south-west Scotland they only reached 14–15 °C. Somewhat colder conditions are envisaged for the more northerly regions of Britain (Coope 1977a). Overall, summer temperatures at least as warm as those of the present day are implied, and these were maintained for around 1000 years.

The decline from the thermal maximum appears to have been less dramatic than the rise, but a marked cooling of perhaps 3–4 °C has been detected at *ca.* 12 000 BP, while at around, or perhaps before (depending upon altitudinal and latitudinal factors) 11 000 BP, a further cooling of at least 4 °C occurred in July temperatures. No data are at present available on winter temperatures but if, as has been suggested (Watson 1977), discontinuous permafrost remained in some areas, then winter temperatures of well below zero are implied for at least part of the period. There is little direct evidence for continued periglacial activity in the Interstadial although the continued inwash of minerogenic sediments into some Scottish lake basins throughout the Lateglacial Interstadial (Walker 1975a; Lowe and Walker 1977) suggests that in some upland areas, climatic conditions were sufficiently severe to prevent a complete vegetation cover from becoming established.

In view of the large number of sites that have now been studied palaeobotanically, more is known about the vegetation cover of the British Isles during the Lateglacial than about any period other than the Flandrian. The data collectively suggest strong regional differentiation in Lateglacial vegetation patterns (Pennington 1977b) due to variations in plant response to climatic change at different latitudes and altitudes, and also to such microscale factors as edaphic conditions, soil development, soil stability, aspect, exposure and so forth. The vegetation of the early part of the Interstadial in most areas was dominated by grasses, sedges and herbaceous plants of low competitive ability that thrive on open ground and disturbed soils. Characteristic taxa were the docks (*Rumex*), the mugworts or wormwoods (*Artemisia*), species of the pinks (Caryophyllaceae) and daisy (Compositae) families and the plantains (*Plantago*) (Fig. 7.14). These pollen assemblages are often found in

Fig. 7.14: Chronostratigraphy, lithostratigraphy and pollen stratigraphy of Lateglacial sediments at Llyn Gwernan, North Wales (for location see Fig. 7.13). Lithological units 1 and 3 are minerogenic, and 2 and 4 are organic lake muds (after Lowe 1981).

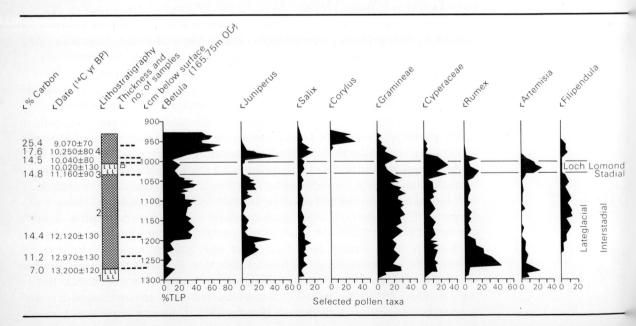

sediments that are minerogenic in nature and with a chemical composition that suggests the inwash into lake basins of unweathered skeletal soils (e.g. Pennington 1970). However, as they also appear to have accumulated at a time when thermophilous beetles were present, it is clear that the deposits can no longer be regarded as indicating cold conditions, which was the initial conclusion based on pollen assemblages. Rather they reflect simply an early stage in plant succession. This phase of open-habitat conditions was succeeded, shortly after 13 000 BP, by a shrub vegetation typically of juniper (*Juniperus communis*) with willows (*Salix*) and, in the more oceanic western regions, the crowberry (*Empetrum*). Birch woodland subsequently developed in many central and southern areas (e.g Fig. 7.14), the dominant tree being the downy birch (*Betula pubescens*) though in some localities the more warmth-loving silver birch (*Betula pendula*) was present (Walker 1966), while Scots pine (*Pinus sylvestris*) was to be found in the woods of south-east England (Godwin 1975). The extent of birch woodland is difficult to establish, although on the basis of macrofossil evidence it is clear that tree birch was present as far north as the Island of Skye and Aberdeenshire during the Lateglacial Interstadial (Vasari and Vasari 1968; H. J. B. Birks 1973b). Birch copses were to be found in valley sides in the south-west of England (Brown 1977), in upland Wales (Moore 1977) and in parts of central and southern Scotland (Gray and Lowe 1977), and a fairly extensive birch canopy appears to have formed in the Lake District lowlands (Pennington 1977b).

By contrast, in the lowland areas of north-east England and eastern Scotland, birch woods appear to have been less well developed (e.g. Newey 1970; Turner and Kershaw 1973), perhaps reflecting exposure to cold easterly winds blowing across the North Sea plain. Severe wind exposure has also been suggested, along with other climatic factors, including low annual temperatures and a short growing season, to explain the rather puzzling absence of birch woodland from Ireland during the Lateglacial Interstadial (Watts 1977). Although macrofossil remains of *Betula pubescens* have been found at a number of Irish sites, birch woodland never developed in the Lateglacial and the landscape was one of open grassland, with locally dense juniper scrub and *Empetrum* heathlands particularly extensive in western areas. An *Empetrum*-dominated heath vegetation was also characteristic of central and northern areas of Scotland during the Lateglacial Interstadial (Pennington *et al*. 1972).

At several sites in Britain, there is evidence for two separate phases of woody plant dominance during the Lateglacial Interstadial, and these apparent vegetational changes are reflected in the sediment stratigraphy by a second phase of minerogenic inwash into lake basins (e.g. Walker and Godwin 1954; Bartley 1962; Walker 1977, 1982a; Caseldine 1980). Renewed minerogenic inwash has also been detected at a number of Irish sites (Watts 1963, 1977). It has been suggested that the biostratigraphic evidence reflects a short period of more severe climatic conditions during the Lateglacial Interstadial, perhaps equivalent to the climatic oscillation of the Bölling/Older Dryas/Alleröd sequence reported from continental north-west Europe (Pennington 1975). At several localities in south-east England, palaeosols found interbedded with soliflucted chalk and loess deposits have also been correlated with the Bölling and Alleröd episodes, partly on the basis of their stratigraphic position, and partly on the evidence of their associated molluscan faunas (Kerney 1963, 1965). There are, however, numerous British sites where no such oscillation has been detected in the pollen record or in other stratigraphic sequences and, moreover, there is only very slight evidence for an oscillation in the coleopteran records (Coope 1975a). Any climatic fluctuation within the Lateglacial Interstadial must therefore have been of relatively short duration and of very low amplitude, and perhaps may only have been local in effect.

The most striking aspect of the vertebrate faunas of the Lateglacial period is the absence of many of the characteristic larger mammals of the Middle and pre-Interstadial Late Devensian, such as mammoth, woolly rhinoceros, bison, musk ox, lion and spotted hyaena (Stuart 1977). This may be seen in the context of the general extinction of many of the large mammals of the northern hemisphere at the close of the last glacial due, it has been suggested, to climatic change and consequent loss of habitat, or to progressive extermination as a result of the improved hunting technology of Palaeolithic man, or to both (Martin and Wright 1967). Certainly sites have been found in Britain where there is evidence of hunting activity, including the elk skeleton and associated barbed points in deposits of Lateglacial Interstadial age at High Furlong in Lancashire (Hallam *et al.* 1973; and photographs in Stuart 1982). Elk appears to have been associated with the spread of birch woodland and has been found at sites in England and Scotland. Other vertebrate finds of Interstadial age on the British mainland are of typical open-habitat animals, including reindeer, horse, arctic fox and also perhaps red deer and brown bear. In Ireland the only significant faunal discoveries have been of the giant deer (*Megaceros giganteus*)[8] and reindeer (Stuart 1982). Significantly, in view of the postulated absence of birch woods from Ireland, elk remains have not yet been recorded from across the Irish Sea (Stuart 1974).

Throughout the Lateglacial, northern Britain was recovering isostatically from the removal of the weight of the Devensian ice sheet. Although precise figures cannot be produced for the amount of glacio-isostatic depression, in Scotland it was probably of the order of 250–300 m (Gray and Lowe 1977). World sea-level at around 14 000 BP was at least 60 m (Fairbridge 1961) and perhaps 90 m (Jelgersma 1979) below that of the present day, and thus although some initial rebound would have occurred by that time, the sea was able to flood in and form shorelines to the heads of the firths and sea lochs (Fig. 7.15). Subsequently these have been raised above sea-level in many areas. In the Paisley area of the Clyde Valley, for example, marine clays occur up to 25 m above present sea-level and the contained shells have yielded radiocarbon dates up to *ca.* 13 000 BP (Bishop and Dickson 1970). Around the coasts of Argyll and west Inverness-shire, raised marine features of Lateglacial age are found up to 40 m above present sea-level, while on the coasts around the Cromarty Firth similar features also occur (Sissons 1976). The most detailed shoreline sequence has, however, been found in the Forth and Tay areas of south-east Scotland (Figs 7.15 and 7.16). The oldest and most steeply

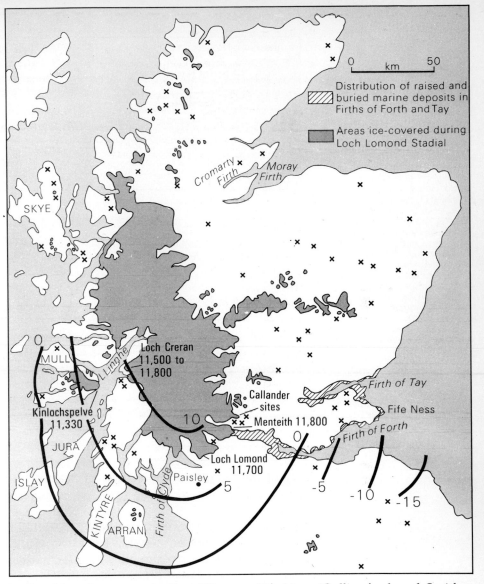

Fig. 7.15: Aspects of landscape evolution in Scotland during the Lateglacial, and the location of sites referred to in the text (based on various sources). A cross indicates a site from which a full Lateglacial pollen stratigraphy has been obtained. Isobases for the Main Lateglacial Shoreline are shown in metres. For further explanation see text.

inclined beaches occur in East Fife near Fife Ness (Cullingford and Smith 1966). These terminate westwards in outwash and appear to have formed very soon after ice wastage began. As such they are strictly speaking older than the lower boundary of the Lateglacial (discussed above). Indeed, on the basis of the gradients and ages of three later shorelines in south-east Scotland, Andrews and Dugdale (1970) have calculated the ages of the East Fife shorelines as between 18 250 and 15 100 BP, although on the basis of more recent data these become 17 600 and 14 750 BP (Sissons 1976).

The most prominent raised marine feature is termed the Main Perth Shoreline and in the Forth Valley it can be traced eastwards for some 70 km from a height of around 40 m near Stirling (Sissons and Smith 1965; Smith *et al.* 1969). No date is available for the age of the shoreline, but it may have formed around 13 500 BP (Cullingford 1977). Following the

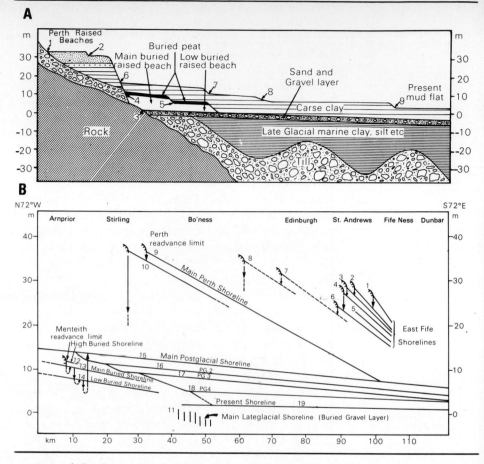

Fig. 7.16: The raised shorelines of the Firth of Forth and adjacent areas of Scotland. A: Schematic diagram of the sequence of deposits in the area around Grangemouth between Edinburgh and Stirling. (1) and (2) are Late Devensian raised beaches; (3) the Main Lateglacial Shoreline (Buried Gravel Layer); (4) and (5) are buried raised beaches of early Flandrian age; (6) to (9) are shorelines related to the early Flandrian sea-level rise (6) and subsequent fall through the mid-Flandrian (after Sissons 1976). B: Shoreline sequence in south-east Scotland related to an axis from Stirling to Dunbar. The numbers indicate relative order of age from oldest (1) to the present shoreline (19). A number of the older shorelines terminate westward in outwash sands and gravels (after Sissons *et al.* 1966; Sissons 1976).

cutting of the Main Perth Shoreline, *relative* sea-level fell rapidly as the land rose isostatically following the final wastage of the Late Devensian ice sheet.

The only other area of Britain where Lateglacial marine features have been found above present day sea-level is in Northern Ireland. On Malin Head in Co. Donegal, for example, Lateglacial marine deposits are present up to 20 m above present sea-level (Mitchell 1977), while in many parts of north-east Ireland, raised shorelines and related deposits of possible Lateglacial Interstadial age occur between 10–20 m OD (Synge 1977). Elsewhere, the evidence lies beneath the sea and only in isolated localities are there indications of Lateglacial Interstadial sea-levels. One such area is Cardigan Bay where estuarine and brackish water silts have been found overlying till of Welsh and Irish Sea origin down to –61 m OD. Microfaunal and microfloral analyses indicate mild climatic conditions, with summer sea temperatures perhaps as high as 12 °C and a Lateglacial Interstadial age has been inferred (Haynes *et al.* 1977). As the southern part of Cardigan Bay lies near the outer limit of the last ice sheet, it is possible that isostatic recovery there may have been minimal and therefore that the figure of –60 m OD represents a realistic estimate for sea-level around that part of the British coasts at some time during the Interstadial. Sea-levels of between –60 and –90 m would mean that Ireland became separated from mainland Britain very soon after the wastage of the Late Devensian ice sheet, although England was connected to the continent of Europe by land bridges across the eastern English Channel and the southern half of the North Sea.

The Loch Lomond Stadial

Towards the end of the Lateglacial Interstadial polar waters began to move southwards once more (Fig. 7.12) and by *ca.* 10 200 BP the oceanic Polar Front had reached its maximum position off the coast of south-west Ireland (Ruddiman *et al.* 1977).[9] With generally falling temperatures, and in response to increased snowfall as the atmospheric Polar Front moved southwards across the country, glaciers began to reform in many upland regions of the British Isles. This phase of renewed glacier activity is generally known as the Loch Lomond Readvance and was the last occasion on which glacier ice existed in the British Isles. The greatest area of ice accumulation was in the Highlands of Scotland (Fig. 7.15). Over Rannoch Moor, one of the major ice-dispersal centres, an ice shed with a surface altitude of 850–900 m developed and the thickness of ice locally exceeded 400 m. In parts of the southern Great Glen ice accumulated to depths of 600 m (Sissons 1979a). Smaller ice caps existed in the Grampian Highlands and on some of the islands of the Inner Hebrides (e.g. Mull and Skye), while numerous cirque and valley glaciers developed in other parts of the Scottish Highlands, the Southern Uplands, and the Lake District (Sissons 1976, 1980; Cornish 1981). In Ireland, small glaciers appeared in the mountain cirques of Kerry, Wicklow, Mayo and Donegal (Synge 1979), while in Wales cirque glaciers existed in the Brecon Beacons (Fig. 7.17), around Cader Idris and, particularly, in Snowdonia (Lewis 1970; Unwin 1975; Gray 1982).

In most of these areas the small size of the glaciers suggests that they formed anew at the beginning of the Loch Lomond Stadial, but in Scotland the very considerable ice volumes that appear to have existed in parts of the western and central Highlands have been taken by some authorities to indicate that ice had remained in many localities throughout the Lateglacial Interstadial (e.g. Peacock 1970; Sugden 1970). Others however have argued for complete deglaciation of Scotland prior to the Loch Lomond Stadial (e.g. Sissons 1972; Sissons and Grant 1972), partly on the nature and distribution of the morphological evidence (see below), and partly on the distribution of radiocarbon-dated pollen sites which, if the radiocarbon dates are correct, indicate that most of Scotland (including areas close to the principal ice-dispersal centres) was ice-free early in the Lateglacial Interstadial. What little ice remained, it is suggested, would be unlikely to have survived until the Loch Lomond Stadial. If this view is sustained it would mean that, strictly speaking, this renewed period of glacier growth should be referred to as the Loch Lomond Advance (Sissons 1977a).

The evidence for this limited glaciation is convincing. Indeed, it was the clarity of the Scottish evidence in particular which was instrumental in persuading British workers in the 1840s and 1850s to accept the views of Agassiz and the early 'glacialists'. The ice limits are marked in many cases by clear end moraines, over 100 of which have been mapped in Scotland

Fig. 7.17: Terminal moraine of Loch Lomond Stadial age in the Brecon Beacons, South Wales.

alone (Sissons 1976), but which most frequently delimit the extent of the smaller cirque glaciers that existed in the hills of Ireland and Wales. In the Lake District and particularly in the Scottish Highlands the former extent of valley glaciers is indicated by the often abrupt downvalley or valley-side termination of a chaotic, highly irregular morainic topography (hummocky moraine – Fig. 2.14), while the former thickness of valley ice can be reconstructed from drift or boulder limits on the hillsides or, where bedrock is outcropping, by the 'trimlines' separating frost-shattered rock above from ice-scoured and ice-moulded bedrock below (Thorp 1981). Radiocarbon dates are available from four sites in Scotland, at Loch Lomond, Menteith, Kinlochspelve and Loch Creran (Fig. 7.15), on marine shells that have been transported or overridden by Loch Lomond Stadial glaciers and these range in age from *ca*. 11 800 to 11 300 years BP in age (Sissons 1967b; Peacock 1971; Gray and Brooks 1972). Similar dates (11 600 ± 200 and 11 500 ± 550 BP) have been obtained from deformed lake clays caught up in a moraine at Lough Nahanagan in the Wicklow Mountains of south-east Ireland (Watts 1977).[10]

Beyond the ice margins a tundra landscape developed once more, with discontinuous permafrost extending from the English Channel to the Scottish border area and possibly continuous permafrost existing further north (Watson 1977). Periglacial activity was widespread and is reflected in present-day landscapes throughout the British Isles. In upland areas many of the extensive summit blockfields, solifluction spreads and lobes (Fig. 2.20), and fossil screes date from this period (Sissons 1973), as do numerous protalus ramparts that developed in association with perennial snow patches (see e.g. Ellis-Gruffydd 1977; Gray 1982) and the remarkable fossil rock glaciers that have been found in Scotland (e.g. Dawson 1977). Widespread ground ice activity is indicated by the occurrence of pingo remains in south-east Ireland (Mitchell 1973), Wales (Watson and Watson 1974) and East Anglia (Sparks *et al.* 1972) and by the distribution of fossil ice and sand wedge pseudomorphs that are found throughout the British Isles (Watson 1977). In southern England, a combination of intermittently frozen substrates, frost shattering and a surface layer soaked by seasonal snow melt resulted in widespread mass movement in chalkland areas (Kerney 1963; Kerney *et al.* 1964), while soil instability and an incomplete vegetation cover are indicated in almost all parts of Britain by the presence of minerogenic sediments (reflecting destroyed Interstadial soils) in many present and former lakes (e.g. Pennington 1970). Many rivers carried increased bed-loads and were characterised by marked channel instability and braiding at that time (Rose *et al.* 1980). Finally, radiocarbon dates of around 11 000 BP beneath coversands in Yorkshire, Lincolnshire and Lancashire indicate that aeolian activity was also an important feature of the periglacial environment of Britain during the Loch Lomond Stadial (Catt 1977).

Biological evidence supports this general environmental reconstruction (Fig. 7.18). Pollen spectra from a large number of sites ranging from Cornwall to northern Scotland (Pennington 1977b) indicate a landscape dominated by grasses and sedges, with representatives of such herbaceous families as Caryophyllaceae, Chenopodiaceae, Cruciferae and Composite. These are typically 'weeds' and many species are characteristic of bare or disturbed soils. Particularly common in the north and west were *Rumex* and *Artemisia* (e.g. Figs 7.14 and 7.18). The latter genus is of particular interest as many species of *Artemisia* (although not all) are associated with dryland habitats and low snow cover (Iversen 1954) and are typical plants of the frost-heaved soils of the periglacial environment. A herbaceous assemblage dominated by *Artemisia* has been found at many Irish sites (Watts 1977), and in areas of central and eastern Scotland (e.g. Walker 1975a; Birks and Mathewes 1978), where low precipitation during the Stadial is indicated (see below). Other characteristic elements in the tundra landscape of northern and western areas were the clubmosses (*Lycopodium* and *Selaginella*) and vegetation associated with perennial snowbeds (e.g. the least willow, *Salix herbacea*).

Very few Stadial sediments have been found to contain macrofossil remains of trees, although pollen analyses from parts of southern and eastern England suggest the limited local presence of birch and pine (Godwin 1975). Elsewhere the birch woodlands of the Lateglacial Interstadial seem to have disappeared from Britain. An open landscape with periglacial conditions is indicated by numerous molluscan assemblages from the chalklands of southern England (Kerney *et al.* 1964; Kerney 1977a, 1977b), while a tundra environment is also implied by the vertebrate remains from sites in Derbyshire and Essex. These include ptarmigan (*Lagopus mutus*), arctic lemming, northern vole, (*Microtus gregalis*), reindeer and horse. A significant absentee from the Stadial faunas is the giant deer, which is not know from deposits younger than the Lateglacial Interstadial.

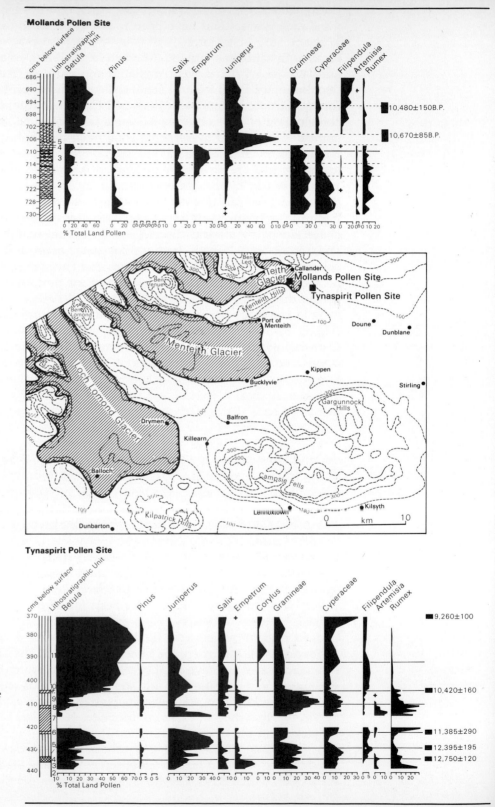

Fig. 7.18: Radiocarbon-dated pollen stratigraphies from infilled kettle-holes close to the Loch Lomond Stadial terminal moraine at Callander, Perthshire, Scotland (based on Lowe 1978). Tynaspirit is situated just outside the arcuate moraine and contains a full sequence of Lateglacial deposits, whereas Mollands, located immediately within the moraine, contains Flandrian sediments only.

Presumably it could not adapt to the more rigorous climatic and vegetational conditions of the Loch Lomond Stadial (Stuart 1977).

Severe frost action along with marine erosion was responsible for the development of an extensive shoreline in northern Britain during the Loch Lomond Stadial. In western Scotland, between the Firth of Clyde and Loch Linnhe (Fig. 7.15) a pronounced rock platform often backed by a cliff occurs above sea-level along much of the present coastline (Fig. 7.19). Known as the Main Lateglacial Shoreline (Sissons 1974b), it attains a maximum altitude of around +11 m OD in the Oban area but, because of isostatic recovery, slopes southwards and westwards eventually passing below present sea-level in southern Kintyre and north-east Islay (Gray 1978; Dawson 1980). In the Forth Valley this shoreline (originally referred to as the Buried Gravel Layer) lies beneath more recent sediments (Fig. 7.16) but can be traced over 80 km to just north of Berwick where it lies at −18 m OD (Sissons 1976). Isobases for elevated platforms in the west and for the Buried Gravel Layer in the east are shown in Fig. 7.15.

Elsewhere around the British coasts evidence relating to sea-level during the Loch Lomond Stadial lies submerged beyond the present coastline, and little can be assigned with certainty to that time period. However, in view of the estimates from other parts of the world which typically place eustatic sea-level towards the end of the Lateglacial at *ca.* −40 to −50 m (e.g. Fairbridge 1961; Shepard and Curray 1967), and assuming that the combination of frost weathering and marine action which clearly proved so effective in Scotland would have operated in a similar fashion in other

Fig. 7.19: A raised rock platform which is part of the Main Lateglacial Shoreline at Grass Point, south-east Mull, Scotland.

areas, it is possible that the pronounced erosional shoreline detected at −40 to −45 m off the coasts of south-west England (Kidson 1977), and the submarine cliff between −33 and −44 m OD in Cardigan Bay (Dobson *et al.* 1971) may be of Loch Lomond Stadial age.

Climatic data for the Loch Lomond Stadial have been derived from a combination of coleopteran and geomorphological evidence. Coleopteran assemblages from many parts of Britain suggest mean July temperatures of around 10 °C in southern England, 9 °C in central England and north Wales, and 8 °C in southern Scotland and the Lake District (Coope 1977a). These figures are in very close agreement with those calculated on the basis of inferred firn-line altitudes for Loch Lomond Stadial glaciers (e.g. Fig. 7.20). This method (outlined in Ch. 2) has produced mean July figures of 6 °C for the south-east Grampians (Sissons 1979b), 7 °C for the south-west Grampians (Sissons and Sutherland 1976), and 8 °C for the Lake District (Sissons 1980). The widespread distribution of permafrost indicated by the periglacial evidence suggests however that mean annual temperatures were rarely above freezing point and may have been as low as, or even lower than, −6° to −8 °C. Hence mean January temperatures of −17° to −20 °C are suggested for much of the British Isles, which represents an annual temperature range of over twice that of the present day.

Generally low levels of precipitation are implied for much of the British Isles during the Stadial with glacier development resulting from a band of snowfall associated with increased cyclonic activity along the Polar Front which affected, in particular, northern and western parts of the country as it moved southwards. Glacier distributions, dimensions and altitudes (Figs 7.15 and 7.20) suggest that the principal snow-bearing winds were south-easterly or south-westerly (Sissons 1979a), and that marked precipitation gradients developed in highland Britain between the maritime margin and inland areas. In Scotland, for example, precipitation levels in the south-west Grampians may have been comparable to those of today (*ca.* 4000 mm per year), whereas in the Cairngorms values had fallen to 500–600 mm. In the adjacent Spey Valley, annual precipitation may have been as low as 200–300 mm (Sissons 1979b). These figures, based on calculations from inferred glacier firn-line altitudes, are clearly highly generalised and provide only a rough estimate of former precipitation levels. Nevertheless, there is a clear inverse relationship between areas of low precipitation indicated by these calculations and the distribution of pollen sites in which the Stadial spectra contain high frequencies of *Artemisia*. In the Spey Valley, for example, 'a rather arid climate' is indicated by the plant macrofossil and pollen analytical evidence (Birks and Mathewes 1978). It is likely that precipitation values as low as, or even lower than, those for the Spey Valley region were the norm for lowland areas of central and southern Britain for much of the Loch Lomond Stadial.

The timing of the onset and ending of the Loch Lomond Stadial is difficult to establish. The conventional time-span is from 10 800 to 10 300 BP, dates which were very much influenced by early radiocarbon age determinations on the boundaries of pollen zone III of the Godwin/Jessen scheme. These dates must now be regarded as only very general

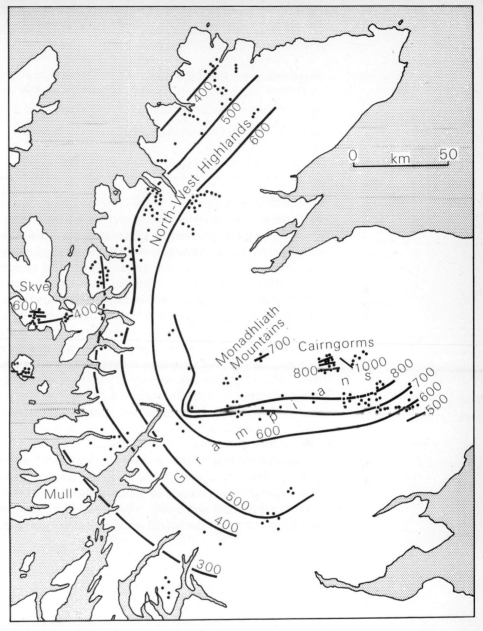

Fig. 7.20: Regional firn-line altitudes (in metres) in the Highlands and Inner Hebrides of Scotland (after Sissons 1980). Dots show those localities for which firn-line altitudes have been calculated.

approximations, however, in view of the markedly time-transgressive nature of climatic change in Britain at the beginning and end of the Loch Lomond Stadial (Gray and Lowe 1977). In some northern areas the onset of severe climatic conditions and the build-up of glacier ice may have begun well before 11 000 BP, while in South Wales and south-east Ireland radiocarbon evidence suggests that the full effects of the periglacial climatic régime of the Stadial may not have been felt until around 10 600 BP (Craig 1978; Walker 1980). Similarly, it is difficult to be precise about the dating of the transition to a more mild climate at the end of the Loch Lomond Stadial. The very considerable number of radiocarbon dates that are now available on this event place the climatic

Site	Date (yrs BP)
Mollands, near Callander, Perthshire	10 670 ± 85 10 480 ± 150
Kingshouse 2, Rannoch Moor, Argyll	10 520 ± 330 10 200 ± 180
Rannoch Station 1, Rannoch Moor, Argyll	10 660 ± 240
Rannoch Station 2, Rannoch Moor, Argyll	10 390 ± 200 10 160 ± 200
Torness, Island of Mull, Argyll	10 170 ± 160

Table 7.6: Basal radiocarbon dates older than 10 000 BP from sites within the Loch Lomond Stadial ice limits in Scotland (after Lowe and Walker 1976, 1977; Walker and Lowe 1979, 1982).

amelioration between *ca.* 10 000 and 10 300 BP, but again there is no spatial consistency in the dates and therefore a firm dating on the close of the Stadial cannot at present be given (see e.g. Lowe and Walker 1980). What is becoming clear, however, is that ice wastage from the Loch Lomond maximum began before the apparent onset of climatic amelioration, the implication being that, as with the Late Devensian ice sheet, glacier retreat was initiated not by thermal improvement but by lack of precipitation. This hypothesis is supported by the occurrence of ice wedges in sediments within the Loch Lomond Readvance limits at a number of sites in Scotland (Sissons 1974a), and by the fact that in the Lake District several glaciers had withdrawn from their end moraines before the end of the *Artemisia* pollen assemblage zone (Pennington 1978). The close of that zone has been radiocarbon-dated to 10 650 ± 170 BP (Pennington 1975). Similar dates have been obtained from basal organic sediments within the Loch Lomond Readvance limits at sites in Scotland, of particular significance being those from Rannoch Moor (Table 7.6), the principal accumulation and dispersal centre in western Scotland (Lowe and Walker 1976; Walker and Lowe 1979). If these dates are correct, they suggest a glacial maximum perhaps around 10 800 BP and glacier wastage on a considerable scale some 400–500 years before the conventional date of 10 000 BP for general climatic amelioration at the start of the Flandrian.

CONCLUSIONS

It is clear from the foregoing that the pieces of the stratigraphic jigsaw puzzle, reflecting environmental changes in Britain over the past one thousand centuries, are beginning to fall into place. However, as was emphasised at the outset, much of the evidence for environmental conditions in Britain during the Devensian can be regarded as equivocal, and it would be quite wrong to create the impression that the palaeogeographic picture outlined above is intended as the definitive account of environments in the British Isles during the last cold stage. What is becoming increasingly apparent, however, is that a proper understanding of environmental changes in Devensian Britain can only be achieved by adopting a broad perspective and by drawing upon evidence from areas outside the British Isles. A recurrent theme throughout this chapter, for example, has been the influence of oceanographic changes in the north-east Atlantic on the British climate and environment. The type of water mass around the coast of the British Isles and adjacent areas of north-west Europe, and in particular the changing position of the Polar Front, appear to have been dominant factors in determining the duration and intensity of sucessive warm and cold phases, and also the timing of glacier build-up and decay. Similarly, the contrasting vegetation cover of the Devensian interstadials may be largely explicable in terms of the position of the European treeline and the locations of arboreal refugia during successive cold stages. Relative sea-level changes and the development of land bridges between Britain and the continent of Europe are further examples of the factors that need to be considered in explaining floral and faunal changes during the Devensian.

Because of the interweaving of environmental components, the detailed study of any one element requires an awareness of the influence of others, and hence solutions to the types of problems considered in this chapter can only be achieved by interdisciplinary and multidisciplinary approaches to palaeoenvironmental reconstruction. This applies, of course, not only to the Devensian, but to the whole of the Quaternary record. Whereas, in the past, Quaternary research has often tended to be fragmented into individual fields of study, it is from investigations arising out of the close co-operation between scientists from a number of different disciplines, perhaps at present best exemplified by the CLIMAP team, that the significant advances of the future are likely to emerge.

NOTES

1. In Ireland, the Lateglacial Interstadial is now referred to as the Woodgrange Interstadial (Mitchell 1976).
2. Grüger (1979) has suggested that Isotope Stages 5a and 5c may be the correlatives of the Odderade and Brørup Interstadials respectively of the continental European sequence.
3. More recent investigations of the Chelford peats suggest that the spruce macrofossils should be referred not to *Picea abies* but to a form of *P. obovata* (Whitehead 1977).
4. Despite the length of the Early Devensian pollen record at Wing, there is no evidence of an interstadial comparable with that at Chelford (Hall 1980).
5. Recent dating of the Teindland deposit has in fact extended the ^{14}C age of the organic layer to *ca*. 40 000 BP (Caseldine and Edwards 1982).
6. The sands and gravels have subsequently been reinterpreted as 'subglacial tunnel esker sediments'. If this is correct, the sediments are clearly not ice-marginal and cannot be used as evidence for an ice limit (Francis 1980).
7. A date of 13 700 $^{+1300}_{-1700}$ BP was obtained from a mommoth tusk underlying glacial till at Kilmaurs, Ayrshire (Fig. 7.1), thus suggesting glaciation after that time (Sissons 1967b). However, a subsequent date of >40 000 (Shotton *et al.* 1970) implies that the faunal remains are

earlier than Late Devensian and therefore the overlying till could well have been deposited by the last ice sheet.

8. Browsing on young shoots by giant deer has been suggested (Watts 1977) as a further factor inhibiting the spread of birch woods in Ireland (cf. the Upton Warren Interstadial in the English Midlands).

9. The dating of this event rests on a single radiocarbon assay of 10 220 ± 280 BP from the interval between 63 and 77 cm in ocean core V23–82 (Sancetta *et al.* 1973). In view of the large standard error and the thickness of sediment used for dating purposes, the age of 10 200 should be regarded as only approximate.

10. In Ireland, the equivalent of the Loch Lomond Stadial is generally referred to as the Lough Nahanagan Stadial.

References

Aaby, B. (1976a) Cyclic climatic variations over the past 5 500 years reflected in raised bogs, *Nature*, **263**, 281–4.

Aaby, B. (1976b) Cyclic palaeoclimatic variations and the future, *Newsl. Stratigr.*, **5**, 66–9.

Aarseth, I. and Mangerud, J. (1974) Younger Dryas end moraines between Hardangerfjorden and Sognefjorden, western Norway, *Boreas*, **3**, 3–22.

Abell, P. I. (1982) Palaeoclimates at Lake Turkana, Kenya, from oxygen isotope ratios of gastropod shells, *Nature*, **297**, 321–3.

Abelson, P. H. (1954) Organic constituents of fossils, *Carnegie Inst. Wash. Yearbook*, **53**, 97–101.

Absolon, A. (1973) Ostracoden aus einigen Profilen spät- und Postglazialer Karbonatablagerungen in Mitteleuropa, *Mitt. Bayer. Staatssammlung Palaont. Hist. Geol.*, **13**, 47–94.

Ahlmann, H. W. son (1948) Glaciological research on the North Atlantic coasts, *Roy. Geogr. Soc. Res. Series*, **1**, (83 pp.).

Aitken, M. J. (1974) *Physics and Archaeology*, Oxford Univ. Press, Oxford.

Allen, T. (1975) *Particle Size Measurement*, Chapman and Hall, London.

American Commission on Stratigraphic Nomenclature (1961) Code of stratigraphic nomenclature, *Amer. Ass. Petrol. Geol. Bull.*, **45**, 645–65.

Andersen, B. G. (1979) The deglaciation of Norway, 15 000–10 000 BP, *Boreas*, **8**, 79–87.

Andersen, B. G. (1980) The deglaciation of Norway after 10 000 BP, *Boreas*, **9**, 211–16.

Andersen, J. L. and Sollid, J. L. (1971) Glacial chronology and glacial geomorphology in the marginal zones of the glaciers Midtolalsbreen and Nigardsbreen, south Norway, *Norsk geogr. Tidsskr.*, **25**, 1–38.

Andersen, L. W. and Andersen, D. S. (1981) Weathering rinds on quartzite clasts as a relative-age indicator and the glacial chronology of Mount Timpanogos, Wasatch Range, Utah, *Arct. Alp. Res.*, **13**, 25–31.

Andrew, R. and West, R. G. (1977) Appendix. Pollen analysis from Four Ashes, Worcs., *Phil. Trans. R. Soc. Lond.*, B 280, 242–6.

Andrews, J. T. (1968) Postglacial rebound in Arctic Canada; similarity and prediction of uplift curves, *Can. J. Earth Sci.*, **5**, 39–47.

Andrews, J. T. (1969) The shoreline relation diagram; physical basis and use in predicting age of relative sea levels: evidence from Arctic Canada, *Arct. Alp. Res.*, **1**, 67–78.

Andrews, J. T. (1970) A geomorphological study of Post-glacial uplift with particular reference to Arctic Canada, *Inst. Brit. Geogr. Spec. Publ.*, **2**.

Andrews, J. T. (1971a) Techniques of till fabric analysis, *Brit. Geomorph. Res. Gp. Tech. Bull.*, **6**.

Andrews, J. T. (1971b) Methods in the analysis of till fabrics, In R. P. Goldthwait (ed.), *Till/A Symposium*, Ohio State Univ. Press, 321–7.

Andrews, J. T. (1973) The Wisconsin Laurentide ice sheet: dispersal centres, problems of rates of retreat, and climatic complications, *Arct. Alp. Res.*, **5**, 185–99.

Andrews, J. T. (1975) *Glacial Systems*, Duxbury Press, Massachusetts.

Andrews, J. T. (1979) The present ice age: Cenozoic, In B. S. John (ed.), *The Winters of the World*, David and Charles, London and North Pomfret (Vt), 173–218.

Andrews, J. T. (1982) On the reconstruction of Pleistocene ice sheets: a review, *Quat. Sci. Rev.*, **1**, 1–30.

Andrews, J. T., Bowen, D. Q. and Kidson, C. (1979) Amino acid ratios and the correlation of raised beach deposits in south-west England and Wales, *Nature*, **281**, 556–8.

Andrews, J. T. and Dugdale, R. E. (1970) Age prediction of glacio-isostatic strandlines based on their gradients, *Bull. Geol. Soc. Amer.*, **81**, 3769–71.

Andrews, J. T. and Ives, J. D. (1978) 'Cockburn' nomenclature and the late Quaternary history of the eastern Canadian Arctic, *Arct. Alp. Res.*, **10**, 617–33.

Andrews, J. T. and Mahaffey, M. A. W. (1976) Growth rate of the Laurentide ice sheet and sea level lowering, *Quat. Res.*, **6**, 167–84.

Andrews, J. T. and Miller, G. H. (1979) Glacial erosion and ice sheet divides, north-eastern Laurentide Ice Sheet, on the basis of the distribution of limestone erratics, *Geology*, **7**, 592–6.

Andrews, J. T. and Miller, G. H. (1980) Dating Quaternary deposits more than 10 000 years old, In R. A. Cullingford, D. A. Davidson and J. Lewin (eds.), *Timescales in Geomorphology*, John Wiley, Chichester and New York, 263–87.

Andrews, J. T. and Shimizu, K. (1966) Three-dimensional vector technique for analysis of till fabrics: discussion and FORTRAN programme, *Geogr. Bull.*, **8**, 151–65.

Andrews, J. T. and Sim, V. W. (1964) The carbonate content of drift of the Foxe Basin, NWT, *Geogr. Bull.*, **21**, 44–53.

Andrews, J. T. and Smith, D. I. (1970) Till fabric analysis: methodology and local and regional variability (with particular reference to the north Yorkshire cliffs), *Q.J. Geol. Soc. Lond.*, **125**, 503–42.

Andrews, J. T. and Smithson, B. N. (1966) Till fabrics of cross-valley moraines of Baffin Island, *Bull. Geol. Soc. Amer.*, **77**, 271–90.

Andrews, J. T. and Webber, P. J. (1964) A lichenometrical study of the northwestern margin of the Barnes Ice Cap: a geomorphological technique, *Geogr. Bull.*, **22**, 80–104.

Andrews, K. W., Dyson, D. J. and Keown, S. R. (1971) Interpretation of Electron Diffraction Patterns (2nd. ed.), Hilger and Watts, London.

Antevs, E. (1953) Geochronology of the deglacial and neothermal ages, *J. Geol.*, **61**, 195–230.

Anthony, R. S. (1977) Iron-rich rhythmically laminated sediments in Lake of the Clouds, north-eastern Minnesota, *Limnol. Oceanogr.*, **22**, 45–54.

Ashley, G. M. (1979) Sedimentology of a tidal lake, Pitt Lake, British Columbia, Canada, In Ch. Schlüchter (ed.), *Moraines and Varves*, Balkema, Rotterdam, 327–45.

Atkinson, T. C., Harmon, R. S., Smart, P. L. and Waltham, A. C. (1978) Palaeoclimatic and geomorphic implications of $^{230}Th/^{234}U$ dates on speleothems from Britain, *Nature*, **272**, 24–8.

Avery, B. W. and Bascombe, E. L. (eds.) (1974) Soil Survey Laboratory Methods, *Soil Surv. Tech. Monograph*, **6**.

Baillie, M. G. L. and Pilcher, J. R. (1973) A simple crossdating program for tree-ring research, *Tree-Ring Bull.*, **33**, 7–14.

Ball, D. F. (1960) Relic-soil on limestone in South Wales, *Nature*, **187**, 497–8.

Bannister, A. and Raymond, S. (1972) *Surveying*, Pitman, London and New York.

Barber, H. G. and Haworth, E. Y. (1981) *A Guide to the Morphology of the Diatom Frustule*, Freshwater Biol. Assoc. Sci. Paper, No. **44**.

Barber, K. E. (1981) *Peat Stratigraphy and Climate Change*, Balkema, Rotterdam.

Barnett, H. F. and Finke, P. G. (1971) Morphometry of landforms: drumlins, *US Army Earth Sci. Lab. Tech. Rep.*, **ES–63**.

Bartley, D. D. (1962) The stratigraphy and pollen analysis of lake deposits near Tadcaster, Yorkshire, *New Phytol.*, **61**, 277–87.

Bate, R. H. and Robinson, J. E. (1978) *A Stratigraphical Index of British Fossil Ostracods*, Seel House Press, Liverpool.

Battarbee, R. W. (1973) Preliminary studies of Lough Neagh sediments, II. Diatom analysis from the uppermost sediment, In H. J. B. Birks and R. G. West (eds.), *Quaternary Plant Ecology*, Blackwell, Oxford, 279–88.

Bé, A. W. H., Damuth, J. E., Lott, L. and Free R. (1976) Late Quaternary climate record in western equatorial Atlantic sediment, *Geol. Soc. Amer. Memoir*, **145**, 165–200.

Beck, R. B., Funnell, B. M. and Lord, A. R. (1972) Correlation of Lower Pleistocene Crag at depth in Suffolk, *Geol. Mag.*, **109**, 137–9.

Becker, B. and Schirmer, W. (1977) Palaeoecological study on the Holocene Valley development of the River Main, southern Germany, *Boreas*, **6**, 303–21.

Belcher, R. and Ingram, G. (1950) A rapid micro-combustion method for the determination of carbon and oxygen, *Anal. Chem, Acta*, **4**, 118–29.

Bell, F. G. (1969) The occurrence of southern, steppe and halophyte elements in Weichselian (full glacial) floras from southern England, *New Phytol.*, **68**, 913–22.

Bell, F. G. (1970) Late Pleistocene flora from Earith, Huntingdonshire, *Phil. Trans. Roy. Soc. Lond.*, **B 258**, 347–78.

Bell, F. G., Coope, G. R., Rice, R. J. and Riley, T. H. (1972) Mid-Weichselian fossil-bearing deposits at Syston, Leicestershire, *Proc. Geol. Assoc.*, **83**, 197–211.

Bell, F. G. and Dickson, C. A. (1971) The Barnwell Station Arctic flora: a reappraisal of some plant identifications, *New Phytol.*, **70**, 627–38.

Bender, M. L. et al. (1979) Uranium-series dating of the Pleistocene reef tracts of Barbados, West Indies, *Geol. Soc. Amer. Bull.*, **90**, 577–94.

Benedict, J. B. (1967) Recent glacial history of an Alpine area in the Colorado Front Range, USA. I. Establishing a lichen growth curve, *J. Glaciol.*, **6**, 817–32.

Benedict, J. B. (1970) Downslope soil movement in a Colorado alpine region: rates, processes, climatic significance, *Arct. Alp. Res.*, **2**, 165–226.

Benson, L. V. (1978) Fluctuations in the level of pluvial Lake Lahontan during the last 40 000 years, *Quat. Res.*, **9**, 300–16.

Benson, L. V. (1981) Palaeoclimatic significance of lake-level fluctuations in the Lahontan Basin, *Quat. Res.*, **16**, 390–403.

Berg, T. E. (1969) Fossil sand wedges at Edmonton, Alberta, *Biul. Peryglac.*, **19**, 325–33.

Berger, W. H., Bé, A. W. H. and Vincent, E. (eds) (1981) Oxygen and carbon isotopes in foraminifera. *Palaeogeogr. Palaeoclimatol. Palaeoecol.*, **33**, 1–276.

Berggren, W. A. et al. (1980) Towards a Quaternary timescale, *Quat. Res.*, **13**, 277–302.

Berglund, B. E. (1979) The deglaciation of southern Sweden 13 500–10 000 BP, *Boreas*, **8**, 89–118.

Berthelsen, A. (1979) Contrasting views on the Weichselian glaciation and deglaciation of Denmark, *Boreas*, **8**, 125–32.

Beschel, R. E. (1973) Lichens as a measure of the age of recent moraines, *Arct. Alp. Res.*, **5**, 303–9.

Bien, G. S., Rakestrw, N. W. and Suess, H. E. (1963) Radiocarbon dating of deep water of the Pacific and Indian Oceans, *Bull. Inst. Oceanogr. Monaco*, **61**, 1278, 1–16.

Bintliff, J. L. (1975) Mediterranean alluviation: new evidence from archaeology, *Proc. Prehist. Soc.*, **41**, 78–84.

Birkeland, P. W. (1974) *Pedology, Weathering and Geomorphological Research*, Oxford Univ. Press, London and New York.

Birkeland, P. W., Crandell, D. R. and Richmond, G. M. (1971) Status of correlation of Quaternary stratigraphic units in the western coterminous United States, *Quat. Res.*, **1**, 208–27.

Birks, H. H. (1972) Studies in the vegetational history of Scotland. III. A radiocarbon-dated pollen diagram from Loch Maree, Ross and Cromarty, *New Phytol.*,**71**, 731–54.

Birks, H. H. (1973) Modern macrofossil assemblages in lake sediments in Minnesota, In H. J. B. Birks and R. G. West (eds.), *Quaternary Plant Ecology*, Blackwell, Oxford, 172–88.

Birks, H. H. and Mathewes, R. W. (1978) Studies in the vegetational history of Scotland. V. Late Devensian and early Flandrian pollen and macrofossil stratigraphy at Abernethy Forest, Inverness-shire, *New Phytol.*, **80**, 455–84.

Birks, H. H., Whiteside, M. C., Stark, D. M. and Bright, R. C. (1976) Recent palaeolimnology of three lakes in north-western Minnesota, *Quat. Res.*, **6**, 249–72.

Birks, H. J. B. (1973a) Modern pollen rain studies in some arctic and alpine environments, In H. J. B. Birks and R. G. West (eds.), *Quaternary Plant Ecology*, Blackwell, Oxford, 143–68.

Birks, H. J. B. (1973b) *The Past and Present Vegetation of the Isle of Skye: a Palaeoecological Study*, Cambridge Univ. Press, Cambridge (1981).

Birks, H. J. B. (1981) The use of pollen analysis in the reconstruction of past climate: a review, In T. M. L. Wigley, M. J. Ingram and G. Former (eds) *Climate and History*, Cambridge, 111–138.

Birks, H. J. B. and Birks, H. H. (1980) *Quaternary Palaeoecology*, Edward Arnold, London.

Bishop, W. W. and Coope, G. R. (1977) Stratigraphical and faunal evidence for Lateglacial and early Flandrian environments in south-west Scotland, In J. M. Gray and J. J. Lowe (eds.), *Studies in the Scottish Lateglacial*, Pergamon, Oxford and New York, 61–88.

Bishop, W. W. and Dickson, J. H. (1970) Radiocarbon dates related to the Scottish Late-glacial sea in the Firth of Clyde, *Nature*, **227**, 480–2.

Black, R. F. (1976) Periglacial features indicative of permafrost: ice and soil wedges, *Quat. Res.*, **6**, 3–26.

Blake, W. Jr (1980) Application of amino acid ratios to studies of Quaternary geology in the High Arctic, In P. E. Hare, T. C. Hoering and K. King Jr (eds.), *Biogeochemistry of Amino Acids*, Wiley, New York and Chichester, 453–61.

Blasing, T. J. and Fritts, H. C. (1976) Reconstructing past climate anomalies in the North Pacific and Western North America from tree-ring data, *Quat. Res.*, **6**, 563–79.

Bleuer, N. K. (1974) Buried till ridges in the Fort Wayne area, Indiana, and their regional significance, *Bull. Geol. Soc. Amer.*, **85**, 917–20.

Bloemendal, J., Oldfield, F. and Thompson, R. (1979) Magnetic measurements used to assess sediment influx at Llyn Goddionduon, *Nature*, **280**, 50–3.

Bloom, A. L. (1971) Glacial-eustatic and isostatic controls of sea level since the Last Glaciation, In K. K. Turekian (ed.), *The Late Cenozoic Glacial Ages*, Yale Univ. Press, New Haven, 355–79.

Bloom, A. L. et al. (1974) Quaternary sea level fluctuations on a tectonic coast: new ^{230}Th/^{234}U dates from the Huon Peninsula, New Guinea, *Quat. Res.*, **4**, 185–205.

Bohinski, R. C. (1979) *Modern Concepts in Biochemistry* (3rd ed.), Allyn and Bacon, London and Boston.

Bonny, A. P. (1976) Recruitment of pollen to the seston and sediment of some English Lake District lakes, *J. Ecol.*, **64**, 859–87.

Boulton, G. S. (1967) The development of a complex supraglacial moraine at the margin of Sørbreen, Ny Friesland, Vestspitsbergen, *J. Glaciol.*, **6**, 717–35.

Boulton, G. S. (1968) Flow tills and related deposits on some Vestspitsbergen glaciers, *J. Glaciol*, **7**, 391–412.

Boulton, G. S. (1970a) On the origin and transport of englacial debris in Svalbard glaciers, *J. Glaciol.*, **9**, 213–29.

Boulton, G. S. (1970b) The deposition of subglacial and melt-out tills at the margin of certain Svalbard glaciers, *J. Glaciol.*, **9**, 231–45.

Boulton, G. S. (1971) Till genesis and fabric in Svalbard, Spitsbergen, In R. P. Goldthwait (ed.), *Till : a Symposium*, Ohio State Univ. Press, 41–72.

Boulton, G. S. (1972a) Modern Arctic glaciers as depositional models for former ice sheets, *Q. J. Geol. Soc. Lond.*, **128**, 361–93.

Boulton, G. S. (1972b) The role of thermal regime in glacial sedimentation, In R. J. Price and D. E. Sugden (eds.), *Polar Geomorphology, Inst. Brit. Geogr. Spec. Publ.*, **4**, 1–19.

Boulton, G. S. (1974) Processes and patterns of glacial erosion, In D. R. Coates (ed.), *Glacial Geomorphology*, State Univ. New York, Binghampton, 41–87.

Boulton, G. S. (1977) A multiple till sequence formed by a Late Devensian\Welsh ice-cap: Glanllynnau, Gwynedd, *Cambria*, **4**, 10–31.

Boulton, G. S. (1980) Classification of till, *Quat. Res. Assoc. (GB), Newsl.*, **31**, 1–12.

Boulton, G. S., Chroston, P. N. and Jarvis, J. (1981) A marine seismic study of late Quaternary sedimentation and inferred glacier fluctuations along western Inverness-shire, Scotland, *Boreas*, **10**, 39–51.

Boulton, G. S., Jones, A. S., Clayton, K. M. and Kenning, M. J. (1977) A British ice sheet model and patterns of glacial erosion and deposition in Britain, In F. W. Shotton (ed.), *British Quaternary Studies – Recent Advances*, Oxford Univ. Press, Oxford, 231–46.

Boulton, G. S. and Worsley, P. (1965) Late Weichselian glaciation of the Cheshire-Shropshire Basin, *Nature*, **207**, 704–6.

Bowen, D. Q. (1973a) The excavation at Minchin Hole 1973, *Gower*, **24**, 12–18.

Bowen, D. Q. (1973b) The Pleistocene succession of the Irish Sea, *Proc. Geol. Assoc.*, **84**, 249–72.

Bowen, D. Q. (1973c) The Pleistocene history of Wales and the borderland, *Geol. J.*, **8**, 207–24.

Bowen, D. Q. (1974) The Quaternary of Wales, In T. R. Owen (ed.), *The Upper Palaeozoic and post-Palaeozoic of Wales*, Cardiff, 373–426.

Bowen, D. Q. (1977a) The coast of Wales, In C. Kidson and M. J. Tooley (eds.), *The Quaternary History of the Irish Sea*, Seel House Press, Liverpool, 223–56.

Bowen, D. Q. (1977b) Late Devensian periglacial environments in upland Britain, *Proc. Xth INQUA Congress, Birmingham, 1977*, p. 49.

Bowen, D. Q. (1978) *Quaternary Geology*, Pergamon, Oxford and New York.

Bowen, D. Q. (1981) The 'South Wales End Moraine': fifty years after, In J. Neale and J. Flenley (eds.), *The Quaternary in Britain*, Pergamon, Oxford and New York, 60–7.

Bowler J. M. (1976) Aridity in Australia: age, origins and expression in aeolian landforms and sediments, *Earth Sci. Rev.*, **12**, 279–310.

Bowler, J. M. and Polach, H. A. (1971) Radiocarbon analysis of soil carbonates: an evaluation from palaeosols in south-eastern Australia, In D. A. Yaalon (ed.), *Palaeopedology*, Internat. Soc. Soil Sci. and Israel Univ. Press Jerusalem, 97–108.

Bowles, F. A. (1975) Palaeoclimatic significance of quartz/illite variations in cores from the eastern equatorial North Atlantic, *Quat. Res.*, **5**, 225–35.

Boydell, A. N. (1970) Late Wisconsin ice near Sundre, Alberta, Unpubl. MSc Thesis, Univ. of Calgary.

Bradbury, J. P. and Waddington, J. C. B. (1973) The impact of European settlement on Shagawa Lake, north-eastern Minnesota, In H. J. B. Birks and R. G. West (eds.), *Quaternary Plant Ecology*, Blackwell, Oxford, 289–307.

Brakenridge, G. R. (1978) Evidence for a cold, dry full-glacial climate in the American south-west, *Quat. Res.*, **9**, 22–40.

Brasier, M. D. (1980) *Microfossils*, George Allen and Unwin, London and Boston.

Brewer. R. (1964) *Fabric and Mineral Analysis of Soils*, Wiley, New York.

Briggs, D. J. (1977) *Sediments*, Butterworths, London and Boston.

Briggs, D. J. and Gilbertson, D. D. (1973) The age of the Hanborough terrace of the River Evenlode, Oxfordshire, *Proc. Geol. Assoc.*, **84**, 155–74.

Briggs, D. J. and Gilbertson, D. D. (1980) Quaternary processes and environments in the upper Thames Valley, *Trans. Inst. Brit. Geogr. (New Series)*, **5**, 53–65.

Broecker, W. S. (1963) A preliminary evaluation of uranium series inequilibrium as a tool for absolute age measurement on marine carbonates, *J. Geophys. Res.*, **68**, 2817–34.

Broecker, W. S. (1965) Isotope geochemistry and the Pleistocene climatic record, In H. E. Wright and D. G. Frey (eds.), *The Quaternary of the United States*, Princeton Univ. Press, Princeton, N.J., 737–53.

Broecker, W. S. and Bender, M. L. (1972) Age determinations on marine strandlines, In W. W. Bishop and J. A. Miller (eds.), *Calibration of Hominid Evolution*, Scottish Academic Press, Edinburgh; Wenner-Gren Found., New York, 19–38.

Broecker, W. S., Gerard, R., Ewing, M. and Heezen, B. C. (1960) Natural radiocarbon in the Atlantic Ocean, *J. Geophys. Res.*, **65**, 2903–31.

Broecker, W. S. and Kaufman, A. (1965) Radiocarbon chronology of Lake Lahontan and Lake Bonneville, Great Basin. II, *Bull. Geol. Soc. Amer.*, **76**, 537–66.

Broecker, W. S. and Ku, T-L. (1969) Caribbean cores P6304–8 and P6304–9: new analysis of absolute chronology, *Science*, **166**, 404–6.

Broecker, W. S. and Thurber, D. L. (1965) Uranium-series dating of corals and oolites from Bahaman and Florida Key limestones, *Science*, **149**, 58–60.

Broecker, W. S. and Van Donk, J. (1970) Insolation changes, ice volumes and the O^{18} record in deep-sea sediments, *Rev. Geophys. and Space Phys.*, **8**, 169–98.

Brown, A. P. (1977) Late Devensian and Flandrian vegetational history of Bodmin Moor, Cornwall, *Phil. Trans. Roy. Soc. Lond.*, B **276**, 251–320.

Brown, C. R., Briggs, D. J. and Gilbertson, D. D. (1980) Depositional environment of late-Pleistocene terrace gravels of the Vale of Bourton, Gloucestershire, *Mercian Geol.*, **7**, 269–78.

Brown, R. J. E. and Péwé, T. L. (1973) Distribution of permafrost in North America and its relationship to the environment: a review, 1963–73. In North American Contribution, Permafrost Second International Conference, Yakutsk, USSR, *Nat. Acad. Sci., Sci. Publ.*, **2115**, 71–100.

Brugam, R. B. (1980) Postglacial diatom stratigraphy of Kirchner Marsh, Minnesota, *Quat. Res.*, **13**, 133–46.

Brunnacker, K. (1975) The Mid-Pleistocene of the Rhine Basin, In K. W. Butzer and G. L. Isaac (eds.), *After the Australopithecines*, Mouton Press, The Hague, 189–224.

Brunskill, G. J. (1969) Fayetteville Green Lake, New York. II. Precipitation and sedimentation of calcite in a meromictic lake with laminated sediments, *Limnol. Oceanogr.*, **14**, 830–47.

Bryson, R. A., Wendland, W. M., Ives, J. D. and Andrews, J. T. (1969) Radiocarbon isochrones on the disintegration of the Laurentide ice sheet, *Arct. Alp. Res.*, **1**, 1–13.

Bucha, V. (1967a) Intensity of the earth's magnetic field during archaeological times in Czechoslovakia, *Archaeometry*, **10**, 12–22.

Bucha, V. (1967b) Archaeomagnetic and palaeomagnetic study of the magnetic field of the earth in the past 600 000 years, *Nature*, **213**, 1005–7.

Bucha, V. (1970) Influence of the earth's magnetic field on radiocarbon dating, In I.U. Olsson (ed.), *Radiocarbon Variations and Absolute Chronology*, Wiley, New York and London, 501–10.

Buchardt, B. and Fritz, P. (1980) Environmental isotopes as environmental climatological indicators, In P. Fritz and J. Ch. Fontes (eds.), *Handbook of Environmental Isotope Geochemistry* (vol. 1), Elsevier, Rotterdam, and New York, 473–504.

Buchsbaum, R. (1948) *Animals Without Backbones* (2 vols.), Pelican, London.

Buckland, P. C. (1976) The use of insect remains in the interpretation of archaeological environments, In D. A. Davidson and M. L. Shackley, *Geoarchaeology*, Duckworth, London. 369–91.

Budd, W. F. and Smith, I. N. (1981) The growth and retreat of ice sheets in response to orbital radiation changes, In *Sea Level, Ice and Climatic Change*, Internat. Assoc. of Hydrol. Sciences, **131**, 369–409.

Bull, P. A. (1976) An electron microscope study of cave sediments from Agen Allwedd, Wales, *Trans. Brit. Cave Res. Assoc.*, **3**, 7–14.

Bull, P. A. (1978a) A study of stream gravels from a cave: Agen Allwedd, *Zeit. für Geomorph.*, **22**, 275–96.

Bull, P. A. (1978b) Observations on small sedimentary quartz particles analysed by scanning electron microscopy, In O. Jahari (ed.), *Scanning Electron Microscopy*, SEM Inc., Illinois, vol. 1, 821–8.

Bull, P. A. (1980) Towards a reconstruction of time-scales and palaeoenvironments from cave sediment studies, In R. A. Cullingford, D. A. Davidson and J. Lewin (eds.), *Timescales in Geomorphology*, Wiley, Chichester and New York, 177–87.

Burckle, E. H. (1978) Marine diatoms, In B. U. Haq and A. Boersma (eds.), *Introduction to Marine Micropalaeontology*, Elsevier, Rotterdam, 245–66.

Burckle E. H. and Biscaye. P. (1971) Sediment transport by Antarctic Bottom Water through the eastern Rio Grande rise, *Geol. Soc. Amer. Ann. Meeting, Abstracts*, 518–19.

Burke, R. M. and Birkeland, P. W. (1979) Reevaluation of multiparameter relative dating techniques and their application to the glacial sequence along the eastern escarpment of the Sierra Nevada, California, *Quat. Res.*, **11**, 21–51.

Burleigh, R. (1972) Carbon-14 dating, with application to dating of remains from caves, *Studies in Speleology*, **2**, 176–190.

Butterfield, B. G. and Meylan, B. A. (1980) *Three-Dimensional Structure of Wood: An Ultrastructural Approach*, Chapman and Hall, London and New York.

Butzer, K. W. (1962) Coastal geomorphology of Majorca, *Ann. Assoc. Amer. Geogr.*, **52**, 192–212.

Butzer, K. W. (1963) Climatic geomorphologic interpretation of Pleistocene sediments in the Eurafrican subtropics, *Viking Fund. Publ. Archaeology*, **36**, 1–27.

Butzer, K. W. (1975) Pleistocene littoral-sedimentary cycles of the Mediterranean Basin: a Mallorquin view, In K. W. Butzer and G. L. Isaac (eds.), *After the Australopithecines*, Mouton Press, The Hague, 45–71.

Butzer, K. W. (1980) Holocene alluvial sequences: problems of dating and correlation, In R. A. Cullingford, D. A. Davidson and J. Lewin (eds.), *Timescales in Geomorphology*, Wiley, Chichester and New York, 131–42.

Butzer, K. W. and Cuerda, J. (1962) Coastal stratigraphy of southern Mallorca and implications for the Pleistocene chronology of the Mediterranean Sea, *J. Geol.*, **70**, 398–416.

Butzer, K. W., Isaac, G. L., Richardson, J. L. and Washbourn-Kamau, C. (1972) Radiocarbon dating of East African lake levels, *Science*, **175**, 1069–76.

Cailleux, A. (1942) Les actions éoliennes périglaciaires en Europe, *Mém. Soc. Géol. Fr.*, **46**, 1–176.

Cailleux, A. (1945) Distinctions des galets marines et fluviatiles, *Bull. Soc. Géol. Fr.*, **5**, 375–404.

Cailleux, A. (1947) L'indice d'émoussé des grains de sable et grès, *Geomorph. Dyn.*, **3**, 78–87.

Campbell, C. A., Paul, E. A., Rennie, D. A. and McCallum, K. J. (1967) Applicability of the carbon dating method to soil humus studies, *Soil Sci.*, **104**, 217–24.

Campbell, J. B. (1977) *The Upper Palaeolithic of Britain* (2 vols.), Oxford Univ. Press.

Carroll, D. (1970) *Clay Minerals: A Guide to their X-ray Identification*, Geol. Soc. Amer., Boulder, Colorado.

Carruthers, R. G. (1947–48) The secret of the glacial drifts, *Proc. Yorks. Geol. Soc.*, **27**, 43–57 and 129–72.

Carruthers, R. G. (1953) *Glacial Drifts and the Undermelt Theory*, Harold Hill, Newcastle-upon-Tyne.

Carver, R. E. (1971a) *Procedures in Sedimentary Petrology*, Wiley, New York.

Carver, R. E. (1971b) Heavy mineral separation, In R. E. Carver (ed.), *Procedures in Sedimentary Petrology*, Wiley, New York, 427–52.

Caseldine, C. J. (1980) A Lateglacial site at Stormont Loch, near Blairgowrie, eastern Scotland, In J. J. Lowe, J. M. Gray and J. E. Robinson (eds.), *Studies in the Lateglacial of North-west Europe*, Pergamon, Oxford and New York, 69–88.

Caseldine, C. J. and Edwards, K. J. (1982) Interstadial and last interglacial deposits covered by till in Scotland: comments and new evidence, *Boreas*, **11**, 119–22.

Catt, J. A. (1977) Loess and coversands, In F. W. Shotton (ed.), *British Quaternary Studies – Recent Advances*, Oxford Univ. Press, Oxford, 221–9.

Catt, J. A. (1978) The contribution of loess to soils in lowland Britain, In S. Limbrey and J. G. Evans (eds.), *The Effects of Man on the Landscape: the lowland zone*, CBA Res. Rep., **21**, 12–20.

Catt, J. A. and Penny, L. F. (1966) The Pleistocene deposits of Holderness, East Yorkshire, *Proc. Yorks. Geol. Soc.*, **35**, 375–420.

Catt, J. A., Weir, A. H. and Madgett, P. A. (1974) The loess of eastern Yorkshire and Lincolnshire, *Proc. Yorks. Geol. Soc.*, **40**, 23–39.

Chamberlin, T. C. (1894) Proposed genetic classification of Pleistocene glacial formations, *J. Geol.*, **2**, 517–38.

Chamberlin, T. C. (1895) Glacial phenomena of North America, In J. Geikie, *The Great Ice Age*, Appleton, New York (2nd ed.), 724–55.

Chaplin, R. E. (1971) *The Study of Animal Bones from Archaeological Sites*, Seminar Press, London.

Chappell, J. M. A. (1974a) Geology of coral terraces, Huon Peninsula, New Guinea: a study of Quaternary tectonic movements and sea level changes, *Bull. Geol. Soc. Amer.*, **85**, 553–70.

Chappell, J. M. A. (1974b) Late Quaternary glacio- and hydro-isostasy on a layered earth, *Quat. Res.*, **4**, 405–28.

Charlesworth, J. K. (1926) The glacial geology of the Southern Uplands, west of Annandale and upper Clydeside, *Trans. Roy. Soc. Edinb.*, **55**, 25–50.

Charlesworth, J. K. (1928) The glacial retreat from central and southern Ireland, *Q.J. Geol. Soc. Lond.*, **84**, 293–344.

Charlesworth, J. K. (1929) The South Wales end-moraine, *Q.J. Geol. Soc. Lond.*, **85**, 335–58.

Chinn, T. J. H. (1981) Use of rock weathering-rind thickness for Holocene absolute age-dating in New Zealand, *Arct. Alp. Res.*, **13**, 33–45.

Chizhov, O. P. (1964) Precipitation, feeding and melting of ice sheets of north-eastern Atlantic in present climatic conditions, *Results of Researches IGY, Glaciol.*, no. **13**, 30–8.

Chorley, R. J. (1959) The shape of drumlins, *J. Glaciol.*, **3**, 339–44.

Chorley, R. J., Dunn, A. J. and Beckinsale, R. P. (1964) *A History of the Study of Landforms*, vol. 1, Methuen, London.

Christensen. L. (1974) Crop-marks revealing large-scale patterned ground structures in cultivated areas, south-western Jutland, Denmark, *Boreas*, **3**, 153–80.

Christiansen, E. A. (1979) The Wisconsin deglaciation of southern Saskatchewan and adjacent areas, *Can. J. Earth Sci.*, **16**, 913–38.

Clapperton, C. M., Gunson, A. R. and Sugden, D. E. (1975) Loch Lomond Readvance in the eastern Cairngorms, *Nature*, **253**, 710–12.

Clapperton, C. M. and Sugden, D. E. (1977) The Late Devensian glaciation of North-East Scotland, In J. M. Gray and J. J. Lowe (eds.), *Studies in the Scottish Lateglacial Environment*, Pergamon, Oxford and New York, 1–13.

Clark, J. A. (1980) The reconstruction of the Laurentide Ice Sheet of North America from sea level data: method and preliminary results, *J. Geophys. Res.*, **85**, 4307–23.

Clark, J. G. D. (1954) *Excavations at Starr Carr, an early Mesolithic site at Seamer, near Scarborough, Yorkshire*, Cambridge Univ. Press, Cambridge.

Clayton, K. M. (1977) River Terraces, In F. W. Shotton (ed.), *British Quaternary Studies – Recent Advances*, Oxford Univ. Press, Oxford, 153–67.

CLIMAP (1976) The surface of ice-age earth, *Science*, **191**, 1131–7.

Cline, R. M. and Hays, J. D. (1976) Investigation of late

Quaternary Palaeooceanography and Palaeoclimatology, *Geol. Soc. Amer. Mem*, **145**.

Colhoun, E. A., Dickson, J. H., McCabe, A. M. and Shotton, F. W. (1972) A Middle Midlandian freshwater series at Derryvree, Maguiresbridge, County Fermanagh, Northern Ireland, *Proc. R. Soc. Lond.*, **B 180**, 273–92.

Colman, S. M. (1982) Chemical weathering of basalts and andesites: evidence from weathering rinds, *US Geol. Surv. Prof. Paper*. 1246.

Colman, S. M. and Pierce, K. L. (1981) Weathering rinds on andesitic stones as a Quaternary age indicator, western United States, *US Geol. Surv. Prof. Paper*, **1210**.

Colquhoun, D. J. and Johnson, H. S. (1968) Tertiary sea level fluctuations in South Carolina, *Palaeogeogr. Palaeoclimatol. Palaeoecol.*, **5**, 105–26.

Conkey, L. (1979) Response of tree-ring density to climate in Maine, USA, *Tree-Ring Bull.*, **39**, 29–38.

Connolly, A. P. (1976) Use of the scanning electron microscope for the identification of seeds with special reference to *Saxifraga* and *Papaver, Folia Quaternaria*, **47**, 29–32.

Cooke, H. B. S. (1981) Age control of Quaternary sedimentary/climatic record from deep boreholes in the Great Hungarian Plain, In W. C. Mahaney (ed.), *Quaternary Palaeoclimate*, GeoAbstracts, Norwich, 1–12.

Cooke, R. U. and Doornkamp, J. C. (1974) *Geomorphology in Environmental Management*, Clarendon Press, Oxford.

Coope, G. R. (1959) A late Pleistocene insect fauna from Chelford, Cheshire, *Proc. R. Soc. Lond.*, **B 151**, 70–86.

Coope, G. R. (1961) On the study of glacial and interglacial faunas, *Proc. Linn. Soc. Lond.*, **172**, 62–5.

Coope, G. R. (1962) A Pleistocene Coleopterous fauna with arctic affinities from Fladbury, Worcestershire, *Q.J. Geol. Soc. Lond.*, **118**, 103–23.

Coope, G. R. (1967) The value of Quaternary insect faunas in the interpretation of ancient ecology and climate, In E. J. Cushing and H. E. Wright (eds.), *Quaternary Palaeoecology*, Yale Univ. Press, New Haven, 359–80.

Coope, G. R. (1968a) An insect fauna from Mid-Weichselian deposits at Brandon, Warwickshire, *Phil. Trans. R. Soc. Lond.*, **B 254**, 425–56.

Coope, G. R. (1968b) Coleoptera from the 'Arctic Bed' at Barnwell Station, Cambridge, *Geol. Mag.*, **105**, 482–6.

Coope, G. R. (1970a) Interpretation of Quaternary insect fossils, *Ann. Rev. Ent.*, **15**, 97–120.

Coope, G. R. (1970b) Climatic interpretation of Late Weichselian coleoptera from the British Isles, *Rev. Géogr. Phys. Geol. Dyn.*, **12**, 149–55.

Coope, G. R. (1974a) Interglacial coleoptera from Bobbitshole, Ipswich, *Q.J. Geol. Soc. Lond.*, **130**, 333–40.

Coope, G. R. (1974b) Tibetan species of dung beetle from Late Pleistocene deposits in England, *Nature*, **245**, 335–6.

Coope, G. R. (1975a) Climatic fluctuations in north-west Europe since the last Interglacial, indicated by fossil assemblages of Coleoptera, In A. E. Wright and F. Moseley (eds.), *Ice Ages: Ancient and Modern*, Seel House Press, Liverpool, 153–68.

Coope, G. R. (1975b) Mid-Weichselian climatic changes in western Europe, reinterpreted from Coleopteran assemblages, In R. P Suggate and M. M. Cresswell (eds.), *Quaternary Studies*, Bull. R. Soc. NZ, **13**, 101–8.

Coope, G. R. (1977a) Fossil coleopteran assemblages as sensitive indicators of climatic changes during the Devensian (last) cold stage, *Phil. Trans. R. Soc. Lond.*, **B 280**, 313–40.

Coope, G. R. (1977b) Quaternary coleoptera as aids in the interpretation of environmental history, In F. W. Shotton (ed.), *British Quaternary Studies – Recent Advances*, Oxford Univ. Press, Oxford, 55–68.

Coope, G. R. and Angus, R. B. (1975) An ecological study of a temperate interlude in the middle of the Last Glaciation, based on fossil Coleoptera from Isleworth, Middlesex, *J. Anim. Ecol.*, **44**, 365–91.

Coope, G. R. and Brophy, J. A. (1972) Late-glacial environmental changes indicated by a coleopteran succession from North Wales, *Boreas*, **1**, 97–142.

Coope, G. R. and Joachim, M. J. (1980) Lateglacial environmental changes interpreted from fossil Coleoptera from St Bees, Cumbria, NW England, In J. J. Lowe, J. M. Gray and J. E. Robinson (eds.), *Studies in the Lateglacial of North-west Europe*, Pergamon, Oxford and New York, 55–68.

Coope, G. R., Morgan, A. and Osborne, P. J. (1971) Fossil coleoptera as indicators of climatic fluctuations during the Last Glaciation in Britain, *Palaeogeogr. Palaeoclimatol. Palaeoecol.*, **10**, 87–101.

Coope, G. R. and Pennington, W. (1977) The Windermere Interstadial of the Late Devensian, *Phil. Trans. R. Soc. Lond.*, **B 280**, 337–9.

Coope, G. R. and Sands, C. H. S. (1966) Insect faunas of the last glaciation from the Tame Valley, Warwickshire, *Proc. R. Soc. Lond.*, **B 165**, 389–412.

Coope, G. R., Shotton, F. W. and Strachan, I. (1961) A late Pleistocene fauna and flora from Upton Warren, Worcestershire, *Phil. Trans. R. Soc. Lond.*, **B 244**, 379–421.

Cornish, R. (1979) The statistical analysis of till fabric data: a review, *Univ. of Edinb., Dept. of Geogr., Res. Disc. Pap.*, **16**.

Cornish, R. (1981) Glaciers of the Loch Lomond Stadial in the Western Southern Uplands of Scotland, *Proc. Geol. Assoc.*, **92**, 105–14.

Cornwall, I. W. (1958) *Soils for the Archaeologist*, Phoenix, London.

Cornwall, I. W. (1974) *Bones for the Archaeologist*, Dent, London.

Cox, A., Doell, R. R. and Dalrymple, G. B. (1963) Geomagnetic polarity epochs and Pleistocene geochronometry, *Nature*, **198**, 1049–51.

Cox, A., Doell, R. R. and Dalrymple, G. B. (1965) Quaternary paleomagnetic stratigraphy, In Wright, H. E. and Frey, D. G. (eds.), *The Quaternary of the United States*, Princeton Univ. Press, Princeton, N. J., 817–30.

Craig, A. J. (1978) Pollen percentage and influx analysis in south-east Ireland: a contribution to the ecological history of the Late-Glacial period, *J. Ecol.*, **66**, 297–324.

Craig, H. (1953) The geochemistry of the stable carbon isotopes, *Geochim. et Cosmochim. Acta*, **3**, 55–92.

Craig, H. (1961) Standard for reporting concentrations of D and ^{18}O in natural waters, *Science*, **133**, 1833–4.

Crofts, R. S. (1974) Detailed geomorphological mapping and land evaluation in Highland Scotland, In E. H. Brown and R. S. Waters (eds.), *Progress in Geomorphology*, Inst. Brit. Geogr. Spec. Publ., **7**, 231–51.

Crofts, R. S. (1981) Mapping techniques in geomorphology, In A. S. Goudie (ed.), *Geomorphological Techniques*, Allen and Unwin, London and Boston, 66–75.

Cronin, T. M. (1980) Biostratigraphic correlation of Pleistocene marine deposits and sea levels, Atlantic coastal plain of the south-eastern United States, *Quat. Res.*, **13**, 213–29.

Cropper, J. P. (1979) Tree-ring skeleton plotting by computer, *Tree-Ring Bull.*, **39**, 47–59.

Cullingford, R. A. (1977) Lateglacial raised shorelines and deglaciation in the Earn-Tay area, In J. M. Gray and J. J. Lowe (eds.), *Studies in the Scottish Lateglacial Environment*, Pergamon, Oxford and New York, 15–32.

Cullingford, R. A., Caseldine, C. J. and Gotts, P. E. (1980) Early Flandrian land and sea-level changes in Lower Strathearn, *Nature*, **284**, 159–61.

Cullingford, R. A. and Smith, D. E. (1966) Late-glacial shorelines in eastern Fife, *Trans. Inst. Brit. Geogr.*, **39**, 31–51.

Cullingford, R. A. and Smith, D. E. (1980) Late Devensian raised shorelines in Angus and Kincardineshire, Scotland, *Boreas*, **9**, 21–38.

Curray, J. R. (1956) The analysis of two-dimensional orientation data, *J. Geol.*, **64**, 117–31.

Cushing, E. J. (1967a) Late Wisconsin pollen stratigraphy and the glacial sequence in Minnesota, In E. J. Cushing and H. E. Wright (eds.), *Quaternary Palaeoecology*, Yale Univ. Press, New Haven, 59–88.

Cushing, E. J. (1967b) Evidence for differential pollen preservation in Late Quaternary sediments in Minnesota, *Rev. Palaeobot. Palynol.*, **4**, 87–101.

Cushing, E. J. and Wright, H. E. (1965) Hand-operated piston corers for lake sediments, *Ecology*, **46**, 380–4.

Czudek, T. and Demek, J. (1970) Thermokarst in Siberia and its influence on the development of lowland relief, *Quat. Res.*, **1**, 103–20.

Dalrymple, G. B. and Lanphere, M. A. (1969) *Potassium-Argon Dating: Principles, Techniques and Applications to Geochronology*, Freeman and Co., San Francisco.

Damon, P. E., Lerman, J. C. and Long, A. (1978) Temporal fluctuations of atmospheric ^{14}C: causal factors and implications, *Ann. Rev. Earth Planet. Sci.*, **6**, 457–94.

Damon, P. E., Long, A. and Wallick, E. J. (1972) Dendrochronologic calibration of the ^{14}C time-scale, *Proc. 8th Internat. Conf. on Radiocarbon Dating*, Roy. Soc. NZ, Wellington, vol. **1**, pp. A28–A43.

Dansgaard, W. (1954) The ^{18}O abundance in fresh water, *Geochim. et Cosmochim. Acta*, **6**, 241–60.

Dansgaard, W. and Duplessy, J. C. (1981) The Eemian Interglacial and its termination, *Boreas*, **10**, 219–28.

Dansgaard, W., Johnsen, S. J., Clausen, H. B. and Langway Jr, C. C. (1971) Climatic record revealed by the Camp Century ice core, In K. K. Turekian (ed.), *The Late Cenozoic Glacial Ages*, Yale Univ. Press, New Haven, 37–56.

Dansgaard, W., Johnsen, S. J., Miller, J. and Langway Jr., C. C. (1969) One thousand centuries of climatic record from the Greenland ice sheet, *Science*, **166**, 371–81.

Dansgaard, W., Johnsen, S. J., et al. (1975) Climatic change, Norsemen and modern man, *Nature*, **255**, 24–8.

Dansgaard, W. and Tauber, H. (1969) Glacier oxygen 18 content and Pleistocene ocean temperatures, *Science*, **166**, 499–502.

Davies, G. L. (1968) *The Earth in Decay*, MacDonald, London.

Davies, G. L. H. and Stephens, N. (1978) *The Geomorphology of the British Isles: Ireland*, Methuen, London.

Davis, M. B. (1976) Erosion rates and land-use history in southern Michigan, *Environ. Conserv.*, **3**, 139–48.

Dawson, A. G. (1977) A fossil lobate rock glacier in Jura, *Scott. J. Geol.*, **13**, 37–42.

Dawson, A. G. (1980) Shore erosion by frost: an example from the Scottish Lateglacial, In J. J. Lowe, J. M. Gray and J. E. Robinson (eds.), *Studies in the Lateglacial of North-west Europe*, Pergamon, Oxford and New York, 45–54.

Deacon, M. (1973) The voyage of HMS Challenger, In R. G. Pirie (ed.), *Oceanography: Contemporary Readings in Ocean Sciences*, Oxford Univ. Press, London and New York, 24–44.

Dean, J. A. (1960) *Flame Photometry*, McGraw-Hill, London.

Deevey, E. S., Gross, M. S., Hutchinson, G. E. and Kraybill, H. L. (1954) The natural ^{14}C contents of materials from hard-water lakes, *Proc. Nat. Acad. Sci.*, **40**, 285–8.

De Geer, G. (1912) A geochronology of the last 12 000 years, *XIth Internat. Geol. Congr. (Stockholm)*, **1**, 241–53.

Delorme, L. D. (1969) Ostracodes as Quaternary palaeoecological indicators, *Can. J. Earth Sci.*, **6**, 1471–6.

Delorme, L. D. (1971) Palaeoecological determinations using Pleistocene freshwater ostracodes, In H. J. Oertli (ed.), *Palaéoécologie Ostracodes Pau, 1970*, *Bull. Centre Rech. Pau SNPA*, **5**, suppl., 341–7.

Delorme, L. D., Zoltai, S. C. and Kalas, L. L. (1976) Freshwater shelled invertebrate indicators of palaeoclimate in north-western Canada during the late glacial, In S. Horie (ed.), *Palaeolimnology of Lake Biwa and the Japanese Pleistocene*, Kyoto Univ., 605–57.

Demek, J. (1969) Cryoplanation terraces, their geographical distribution, genesis and development, *Rozpr. csl. Akad. Ved. rada MPV*, **79**, 1–80.

Demorest, M. (1938) Ice flowage as revealed by glacial striae, *J. Geol.*, **46**, 700–25.

Denton, G. H. and Hughes, T. J. (1981) *The Last Great Ice Sheets*, Wiley, New York.

Denton, G. H. and Karlén, W. (1973a) Holocene climatic variations – their pattern and possible cause, *Quat. Res.*, **3**, 155–205.

Denton, G. H. and Karlén, W. (1973b) Lichenometry: its application to Holocene moraine studies in southern Alaska and Swedish Lapland, *Arct. Alp. Res.*, **5**, 347–72.

Denton, G. H. and Karlén, W. (1977) Holocene glacial and tree-line variations in the White River Valley and Skolai Pass, Alaska and Yukon Territory, *Quat. Res.*, **7**, 63–111.

Derbyshire, E. (1972) Tors, rock weathering and climate in southern Victoria Land, Antarctica, In R. J. Price and D. E. Sugden (eds.), *Polar Geomorphology*, Inst. Brit. Geogr. Spec. Publ., **4**, 93–105.

Devoy, R. J. N. (1977) Flandrian sea-level changes in the Thames Estuary and the implications for land subsidence in England and Wales, *Nature*, **220**, 712–15.

De Vries, H1. (1958) Variation in concentration of radiocarbon with time and location on earth, *Koninkl. Ned. Akad. Wetenschap. Proc.*, **B6**, 94–102.

Dickson, C. A. (1970) The study of plant macrofossils in British Quaternary deposits, In D. Walker and R. G. West (eds.), *Studies in the Vegetational History of the British Isles*, Cambridge Univ. Press, 233–54.

Dickson, J. H. et al. (1978) Palynology, palaeomagnetism and radiometric dating of Flandrian marine and freshwater sediments of Loch Lomond, *Nature*, **274**, 548–53.

Diebel, K. and Pietrzeniuk, E. (1977) Ostracoder aus dem

Travertin von Taubuch bei Weimar, *Quatärpalontologie*, **2**, 119–37.

Dimbleby, G. W. (1957) Pollen analysis of terrestrial soils, *New Phytol.*, **56**, 12–28.

Dimbleby, G. W. (1961) Soil pollen analysis, *J. Soil Sci.*, **12**, 1–11.

Dobson, M. R., Evans, W. E. and James, K. H. (1971) The sediment of the floor of the Southern Irish Sea, *Marine Geol.*, **11**, 27–69.

D'Olier, B. (1975) Some aspects of Late Pleistocene-Holocene drainage of the River Thames in the eastern part of the London Basin, *Phil. Trans. R. Soc. Lond.*, **A 279**, 269–77.

D'Olier, B. and Madrell, R. J. (1970) Buried channels of the Thames Estuary, *Nature*, **226**, 347–8.

Donner, J. J. (1979) The Early or Middle Devensian peat at Burn of Benholm, Kincardineshire, *Scott. J. Geol.*, **15**, 247–50.

Doornkamp, J. C. and King, C. A. M. (1971) *Numerical Analysis in Geomorphology*, Edward Arnold, London.

Dormaar, J. F. and Lutwick, L. E. (1969) Infra-red spectra of humic acids and opal phytoliths as indicators of palaeosols, *Can. J. Soil Sci.*, **49**, 29–37.

Doucas, G. *et al.*, (1978) Detection of ^{14}C using a small van de Graaff accelerator, *Nature*, **276**, 253–5.

Douglass, R. C. (1965) Larger foraminifera, In B. Kummel and D. Raup (eds.), *Handbook of Palaeontological Techniques*, Freeman and Co., San Francisco.

Drake, L. D. (1977) Human factor in till-fabric analysis, *Geology*, **5**, 180–4.

Dreimanis, A., Hutt, G., Raukas, A. and Whippery, P. W. (1979) Dating methods of Pleistocene deposits and their problems. I. Thermo-luminescent dating, *Geoscience (Canada)*, **5**, 55–60.

Dreimanis, A. and Raukas, A. (1975) Did Middle Wisconsin, Middle Weichselian and their equivalents represent an Interstadial Complex in the Northern Hemisphere?, In R. P. Suggate and M. M. Cresswell (eds.), *Quaternary Studies*, Roy. Soc. NZ Bull., **13**, 109–20.

Dumanski, J. (1969) Micromorphology as a tool in the Quaternary record, In S. Pawluk (ed.), *Pedology and Quaternary Research*, Univ. Alberta Press, Edmonton, 39–52.

Du Saar, A. (1978) Diatom investigation of a sediment core: Downholland Moss-15, In M. J. Tooley (ed.), *Sea-Level Changes in North-West England during the Flandrian Stage*, Clarendon Press, Oxford, 203–8.

Eardley, A. J. *et al.* (1973) Lake cycles in the Bonneville Basin, Utah, *Bull. Geol. Soc. Amer.*, **84**, 211–16.

Ebert, J. I. and Hitchcock, R. K. (1978) Ancient Lake MakGadikgadi, Botswana: mapping, measurement and palaeoclimatic significance, In E. M. Van Zinderen Bakker and J. A. Coetzee (eds.), *Palaeoecology of Africa and the Surrounding Islands*, Balkema, Rotterdam, 47–56.

Edwards, K. J., Caseldine, C. J. and Chester, D. K. (1976) Possible interstadial and interglacial pollen floras from Teindland, Scotland, *Nature*, **264**, 742–4.

Edwards, K. J. and Rowntree, K. M. (1980) Radiocarbon and palaeoenvironmental evidence for changing rates of erosion at a Flandrian stage site in Scotland, In R. A. Cullingford, D. A. Davidson and J. Lewin (eds.), *Timescales in Geomorphology*, Wiley, Chichester and New York, 207–23.

Ellis, A. E. (1978) British Freshwater Bivalve Mollusca, *Linnaean Synopses of the British Fauna, New Series*, **11**, 1–109.

Ellis-Gruffydd, I. D. (1977) Late Devensian glaciation in the Upper Usk basin, *Cambria*, **4**, 46–55.

Embleton, C. and King, C. A. M. (1975a) *Glacial Geomorphology*, Edward Arnold, London; Halstead, New York, 2nd ed.

Embleton, C. and King, C. A. M. (1975b) *Periglacial Geomorphology*, Edward Arnold, London; Halstead, New York, 2nd ed.

Emiliani, C. (1955) Pleistocene temperatures, *J. Geol.*, **63**, 538–75.

Emiliani, C. (1966) Isotopic palaeotemperatures, *Science*, **154**, 851–7.

Emiliani, C. (1971) The amplitude of Pleistocene climatic cycles at low latitudes and the isotopic composition of glacial ice, In K. K. Turekian (ed.), *The Late Cenozoic Glacial Ages*, Yale Univ. Press, New Haven, 183–97.

Enjalbert, H. (1968) La genèse des reliefs karstiques dans les pays tempérés et dans les pays tropicaux, In P. Fénelon (ed.), *Phénomènes Karstiques, CNRS*, **4**, 295–327.

Epstein, S., Sharp, R. P. and Gow, A. J. (1970) Antarctic ice sheet: stable isotope analyses of Byrd Station cores and interhemispheric climatic implications, *Science*, **168**, 1570–72.

Ericson, D. B., Broecker, W. S., Kulp, J. L. and Wollin, G. (1956) Late Pleistocene climate and deep-sea sediments, *Science*, **124**, 385–9.

Ericson, D. B., Ewing, M., Wollin, G. and Heezen, B. C. (1961) Atlantic deep-sea sediment cores, *Bull. Geol. Soc. Amer.*, **72**, 193–286.

Ericson, D. B. and Wollin, G. (1968) Pleistocene climates and chronology in deep-sea sediments, *Science*, **162**, 1227–34.

Evans, G. H. (1970) Pollen and diatom analysis of Late Quaternary deposits in the Blelham Basin, north Lancashire, *New Phytol.*, **69**, 821–74.

Evans, G. H. and Walker, R. (1977) The Late Quaternary history of the diatom flora of Llyn Clyd and Llyn Glas, two small oligotrophic high mountain tarns in Snowdonia, Wales, *New Phytol.*, **78**, 221–36.

Evans, J. G. (1972) *Land Snails in Archaeology*, Seminar Press, London.

Evans, J. G., French, C. and Leighton, D. (1978) Habitat change in two late-glacial and post-glacial sites in southern Britain: the molluscan evidence, In S. Limbrey and J. G. Evans (eds.), *The Effect of Man on the Landscape: the Lowland Zone*, C. B. A. Res., Rep. **21**, 63–75.

Evans, P. (1971) Towards a Pleistocene time-scale, In *The Phanerozoic Time-scale: A supplement*, Geol. Soc. Lond. Spec. Publ. **5**, 123–351.

Eyles, N. and Slatt, R. M. (1977) Ice marginal sedimentary glacitectonic and morphological features of Pleistocene drift: an example from Newfoundland, *Quat. Res.*, **8**, 267–81.

Faegri, K. and Iversen, J. (1975) *Textbook of Pollen Analysis*, (3rd ed.), Munksgaard, Copenhagen.

Fairbridge, R. W. (1961) Eustatic changes of sea level, *Phys. Chem. of the Earth*, **5**, 99–185.

Fairbridge, R. W. (1968) Indicator boulder, In R. W. Fairbridge (ed.), *Encyclopaedia of Geomorphology*, Reinhold, New York, 550–2.

Falconer, G., Ives, J. D., Loken, O. H. and Andrews, J. T. (1965) Major end moraines in eastern and central Arctic Canada, *Geogr. Bull.*, **7**, 137–53.

Farrand, W. R. (1975) Sediment analysis of a Pleistocene rock shelter: the Abri Pataud, *Quat. Res.*, **5**, 1–26.

Ferguson, C. W. (1970) Dendrochronology of bristlecone

pine, *Pinus aristata*: establishment of a 7484–year chronology in the White Mountains of Eastern California, USA, In I. U. Olsson (ed.), *Radiocarbon Variations and Absolute Chronology*, Wiley, New York and London, 237–59.

Feyling-Hanssen, R. W. (1964) Foraminifera in Late Quaternary deposits from the Oslofjord area, *Nor. Geol. Unders.*, **225**, 1–383.

Feyling-Hanssen, R. W., Jørgensen, J. A., Knudsen, K. L. and Andersen, A.-L. L. (1971) Late Quaternary foraminifera from Vendsyssel, Denmark and Sandnes, Norway, *Geol. Soc. Denm. Bull.*, **21**, 67–317.

Fink, J. and Kukla, G. J. (1977) Pleistocene climates in central Europe: at least 17 interglacials after the Olduvai event, *Quat. Res.*, **7**, 363–71.

Firbas, F. (1949) *Spät- und Nacheiszeitliche Waldegeschichte Mitteleuropas nördlich der Alpen. Bd. 1. Allgemeine Waldgeschichte*, Gustav Fischer, Jena.

Fitch, J. F., Hooker, P. J. and Miller, J. A. (1976) Argon-40/argon-39 dating of the KBS tuff in Koobi Fora Formation, East Rudolph, Kenya, *Nature*, **263**, 740–4.

FitzPatrick, E. A. (1965) An interglacial soil at Teindland, Morayshire, *Nature*, **207**, 621–2.

Fleischer, R. L., Price, P. B. and Walker, R. M. (1965) Effects of temperature, pressure and ionization of the formation and stability of fission tracks in minerals and glasses, *J. Geophys. Res.*, **70**, 1497–502.

Fleischer, R. L., Price, P. B. and Walker, R. M. (1969) Quaternary dating by the fission-track technique, In D. Brothwell and E. Higgs (eds.), *Science in Archaeology*, Thames and Hudson, Bristol, 58–61.

Flint, R. F. (1943) Growth of the North American ice sheet during the Wisconsin age, *Bull. Geol. Soc. Amer.*, **54**, 325–62.

Flint, R. F. (1965a) The Pliocene-Pleistocene boundary, *Geol. Soc. Amer. Spec. Paper*, **84**, 497–533.

Flint, R. F. (1965b) Introduction: historical perspectives, In H. E. Wright and D. G. Frey (eds.), *The Quaternary of the United States*, Princeton Univ. Press, Princeton, NJ, 3–11.

Flint, R. F. (1971) *Glacial and Quaternary Geology*, Wiley, New York and London.

Flint, R. F. and Gebert, J. A. (1976) Latest Laurentide ice sheet: new evidence from southern New England, *Bull. Geol. Soc. Amer.*, **87**, 182–8.

Folger, D. W. (1970) Wind transport of land-derived minerogenic, biogenic and industrial matter over the north Atlantic, *Deep Sea Res.*, **17**, 337–52.

Folk, R. L. (1974) *Petrology of Sedimentary Rocks*, Hemphill Publ. Co., Austin, Texas.

Folk, R. L. and Ward, W. (1957) Brazos River Bar: a study of the significance of grain size parameters, *J. Sed. Petrol.*, **27**, 3–26.

Follmer, L. R. (1978) The Sangamon Soil in the type area – a review, In W. C. Mahaney (ed.), *Quaternary Soils*, GeoAbstracts, Norwich, England, 125–66.

Forbes, E. (1846) On the connexion between the distribution of the existing fauna and flora of the British Isles, and the geological changes which have affected their area, especially during the epoch of the Northern Drift, *Mem. Geol. Surv. GB*, **1**, 336–432.

Ford, D. C., Thompson, P. and Schwarcz, H. P. (1972) Principles of uranium-thorium radiometric dating methods, In E. Yatsu and A. Falconer (eds.), *Research Methods in Geomorphology*, Univ. of Guelph, Ontario, and GeoAbstracts, Norwich, 247–55.

Ford, T. D. (1975) Sediments in caves, *Trans, Brit. Cave Res. Assoc.*, **2**, 41–6.

Foster, H. D. (1970) Sarn Badrig, a submarine moraine in Cardigan Bay, North Wales, *Zeit. für Geom.*, **14**, 475–86.

Francis, E. A. (1975) Glacial sediments: a selective review, In A. E. Wright and F. Moseley (eds.), *Ice Ages: Ancient and Modern*, Seel House Press, Liverpool, 43–68.

Francis, E. A. (1980) The limit of the last glaciation in England: a consideration of its definition with special reference to the West Midlands, the South West Pennines and the Vale of York, *Quat. Res. Assoc. (GB) Newsl.*, **30**, 1–4.

Fredskild, B. (1973) Studies in the vegetational history of Greenland, *Meddr. om Grønland*, **198**, 1–245.

French, H. M. (1976) *The Periglacial Environment*, Longman, London and New York.

Frenzel, B. (1964) Zur pollenanalyse von lössen, *Eiszeit. und Gegenw.*, **15**, 5–39.

Frey, D. G. (1964) Remains of animals in Quaternary lake and bog sediments and their interpretation, *Arch. Hydrobiol. Beih.*, **2**, 1–114.

Friedman, I. (1968) Hydration rate dates rhyolite flows, *Science*, **159**, 878–80.

Friedman, I. and Long, W. (1976) Hydration rate of obsidian, *Science*, **191**, 347–52.

Friedman, I., Smith, R. L. and Clark, D. (1969) Obsidian hydration, In D. Brothwell and E. Higgs (eds.), *Science in Archaeology*, Thames and Hudson, Bristol, 62–75.

Fritts, H. C. (1976) *Tree Rings and Climate*, Academic Press, London and New York.

Fromm, E. (1970) An estimation of errors in the Swedish varve chronology, In I. U. Olsson (ed.), *Radiocarbon Variations and Absolute Chronology*, Wiley, New York and London, 163–72.

Fryberger, S. G. (1980) Dune forms and wind regime, Mauritania, West Africa: implications for past climate, In E. M. Van Zinderen Bakker and J. A. Coetzee (eds.), *Palaeoecology of Africa*, Balkema, Rotterdam, 79–96.

Frye, J. C., Glass, H. D. and Willman, H. B. (1962) Stratigraphy and mineralogy of the Wisconsinan loesses of Illinois, *Circ. Illinois Geol. Surv.*, **334**.

Funnell, B. H. (1961) The Palaeogene and Early Pleistocene of Norfolk, *Trans. Norfolk Norwich Nat. Soc.*, **19**, 340–56.

Funnel, B. M. and Riedel, W. R. (eds.) (1971) *The Micropalaeontology of Oceans*, Cambridge Univ. Press.

Funnell, B. M. and West, R. G. (1962) The Early Pleistocene of Easton Bavents, Suffolk, *Q.J. Geol. Soc. Lond.*, **117**, 125–41.

Funnell, B. M. and West, R. G. (1977) Preglacial Pleistocene deposits of East Anglia, In F. W. Shotton (ed.), *British Quaternary Studies – Recent Advances*, Oxford Univ. Press, 247–65.

Galehouse, J. S. (1971) Sedimentation analysis, In R. E. Carver (ed.), *Procedures in Sedimentary Petrology*, Wiley, New York, 69–94.

Galloway, R. W. (1970) The full glacial climate in the south-western United States, *Ann. Assoc. Amer. Geogr.*, **60**, 245–56.

Garlick, J. D. (1969) Buried bone: the experimental approach in the study of nitrogen content and blood group activity, In D. Brothwell and E. Higgs (eds.), *Science in Archaeology*, Thames and Hudson, Bristol, 503–12.

Garrard, R. A. (1977) The sediments of the South Irish Sea and Nymphe Bank area of the Celtic Sea, In C. Kidson and M. J. Tooley (eds.), *The Quaternary History of the Irish Sea*, Seel House Press, Liverpool, 69–92.

Garrard, R. A. and Dobson, M. R. (1974) The nature and maximum extent of glacial sediments off the west coast of Wales, *Marine Geol.*, **16**, 31–44.

Garrett, P. (1970) Phanerozoic stromatolites: non-competitive ecologic restriction by grazing and burrowing animals, *Science*, **169**, 171–3.

Garrison, E. G., Rowlett, R. M., Cowan, D. L. and Holroyd, L. V. (1981) ESR dating of ancient flints, *Nature*, **290**, 44–5.

Gascoyne, M., Schwarcz, H. P. and Ford, D. C. (1978) Uranium series dating and stable isotope studies of speleothems. Part I. Theory and techniques, *Trans. Brit. Cave Res. Assoc.*, **5**, 91–111.

Gaunt, G. D. (1974) A radiocarbon date relating to Lake Humber, *Proc. Yorks. Geol. Soc.*, **40**, 195–7.

Gaunt, G. D. (1976) The Devensian maximum ice limit in the Vale of York, *Proc. Yorks. Geol. Soc.*, **40**, 631–7.

Gaunt, G. D., Coope, G. R. and Franks, J. W. (1970) Quaternary deposits at Oxbow opencast Coal Site in the Aire Valley, Yorks., *Proc. Yorks. Geol. Soc.*, **38**, 175–200.

Gentner, W., Glass, B. P., Storzer, D. and Wagner, G. A. (1970) Fission track ages and ages of deposition of deep-sea microtektites, *Science*, **168**, 359–61

Geol. Soc. Amer. (1958) Glacial Map of the United States east of the Rocky Mountains, 1 : 1, 750 000 (2 sheets).

Geyh, M. A., Benzler, J. H. and Roeschmann, G. (1971) Problems of dating Pleistocene and Holocene soils by radiometric methods, In D. A. Yaalon (ed.), *Palaeopedology*, Internal. Soc. Soil Sci. and Israel Univ. Press Jerusalem, 63–75

Gibbard, P. L. (1977) Pleistocene history of the Vale of St Albans, *Phil. Trans. R. Soc. Lond.*, **B 280**, 445–83.

Gibbard, P. L. *et al.* (1982) Middle Devensian deposits beneath 'Upper Floodplain' terrace of the River Thames at Kempton Park, Sunbury, England, *Proc. Geol. Assoc.*, **93**, 275–90.

Gibbard, P. L. and Hall, A. R. (1982) Late Devensian river deposits in the lower Colne Valley, West London, England, *Proc. Geol. Assoc.*, **93**, 291–9.

Girling, M. A. (1974) Evidence from Lincolnshire of the age and intensity of the mid-Devensian temperate episode, *Nature*, **250**, 270.

Gladfelter, B. G. (1972) Pleistocene terraces of the Alto Henares (Guadalajara), Spain, *Quat. Res.*, **2**, 473–86.

Glantz, M. H. (ed.) (1977) *Desertification: Environmental Degradation in and around Arid Lands*, Westview Press, Boulder, Colorado.

Glaser, P. H. (1981) Transport and deposition of leaves and seeds on tundra: a late-glacial analog, *Arct. Alp. Res.*, **13**, 173–82.

Glen, J. W., Donner, J. J. and West, R. G. (1957) On the mechanism by which stones in till become orientated, *Amer. J. Sci.*, **255**, 194–205.

Glob, P. V. (1969) *The Bog People*, Faber and Faber, London.

Godwin, H. (1940) Pollen analysis and forest history in England and Wales, *New Phytol.*, **39**, 370–400.

Godwin, H. (1956) *The History of the British Flora*, Cambridge Univ. Press, 1st ed.

Godwin, H. (1962) Half-life of radiocarbon, *Nature*, **195**, 984.

Godwin, H. (1964) Late-Weichselian conditions in south-eastern Britain: organic deposits at Colney Heath, Herts., *Proc. R. Soc. Lond.*, **B 150**, 199–215.

Godwin, H. (1975) *The History of the British Flora* (2nd ed.), Cambridge Univ. Press.

Goh, K. M. (1972) Amino acid levels as indicators of paleosols in New Zealand soil profiles, *Geoderma*, **7**, 33–47.

Gold, L. W. and Lachenbruch, A. H. (1973) Thermal variations in permafrost – a review of North American literature, North American Contribution, *Permafrost Second Internat. Conf., Yakutsk, USSR, Nat. Acad. Sci., Washington*, 3–23.

Goldthwait, R. P. (1971) Introduction to till, today, In R. P. Goldthwait (ed.), *Till: a Symposium*, Ohio State Univ. Press, 3–26.

Gooding, A. M. (1971) Postglacial alluvial history in the upper Whitewater Basin, south-eastern Indiana, and possible regional relationships, *Amer. J. Sci.*, **271**, 389–401.

Gordon, A. D. and Birks, H. J. B. (1972) Numerical methods in Quaternary palaeoecology. I. Zonation of pollen diagrams, *New Phytol.*, **71**, 961–79.

Gordon, A. D. and Birks, H. J. B. (1974) Numerical methods in Quaternary palaeoecology. II. Comparison of pollen diagrams, *New Phytol.*, **73**, 221–49.

Gordon, J. E. (1981) Glacier margin fluctuations during the 19th and 20th centuries in the Ikamiut Kangerdluarssuat area, west Greenland, *Arct. Alp. Res.*, **13**, 47–62.

Goudie, A. S. (1973) *Duricrusts in Tropical and Subtropical Landscapes*, Clarendon Press, Oxford.

Goudie, A. S. (1977) *Environmental Change*, Clarendon Press, Oxford.

Goudie, A. S. (1981) *Geomorphological Techniques*, Allen and Unwin, London and Boston.

Goudie, A. S., Allchin, B. and Hegde, K. T. M. (1973) The former extensions of the Great Indian Sand Desert, *Geogr. J.*, **139**, 243–57.

Granlund, E. (1932) De svenska högmossarnas geologi, *Sveriges Geol. Unders., Scr. C26*, **no. 373**.

Gray, J. M. (1975) Measurement and analysis of Scottish raised shoreline altitudes, *Queen Mary Coll., Univ. of London. Dept. of Geogr., Occas. Pap.*, **2**.

Gray, J. M. (1978) Low-level shore platforms in the south-west Scottish Highlands: altitude, age and correlation, *Trans. Inst. Brit. Geogr.*, **3**, 151–64.

Gray, J. M. (1981) Large-scale geomorphological field mapping: teaching the first stage, *J. Geogr. Higher Ed.*, **5**, 37–44.

Gray, J. M. (1982) The last glaciers (Loch Lomond Advance) in Snowdonia, N. Wales, *Geol. J.*, **17**, 111–33.

Gray, J. M. and Brooks, C. L. (1972) The Loch Lomond Readvance moraines of Mull and Menteith, *Scott. J. Geol.*, **8**, 95–103.

Gray, J. M. and Lowe, J. J. (1977) The Scottish Lateglacial environment: a synthesis, In J. M. Gray and J. J. Lowe (eds.), *Studies in the Scottish Lateglacial Environment*, Pergamon, Oxford and New York, 163–81.

Gray, J. M. and Lowe, J. J. (1982) Problems in the interpretation of small-scale erosional forms on glaciated bedrock surfaces: examples from Snowdonia, North Wales, *Proc. Geol. Assoc.*, **93**, 403–414.

Green, P. (1979) Tracking down the past, *New Scientist* (22 Nov. 1979), 624–6.

Gregory, K. J. and Cullingford, R. A. (1974) Lateral

variation in pebble shape in north-west Yorkshire, *Sedimentary Geol.*, **12**, 237–48.

Griffey, N. J. and Matthews, J. A. (1978) Major Neoglacial glacier expansion episodes in southern Norway: evidence from moraine ridge stratigraphy with ^{14}C dates on buried palaeosols and moss layers, *Geogr. Annlr.*, **60A**, 73–90.

Griffin, G. M. (1971) Interpretation of X-ray diffraction data, In R. E. Carver (ed.), *Procedures in Sedimentary Petrology*, Wiley, New York, 541–69.

Grootes, P. M. (1978) Carbon-14 timescale extended: comparison of chronologies, *Science*, **200**, 11–15.

Grove, A. T. (1969) Landforms and climatic change in the Kalahari and Ngamiland, *Geogr. J.*, **135**, 191–212.

Grove, A. T. and Pullan, R. A. (1963) Some aspects of the Pleistocene palaeogeography of the Chad Basin, *Viking Fund Publ. Anthropol.*, **36**, 230–45.

Grove, A. T., Street, F. A. and Goudie, A. S. (1975) Former lake levels and climatic change in the rift valley of southern Ethiopia, *Geogr. J.*, **141**, 177–202.

Grove, A. T. and Warren, A. (1968) Quaternary landforms and climate on the south side of the Sahara, *Geogr. J.*, **134**, 194–208.

Grüger, E. (1979) Die Seeablagerungen vom Samerberg/Obb. und ihre Stellung im Jungpleistozan, *Eiszeit. und Gegenw.*, **29**, 23–34.

Gruhn, R., Bryan, A. L. and Moss, A. J. (1974) A contribution to the Quaternary of south-east Essex, England, *Quat. Res.*, **4**, 53–75.

Guilcher, A. (1969) Pleistocene and Holocene sea level changes, *Earth Sci. Rev.*, **5**, 69–97.

Hall, A. R. (1980) Late Pleistocene deposits at Wing, Rutland, *Phil. Trans. R. Soc. Lond.*, **B 289**, 135–64.

Hallam, J. S., Edwards, B. J. N., Barnes, B. and Stuart, A. J. (1973) The remains of a Late Glacial elk associated with barbed points from High Furlong near Blackpool, Lancashire, *Proc. Prehist. Soc.*, **39**, 100–28.

Hammer, C. U. *et al.* (1978) Dating of Greenland ice cores by flow models, isotopes, volcanic debris and continental dust, *J. Glaciol.*, **20**, 3–26.

Hansen, S. (1940) Varvity in Danish and Scanian late glacial deposits, *Danm. Geol. Unders.*, **II Raekke 63**, 478 pp.

Haq, B. U. (1978) Calcareous nannoplankton, In B. U. Haq and A. Boersma (eds.), *Introduction to Marine Micropalaeontology*, Elsevier, Amsterdam, 79–107.

Haq, B. U. and Boersma, A. (eds.) (1978) *Introduction to Marine Micropalaeontology*, Elsevier, Amsterdam.

Hare, F. K. (1947) The geomorphology of parts of the Middle Thames, *Proc. Geol. Assoc.*, **58**, 294–339.

Haring, A., de Vries, A. E. and de Vries, H. (1958) Radiocarbon dating up to 70 000 years by isotopic enrichment, *Science*, **128**, 472–3.

Harkness, D. D. (1975) The role of the archaeologist in C-14 measurement, In T. Watkins (ed.), *Radiocarbon: Calibration and Prehistory*, Edinb. Univ. Press, Edinburgh, 128–35.

Harkness, D. D. (1979) Radiocarbon dates from Antarctica, *Br. Antarct. Surv. Bull.*, **47**, 43–59.

Harkness, D. D. and Burleigh, R. (1974) Possible carbon-14 enrichment in high altitude wood, *Archaeometry*, **16**, 121–7.

Harmon, R. S. (1977) [230]Th/[234]U dating of speleothems and a glacial chronology for alpine karst areas of western North America, *Xth INQUA Congress, Birmingham, 1977, Abstracts*, p. 195.

Harmon, R. S. *et al.* (1981) Bermuda sea level during the last interglacial, *Nature*, **289**, 481–3.

Harmon, R. S., Schwarcz, H. P. and Ford, D. C. (1978) Late Pleistocene sea level history of Bermuda, *Quat. Res.*, **9**, 205–18.

Harmon, R. S., Thompson, P., Schwarcz, H. P. and Ford, D. C. (1978) Late Pleistocene palaeoclimates of North America as inferred from stable isotope studies of speleothems, *Quat. Res.*, **9**, 54–70.

Harrison, P. W. (1957a) A clay till fabric: its character and origin, *J. Geol.*, **65**, 275–308.

Harrison, P. W. (1957b) New technique for three-dimensional fabric analysis of till and englacial debris containing particles from 3 to 40 mm in size, *J. Geol.*, **65**, 98–105.

Harrod, T. M., Catt, J. A. and Weir, A. H. (1973) Loess in Devon, *Proc. Ussher Soc.*, **2**, 554–64.

Havinga, A. J. (1964) Investigation into the differential corrosion susceptibility of pollen and spores, *Pollen Spores*, **4**, 621–35.

Havinga, A. J. (1967) Palynology and pollen preservation, *Rev. Palaeobotan. Palynol.*, **2**, 81–98.

Havinga, A. J. (1974) Problems in the interpretation of pollen diagrams from mineral soils, *Geol. en Mijn.*, **53**, 449–53.

Haworth, E. Y. (1976) Two Late-glacial diatom assemblage profiles from northern Scotland, *New Phytol.*, **77**, 227–56.

Hay, R. L. (1967) Revised stratigraphy of Olduvai Gorge, In W. W. Bishop and J. D. Clark (eds.), *Background to Evolution in Africa*, Univ. Chicago Press, 221–8.

Hay, R. L. (1973) Lithofacies and environments of Bed I, Olduvai Gorge, Tanzania, *Quat. Res.*, **3**, 541–60.

Haynes, J., Kiteley, R. J., Whatley, R. C. and Wilks, P. J. (1977) Microfaunas, microfloras and the environmental stratigraphy of the Late Glacial and Holocene in Cardigan Bay, *Geol. J.*, **12**, 129–58.

Hays, J. D., Imbrie, J. and Shackleton, N. J. (1976) Variations in the earth's orbit: pacemaker of the ice ages, *Science*, **194**, 1121–32.

Healy, T. (1981) Submarine terraces and morphology in the Kieler Bucht, western Baltic, and their relation to Quaternary events, *Boreas*, **10**, 209–17.

Hecht, A. D. (1973) Faunal and oxygen isotopic palaeotemperatures, and the amplitude of glacial/interglacial temperature changes in the equatorial Atlantic, Caribbean Sea, and Gulf of Mexico, *Quat. Res.*, **3**, 671–90.

Hedberg, H. D. (1976) *International Stratigraphic Guide*, Wiley, New York.

Hedges, R. E. M. and Moore, C. B. (1978) Enrichment of ^{14}C for radiocarbon dating, *Nature*, **276**, 255–7.

Heer, O. (1865) *Die Urwelt der Schweiz*, F. Schulthess, Zurich.

Hendy, C. H. and Wilson, A. T. (1968) Palaeoclimatic data from speleothems, *Nature*, **216**, 48–51.

Heusser, L. and Balsam, W. L. (1977) Pollen distribution in the north-east Pacific Ocean, *Quat. Res.*, **7**, 45–62.

Hey, R. W. (1978) Horizontal Quaternary shorelines of the Mediterranean, *Quat. Res.*, **10**, 197–203.

Hibbert, F. A. and Switsur, V. R. (1976) Radiocarbon dating of Flandrian pollen zones in Wales and northern England, *New Phytol.*, **77**, 793–807.

Higgins, A. L. (1970) *Elementary Surveying* (3rd ed.), Longman, London.

Hill, A. R. (1968) An experimental test of the field technique of till macrofabric analysis, *Trans. Inst. Brit. Geogr.*, **45**, 93–105.

Hill, A. R. and Prior, D. B. (1968) Directions of ice movement in north-east Ireland, *Proc. R. Irish Acad.*, **B 66**, 71–84.

Hillefors, A. (1979) Deglaciation models from the Swedish west coast, *Boreas*, **8**, 153–69.

Hjort, C. (1979) Glaciation in northern East Greenland during the Late Weichselian and Early Flandrian, *Boreas*, **8**, 281–96.

Hjort, C. and Funder, S. (1974) The subfossil occurrence of *Mytilus edulis* L. in central East Greenland, *Boreas*, **3**, 23–33.

Hollin, J. T. (1977) Thames interglacial sites, Ipswichian sea levels and Antarctic ice surges, *Boreas*, **6**, 38–52.

Holmes, C. D. (1941) Till fabric, *Bull. Geol. Soc. Amer.*, **52**, 1299–354.

Holmes, G. E., Hopkins, D. M. and Foster, H. L. (1968) Pingos in central Alaska, *US Geol. Surv. Bull.*, **1241-H**.

Holtedahl, H. and Sellevoll, M. (1972) Notes on the influence of glaciation on the Norwegian continental shelf bordering on the Norwegian Sea, In E. Dahl, J–O. Stromberg and O. G. Tandberg (eds.), *The Norwegian Sea Region: its hydrography, glacial and biological history*, Ambio Spec. Rep., **2**, 31–8.

Hopkins, D. M. (1967) *The Bering Land Bridge*, Stanford Univ. Press.

Horie, S. (ed.) (1976) *Palaeolimnology of Lake Biwa and the Japanese Pleistocene* (vol. 4), Contribution to the 'Paleolimnology of Lake Biwa and the Japanese Pleistocene', No. **155**, Kyoto Univ., Otsu.

Horie, S. (ed.) (1979) *International Project on Palaeolimnology and late Cenozoic Climate*, No. 2, Contribution to the 'Palaeolimnology of Lake Biwa and the Japanese Pleistocene', No. **263**, Kyoto Univ., Otsu.

Howell, F. C. et al. (1972) Uranium-series dating of bone from the Isimila prehistoric site, Tanzania, *Nature*, **237**, 51–2.

Huddart, D. (1971) Textural distinction between Main Glaciation and Scottish Readvance tills in the Cumberland Lowland, *Geol. Mag.*, **108**, 317–24.

Huddart, D., Tooley, M. J. and Carter, P. A. (1977) The coasts of north-west England, In C. Kidson and M. J. Tooley (eds.), *The Quaternary History of the Irish Sea*, Seel House Press, Liverpool, 119–54.

Hurford, A. J., Gleadow, A. J. W. and Naeser, C. W. (1976) Fission-track dating of pumice from the KBS tuffs, East Rudolph, Kenya, *Nature*, **263**, 738–40.

Hutchinson, C. S. (1974) *Laboratory Handbook of Petrographic Techniques*, Wiley, New York.

Hyvarinen, H. (1973) The deglacial history of eastern Fennoscandia – recent data from Finland, *Boreas*, **2**, 85–102.

Ignatius, H., Korpela, K. and Kajunsuu, R. (1980) The deglaciation of Finland after 10 000 BP, *Boreas*, **9**, 217–28.

Ikeya, M. (1975) Dating a stalactite by electron paramagnetic resonance, *Nature*, **255**, 48–50.

Imbrie, J. and Imbrie, K. P. (1979) *Ice Ages: Solving the Mystery*, MacMillan, London and Basingstoke.

Imbrie, J. and Kipp, N. (1971) A new micropalaeontological method of quantitative palaeoclimatology: application to a Late Pleistocene Caribbean core, In K. K. Turekian (ed.), *The Late Cenozoic Glacial Ages*, Yale Univ. Press, New Haven, 71–181.

Imbrie, J., Van Donk, J. and Kipp, N. G. (1973) Palaeoclimatic investigations of a late Pleistocene Caribbean deep-sea core: comparisons of isotopic and faunal methods, *Quat. Res.*, **3**, 10–38.

Ince, J. (1981) Pollen analysis and radiocarbon dating of Lateglacial and early Flandrian deposits in Snowdonia, N. Wales, Unpubl. PhD Thesis, City of London Polytechnic.

Institute of Geological Sciences (1977) Maps of the Quaternary deposits of the British Isles, 1:625 000 (2 sheets).

Iversen, J. (1954) The lateglacial flora of Denmark and its relation to climate and soil, *Danm. Geol. Unders., ser II*, **80**, 87–119.

Ives, J. D. (1976) The Saglek moraines of northern Labrador: a commentary, *Arct. Alp. Res.*, **8**, 403–8.

Ives, J. D., Andrews, J. T. and Barry, R. G. (1975) Growth and decay of the Laurentide Ice Sheet and comparisons with Fenno-Scandinavia, *Naturwissenschaften*, **62**, 118–25.

Izett, G. A., Wilcox, R. E. and Borchardt, G. A. (1972) Correlation of a volcanic ash band in Pleistocene deposits near Mount Blanco, Texas, and the Guaje Pumice bed of the Jemez Mountains, New Mexico, *Quat. Res.*, **2**, 554–78.

Jäger, E. and Hunziker, J. C. (1979) *Lectures in Isotope Geology*, Springer-Verlag, Berlin and New York.

Jane, F. W. (1970) *The Structure of Wood*, A. and C. Black, London.

Jardine, W. G. (1981) Status and relationships of the Loch Lomond Readvance and its stratigraphical correlatives, In J. Neale and J. Flenley (eds.), *The Quaternary in Britain*, Pergamon, Oxford and New York, 168–73.

Jauhiainen, E. (1975) Morphometric analysis of drumlin fields in northern central Europe, *Boreas*, **4**, 219–30.

Jelgersma, S. (1966) Sea level changes in the last 10 000 years, In *International Symposium on World Climate from 8000 – 0 BC*, Roy. Met. Soc., 54–69.

Jelgersma, S. (1979) Sea level changes in the North Sea basin, In F. Oele, R. T. E. Schüttenholm and A. J. Wiggers (eds.), *The Quaternary History of the North Sea*, Almqvist and Wiksell, Stockholm, 233–48.

Jochimsen, M. (1973) Does the size of lichen thalli really constitute a valid measure for dating glacier surfaces?, *Arct. Alp. Res.*, **5**, 417–24.

John, B. S. (1970) Pembrokeshire, In C. A. Lewis (ed.), *The Glaciations of Wales and Adjoining Region's*, Longman, London, 229–65.

Johnsen, S. J., Dansgaard, W., Clausen, H. B. and Langway Jr., C. C. (1972) Oxygen isotope profiles through the Antarctic and Greenland ice sheets, *Nature*, **235**, 429–34.

Johnson, R. G. (1982) Brunhes-Matuyama reversal dated at 790 000 yr BP by marine-astronomical correlations, *Quat. Res.*, **17**, 135–47.

Jones, D. K. C. (1974) The influence of the Calabrian transgression on the drainage evolution of south-east England, In E. H. Brown and R. S. Waters (eds.), *Progress in Geomorphology, Inst. Brit. Geogr. Spec. Publ.*, **7**, 139–58.

Jones, D. K. C. (1981) *South-east and Southern England*, Methuen, London and New York.

Jones, R. L. and Cundill, P. R. (1978) Introduction to pollen analysis, *Brit. Geomorph. Res. Group Tech. Bull.*, **22**.

Jørgensen, P. (1977) Some properties of Norwegian tills, *Boreas*, **6**, 149–57.

Kahn, M. I., Oba, T. and Ku, T-L. (1981) Palaeotemperatures and the glacially-induced changes in the oxygen isotope composition of sea water during Late Pleistocene and Holocene time in the Tanner Basin, California, *Geology*, **9**, 485–90.

Karlén, W. (1973) Holocene glacier and climatic variations, Kebnekaise Mountains, Swedish Lapland, *Geogr. Annlr.*, **55A**, 29–63.

Karlén, W. and Denton, G. H. (1976) Holocene glacier variations in Sarek National Park, northern Sweden, *Boreas*, **5**, 25–56.

Karte, J. and Liedtke, H. (1981) The theoretical and practical definition of the term 'Periglacial' in its geographical and geological meaning, *Biul. Peryglac.*, **28**, 123–35.

Kaufman, A. (1971) U-series dating of Dead Sea Basin carbonates, *Geochim. et Cosmochim. Acta*, **35** 1269–81.

Kaufman, A. and Broecker, W. S. (1965) Comparison of ^{230}Th and ^{14}C ages for carbonate materials from Lakes Lahontan and Bonneville, *J. Geophys. Res.*, **70**, 4039–54.

Kaufman, A., Broecker, W. S., Ku, T-L., and Thurber, D. L. (1971) The status of U-series methods of mollusk dating, *Geochim. et Cosmochim. Acta*, **35**, 1155–83.

Kellogg, T. B. (1976) Late Quaternary climatic changes: evidence from deep-sea cores of Norwegian and Greenland seas, *Geol. Soc. Amer. Mem.*, **145**.

Kendall, R. L. (1969) An ecological history of the Lake Victoria basin, *Ecol. Monogr.*, **39**, 121–76.

Kennard, A. S. (1944) The Crayford brickearths, *Proc. Geol. Assoc.*, **55**, 121–69.

Kenward, H. K. (1975a) Pitfalls in the environmental interpretation of death assemblages. *J. Arch. Sci.*, **2**, 85–94.

Kenward, H. K. (1975b) The biological and archaeological implications of the beetle *Aglenus brunneus* (Gyllenhall) in ancient faunas, *J. Arch. Sci.*, **2**, 63–9.

Kenward, H. K. (1976) Reconstructing ancient ecological conditions from insect remains: some problems and an experimental approach, *Ecol. Entomol.*, **1**, 7–17.

Kerney, M. P. (1963) Late-glacial deposits in the Chalk of south-east England, *Phil. Trans. R. Soc. Lond.*, **B 246**, 203–54.

Kerney, M. P. (1965) Weichselian deposits on the Isle of Thanet, East Kent, *Proc. Geol. Assoc.*, **76**, 269–74.

Kerney, M. P. (1968) Britain's fauna of land mollusca and its relation to the post-glacial thermal optimum, *Symp. Zool. Soc. Lond.*, **22**, 273–91.

Kerney, M. P. (1971a) A Middle Weichselian deposit at Halling, Kent, *Proc. Geol. Assoc.*, **82**, 1–11.

Kerney, M. P. (1971b) Interglacial deposits in Barnfield Pit, Swanscombe, and their molluscan fauna, *Q. J. Geol. Soc., Lond.*, **127**, 69–73.

Kerney, M. P. (1976) *Atlas of the Non-Marine Mollusca of the British Isles*, Conch. Soc. GB and Ireland, and Nat. Env. Res. Council.

Kerney, M. P. (1977a) British Quaternary non-marine mollusca: a brief review, In F. W. Shotton (ed.), *British Quaternary Studies – Recent Advances*, Oxford Univ. Press, 31–42.

Kerney, M. P. (1977b) A proposed zonation scheme for late-glacial and post-glacial deposits using land mollusca, *J. Arch. Sci.*, **4**, 387–90.

Kerney, M. P., Brown, E. H. and Chandler, T. J. (1964)

The late-glacial and Postglacial history of the Chalk escarpment near Brook, Kent, *Phil. Trans. R. Soc. Lond.*, **B 248**, 135–204.

Kerney, M. P. and Cameron, R. A. D. (1979) *A Field Guide to the Land Snails of Britain and North-west Europe*, Collins, London.

Kerney, M. P., Preece, R. C. and Turner, C. (1980) Molluscan and plant biostratigraphy of some Late Devensian and Flandrian deposits, Kent, *Phil. Trans. R. Soc. Lond.*, **B 291**, 1–43.

Kerschner, H. (1978) Palaeoclimatic inferences from late Wurm rock glaciers, western Tyrol, Austria, *Arct. Alp. Res.*, **10**, 635–44.

Kesling, R. V. (1965) Ostracod investigations, *Nat. Sci. Foundation, Rep.*, **1**.

Kidson, C. (1977) The coast of south-west England, In C. Kidson and M. J. Tooley (eds.), *The Quaternary History of the Irish Sea*, Seel House Press, Liverpool, 257–98.

Kidson, C., Gilberston, D. D. et al. (1978) Interglacial marine deposits of the Somerset Levels, south-west England, *Boreas*, **7**, 215–28.

Kidson, C. and Heyworth, A. (1973) The Flandrian sea level rise in the Bristol Channel, *Proc. Ussher Soc.*, **2**, 565–84.

Kilford, W. K. (1963) *Elementary Air Survey*, Pitman, London.

King, C. A. M. (1966) *Techniques in Geomorphology*, Edward Arnold, London.

King, C. A. M. and Buckley, J. (1968) Analysis of stone size and shape in Arctic environments, *J. Sed. Petrol.*, **38**, 200–14.

King, S. A. (1978) Radiolaria, In B. U. Haq and A. Boersma (eds.), *Introduction to Marine Micropalaeontology*, Elsevier, Amsterdam, 203–44.

King, W. B. R. and Oakley, K P. (1936) The Pleistocene succession in the lower part of the Thames Valley, *Proc. Prehist. Soc.*, **2**, 52–76.

Kipp, N. G. (1976) New transfer function for estimating past sea-surface conditions from sea bed distribution of planktonic foraminiferal assemblages in the north Atlantic, *Geol. Soc. Amer. Mem.* **145**, 3–41.

Kirk, W. and Godwin, H. (1963) A Lateglacial site at Loch Droma, Ross and Cromarty, *Trans. Roy. Soc. Edinb.*, **65**, 225–49.

Knudsen, K. L. (1977) Foraminiferal faunas of the Quaternary Hostrup clay from northern Jutland, *Boreas*, **6**, 229–45.

Kostyaev, A. G. (1969) Wedge and fold-like diagenetic disturbances in Quaternary sediments and their palaeogeographic significance, *Biul. Peryglac.*, **19**, 231–70.

Krenke, A. N. and Khodakov, V. G. (1966) Connection between the surface melting of glaciers and the air temperature, *Materials Glaciol. Res.*, A. K. Nauk, Inst. Geogr. Chronik, Moscow, **12**, 153–64.

Krinsley, D. H. and Doornkamp, J. C. (1973) *An Atlas of Quartz Sand Surface Textures*, Cambridge Univ. Press, Cambridge.

Krinsley, D. H. and Funnell, B. M. (1965) Environmental history of quartz sand grains from the Lower and Middle Pleistocene of Norfolk, England, *Q. J. Geol. Soc. Lond.*, **124**, 435–61.

Krumbein, W. C. (1939) Preferred orientations of pebbles in sedimentary deposits, *J. Geol.*, **47**, 672–706.

Krumbein, W. C. (1941) Measurement and geological significance of shape and roundess of sedimentary particles, *J. Sed. Petrol.*, **11**, 64–72.

Krumbein, W. C. and Pettijohn, F. J. (1938) *Manual of Sedimentary Petrography*, Appleton-Century Crofts, New York.

Ku, T-L. (1976) The uranium-series methods of age determination, *Ann. Rev. Earth Planet. Sci.*, **4**, 347–79.

Kukla, G. J. (1970) Correlation between loesses and deep-sea sediments, *Geol. Fören. Stockh. Förh.*, **92**, 148–80.

Kukla, G. J. (1975) Loess stratigraphy of central Europe, In K. W. Butzer and G. L. Isaac (eds.), *After the Australopithecines*, Mouton Press, The Hague, 99–188.

Kullenberg, B. (1947) The piston core sampler, *Svenska Hydro-Biol. Komm. Skr. Ser.* **3**, 1–46.

Kullenberg, B. (1955) Deep-sea coring. *Rep. Swedish Deep-Sea Expeditions*, **4**, 35–96.

Kummel, B. and Raup, D. (1965) *Handbook of Palaeontological Techniques*, Freeman, San Francisco.

Kurten, B. (1968) *Pleistocene Mammals of Europe*, Aldine Publ. Co., Chicago.

Kutzbach, J. E. (1980) Estimates of past climate at palaeolake Chad, North Africa, based on a hydrological and energy balance model, *Quat. Res.*, **14**, 210–23.

Kvenvolden, K. A. (1975) Advances in the geochemistry of amino acids, *Ann. Rev. Earth Planet. Sci.*, **3** 183–212.

Lachenbruch, A. H. (1966) Contraction theory of ice wedge polygons: a qualitative discussion, In *Proc. 1st Internat. Conf.*, Nat. Acad. Sci., Nat. Res. Council, Canada, Publ. **1237** 63–71.

Lajoie, K. R., Peterson, E. and Gerow, B. A. (1980) Amino acid bone dating: a feasibility study, South San Francisco Bay Region, California, In P. E. Hare, T. C. Hoering and K. King Jr (eds.), *Biogeochemistry of Amino Acids*, Wiley, New York and Chichester, 477–89.

LaMarche Jr, V. C. (1970) Frost-damage rings in subalpine conifers and their application to tree-ring dating problems, In J. H. G. Smith and J. Worrall (eds.), *Tree-Ring Analysis with special reference to North-west America*, Univ. British Columbia, Faculty of Forestry, Bull. **7**.

LaMarche Jr, V. C. and Fritts, H. C. (1971) Anomaly patterns of climate over the Western United States, 1700–1930, derived from Principal Components Analysis of tree-ring data, *Monthly Weather Rev.*, **99**, 138–42.

Lamb, H. H. (1977a) Climatic analysis, *Phil. Trans. R. Soc. Lond.*, **B 280**, 341–50.

Lamb, H. H. (1977b) The Late Quaternary history of the climate of the British Isles, In F. W. Shotton (ed.), *British Quaternary Studies – Recent Advances*, Oxford Univ. Press, 283–98.

Lamb, H. H. and Woodroffe, A. (1970) Atmospheric circulation during the last ice age, *Quat. Res.*, **1**, 29–58.

Landmesser, C. W., Johnson, T. C. and Wold, R. J. (1982) Seismic reflection study of recessional moraines beneath Lake Superior and their relationship to regional deglaciation, *Quat. Res.*, **17**, 173–90.

Langway Jr, C. C. (1970) Stratigraphic analysis of a deep ice core from Greenland, *Geol. Soc. Amer. Spec. Pap.*, **125**.

Laville, H. (1976) Deposits in calcareous rock shelters: analytical methods and climatic interpretation, In D. A. Davidson and M. L. Shackley (eds.), *Geoarchaeology*, Duckworths, London, 137–55.

Lemke, R. W. (1958) Narrow linear drumlins near Velva, North Dakota, *Amer. J. Sci.*, **256**, 270–4.

Leonard, A. B. and Frye, J. C. (1954) Ecological conditions accompanying loess deposition in the Great Plains region, *J. Geol.*, **62**, 399–404.

Lerman, J. C. (1972) Carbon 14 dating: origin and correction of isotope fractionation errors in terrestrial living matter, *Proc. 8th Internat. Conf. on Radiocarbon Dating, Roy. Soc. New Zealand, Wellington*, pp. H17–H28.

Lewis, C. A. (ed.) (1970) *The Glaciations of Wales and Adjoining Regions*, Longman, London.

Libby, W. F. (1955) *Radiocarbon Dating*, Univ. Chicago Press (2nd ed.).

Lichti-Federovich, S. and Ritchie, J. C. (1968) Recent pollen assemblages from the western interior of Canada, *Rev. Palaeobotan. Palynol.*, **7**, 297–344.

Liestøl, O. (1967) *Storbreen glacier in Jotunheimen, Norway*, Norsk Polarinstitutt. Skrifter, Nr. 141.

Lill, G. O. and Smalley, I. J. (1978) Distribution of loess in Britain, *Proc. Geol. Assoc.*, **88**, 57–65.

Linton, D. L. (1957) Radiating valleys in glaciated lands, In C. Embleton (ed.), *Glaciers and Glacial Erosion*, MacMillan, London, 130–48.

Linton, D. L. (1963) The forms of glacial erosion, *Trans. Inst. Brit. Geogr.*, **33**, 1–28.

Lisitzin, E. (1974) *Sea-Level Changes*, Elsevier, Amsterdam and New York.

Livingstone, D. A. (1955) A lightweight piston sampler for lake deposits, *Ecology*, **36**, 137–9.

Lock, W. W., Andrews, J. T. and Webber, P. J. (1980) *Manual for Lichenometry*, Brit. Geomorph. Res. Group, Tech. Bull., **26**.

Lockwood, J. G. (1979) Water balance of Britain 50 000 to the present day, *Quat. Res.*, **12**, 297–310.

Loewe, F. (1971) Consideration on the origin of the Quaternary ice sheet of North America, *Arct. Alp. Res.*, **3**, 331–44.

Löffler, H. and Danielpol, D. (eds.) (1977) *Aspects of Ecology and Zoogeography of Recent and Fossil Ostracoda*, W. Junk, The Hague.

Lord, A. R. (1980) Interpretation of Lateglacial marine environment of N.W. Europe by means of foraminifera, In J. J. Lowe, J. M. Gray and J. E. Robinson (eds.), *Studies in the Lateglacial of North-west Europe*, Pergamon, Oxford and New York, 103–14.

Løvlie, R. and Larsen E. (1981) Palaeomagnetism and magnetostratigraphy of a Holocene lake sediment from Vagsøy, western Norway, *Phys. Earth Planet, Interiors*, **27**, 143–50.

Lowe, J. J. (1978) Radiocarbon-dated Lateglacial and early Flandrian pollen profiles from the Teith Valley, Perthshire, Scotland, *Pollen Spores*, **20**, 367–97.

Lowe, J. J. (1982) Three Flandrian pollen profiles from the Teith Valley, Perthshire Scotland. II. Analysis of deteriorated pollen, *New Phytol.*, **90**, 371–85.

Lowe, J. J. and Gray, J. M. (1980) The stratigraphic subdivision of the Lateglacial of North-west Europe, In J. J. Lowe, J. M. Gray and J. E. Robinson (eds.), *Studies in the Lateglacial of North-west Europe*, Pergamon, Oxford and New York, 157–75.

Lowe, J. J. and Walker, M. J. C. (1976) Radiocarbon dates and the deglaciation of Rannoch Moor, Scotland, *Nature*, **246**, 632–3.

Lowe, J. J. and Walker, M. J. C. (1977) The reconstruction of the Lateglacial environment in the southern and eastern Grampian Highlands, In J. M. Gray and J. J. Lowe (eds.). *Studies in the Scottish Lateglacial Environment*, Pergamon, Oxford and New York, 101–18.

Lowe, J. J. and Walker, M. J. C. (1980) Problems associated with radiocarbon dating the close of the

Lateglacial period in the Rannoch Moor area, Scotland, In J. J. Lowe, J. M. Gray and J. E. Robinson (eds), *Studies in the Lateglacial of North-west Europe*, Pergamon, Oxford and New York. 123–37.

Lowe, J. J. and Walker, M. J. C. (1981) The early Postglacial environment of Scotland: evidence from a site near Tyndrum, Perthshire, *Boreas*, **10**, 281–94.

Lowe, S. (1981) Radiocarbon dating and stratigraphic resolution in Welsh lateglacial chronology, *Nature*, **293**, 210–12.

Lozeck, V. (1964) Quätamolluskan der Tschecoslowakei, *Rozpravy Ústredniho Ustavu. Geologického*, **31**, Praha.

Lozeck, V. (1972) Holocene interglacial in central Europe and its land snails, *Quat. Res.*, **2**, 327–34.

Luckman, B. H. (1970) The Hereford Basin, In C. A. Lewis (ed.), *The Glaciations of Wales and Adjoining Regions*, Longman, London, 175–96.

Ludlam, S. D. (1979) Rhythmite deposition in lakes of the north-eastern United States, In Ch. Schlüchter (ed.), *Moraines and Varves*, Balkema, Rotterdam, 295–302.

Lundelius, E. L. (1976) Vertebrate palaeontology of the Pleistocene: an overview, *Geoscience and Man*, **13**, 45–59.

Lundqvist, G. (1962) Geological radiocarbon datings from the Swedish station, *Sveriges Geol. Unders. Arbok.*, **56**, 1–23.

Lundqvist, J. (1975) Ice recession in central Sweden, and the Swedish Time Scale, *Boreas*, **4**, 47–54.

Lundqvist, J. (1980) The deglaciation of Sweden after 10 000 BP, *Boreas*, **9**, 229–38.

Lutwick, L. E. (1969) Identification of phytoliths in soils, In S. W. Pawluk (ed.), *Pedology and Quaternary Research*, Univ. of Alberta Press, Edmonton, 77–82.

Maarleveld, G. C. (1960) Wind directions and cover sands in the Netherlands, *Biul. Peryglac.*, **8**, 49–58.

Maarleveld, G. C. (1964) Periglacial phenomena in the Netherlands during different parts of the Wurm time, *Biul. Peryglac.*, **14**, 251–6.

Maarleveld, G. C. (1976) Periglacial phenomena and mean annual temperatures during the last glacial time in the Netherlands, *Biul. Peryglac.*, **26**, 57–78.

Mabbutt, J. A. (1977) *Desert Landforms*, MIT Press, Cambridge, Mass.

McCann, S. B., Howarth, P. J. and Cogley, J. G. (1972) Fluvial processes in a periglacial environment, *Trans. Inst. Brit. Geogr.*, **55**, 69–82.

McCave, I. N., Caston, V. N. D. and Fannin, N. G. T. (1977) The Quaternary of the North Sea, In F. W. Shotton (ed.), *British Quaternary Studies – Recent Advances*, Oxford Univ. Press, 187–204.

McCave, I. N. and Jarvis, J. (1973) Use of the Model T Coulter Counter in size analysis of fine to coarse sand, *Sedimentol.*, **20**, 305–15.

McFarlane, M. J. (1977) *Laterite and Landscape*, Academic Press, New York.

McGraw, J. D. (1975) Quaternary airfall deposits of New Zealand, In R. P. Suggate and M. M. Cresswell (eds.), *Quaternary Studies, Roy. Soc. New Zealand Bull.*, **13**, 35–44.

McGregor, D. F. M. and Green, C. P. (1978) Gravels of the River Thames as a guide to Pleistocene catchment changes, *Boreas*, **7**, 197–203.

McHenry, J. R., Ritchie, J. C. and Gill, A. C. (1973) Accumulation of fallout caesium-137 in soils and sediments in selected watersheds, *Water Resource Res.*, **9**, 676–86.

McIntyre, A., Kipp, N. G. *et al.* (1976) Glacial North

Atlantic 18 000 years ago: A CLIMAP reconstruction, *Geol. Soc. Amer. Mem.*, **145**, 43–76.

McIntyre, A. and McIntyre, R. (1971) Coccolith concentrations and differential solution in oceanic sediments, In B. M. Funnell and W. R. Riedel (eds.), *The Micropalaeontology of Oceans*, Cambridge Univ. Press. 253–61.

McIntyre, A. and Ruddiman, W. F. (1972) North-east Atlantic post-Eemian palaeoceanography: a predictive analog for the future, *Quat. Res.*, **2**, 350–4.

McIntyre, A., Ruddiman, W. F. and Jantzen, R. (1972) Southward penetration of the North Atlantic Polar Front and faunal and floral evidence of large scale surface water mass movements over the past 225 000 years, *Deep Sea Res.*, **19**, 61–77.

Mackay, J. R. (1962) Pingos of the Pleistocene Mackenzie River delta area, *Geogr. Bull.*, **18**, 21–63.

Mackereth, F. J. H. (1958) A portable piston sampler for lake deposits, *Limnol. Oceanogr.*, **3**, 181–91.

Mackereth, F. J. H. (1965) Chemical investigation of lake sediments and their interpretation, *Proc. R. Soc. Lond.*, **B 161**, 295–309.

Mackereth, F. J. H. (1966) Some chemical observations on post-glacial lake sediments, *Phil. Trans. R. Soc. Lond.*, **B 250**, 165–213.

McVean, D. N. (1953) Biological flora of the British Isles: *Alnus* Mill., *J. Ecol.*, **41**, 447–66.

Madgett, P. A. and Catt, J. A. (1978) Petrography, stratigraphy and weathering of Late Pleistocene tills in east Yorkshire, Lincolnshire and north Norfolk, *Proc. Yorks. Geol. Soc.*, **42**, 55–108.

Mahaney, W. C. (ed.) (1976) *Quaternary Stratigraphy of North America*, Dowden, Hutchinson and Ross, Stroudsberg, Pennsylvania.

Mahaney, W. C. and Fahey, B. D. (1976) Quaternary soil stratigraphy in the Front Range, Colorado, In W. C. Mahaney (ed.), *Quaternary Stratigraphy of North America*, Dowden, Hutchinson and Ross, Stroudsberg, Penn., 319–52.

Mangerud, J. (1972) Radiocarbon dating of marine shells, including a discussion of apparent age of Recent shells from Norway, *Boreas*, **1**, 143–72.

Mangerud, J. (1977) Late Weichselian marine sediments containing shells, foraminifera and pollen at Ågotnes, western Norway, *Norsk. Geol. Tidsskr.*, **57**, 23–54.

Mangerud, J. (1980) Ice-front variations of different parts of the Scandinavian ice sheet, 13 000 – 10 000 years BP, In J. J. Lowe, J. M. Gray and J. E. Robinson (eds.), *Studies in the Lateglacial of North-west Europe*, Pergamon, Oxford and New York, 23–30.

Mangerud, J., Andersen, S. Th., Berglund, B. E. and Donner, J. J. (1974) Quaternary stratigraphy of Norden, a proposal for terminology and classification, *Boreas*, **3**, 109–26.

Mangerud, J. and Gulliksen, S. (1975) Apparent radiocarbon ages of Recent marine shells from Norway, Spitsbergen and Arctic Canada, *Quat. Res.*, **5**, 263–73.

Mangerud, J., Sønstegaard, E. and Sejrup, H-P., (1979) The correlation of the Eemian (interglacial) Stage and the deep-sea oxygen-isotope stratigraphy, *Nature*, **277**, 189–92.

Mankinen, E. A. and Dalrymple, G. B. (1979) Revised geomagnetic polarity time-scale for the interval 0–5 my BP, *J. Geophys. Res.*, **84**, 615–26.

Manley, G. (1975) Fluctuations of snowfall and persistence of snow cover in marginal-oceanic climates, In *Proc.*

WMO/IAMAP Symp. on Long-term Climatic Fluctuations, WMO, Geneva, 183–8.

Mark, D. M. (1973) Analysis of axial orientation data, including till fabrics, *Bull. Geol. Soc. Amer.*, **84**, 1369–74.

Marshall, J. F. and Launay, J. (1978) Uplift rates of the Loyalty Islands as determined by ^{230}Th/^{234}U dating of raised coral terraces, *Quat. Res.*, **9**, 186–92.

Martin, P. S. and Wright, H. E. (eds.) (1967) *Pleistocene Extinctions: the Search for a Cause*, Yale Univ. Press.

Matthews, B. (1970) Age and origin of aeolian sand in the Vale of York, *Nature*, **227**, 1234–6.

Matthews, J. (1974) Quaternary environments at Cape Deceit (Seward Peninsula, Alaska); evolution of a tundra ecosystem, *Bull. Geol. Soc. Amer.*, **85**, 1353–84.

Matthews, J. (1976) Evolution of the subgenus *Cyphelophorus* (genus *Helophorus*, Hydrophilidae, Coleoptera): description of two new fossil species and discussion of *Helophorus tuberculatus* Gyll., *Can. J. Zool.*, **54**, 652–73.

Matthews, J. A. (1973) Lichen growth on an active medial moraine, Jotunheim, Norway, *J. Glaciol.*, **65**, 305–13.

Matthews, J. A. (1974) Families of lichenometric dating curves from the Storbreen gletschervorfeld, Jotunheimen, Norway, *Norsk Geogr. Tidsskr.*, **28** 215–35.

Matthews, J. A. (1975) Experiments on the reproducibility and reliability of lichenometric dates, Storbreen gletschervorfeld, Jotunheim, Norway, *Norsk Geogr. Tidsskr.*, **29**, 97–109.

Matthews, J. A. (1978) Plant colonisation patterns on a gletschervorfeld, southern Norway: a meso-scale geographical approach to vegetation change and phytometric dating, *Boreas*, **7**, 155–78.

Matthews, J. A. (1980) Some problems and implications of ^{14}C dates from a podzol buried beneath an end moraine at Haugabreen, southern Norway, *Geogr. Annlr.*, **62A**, 185–208.

Mayhew, D. F. (1977) Avian predators as accumulators of fossil mammal material, *Boreas*, **6**, 25–31.

Melcher, C. L. (1981) Thermoluminescence of meteorites and their terrestrial ages, *Geochim. et Cosmochim. Acta*, **45**, 615–26.

Menzies, J. (1978) A review of the literature on the formation and location of drumlins, *Earth-Science Reviews*, **14**, 315–50.

Mercer, J. H. (1968) The discontinuous glacio-eustatic fall in Tertiary sea level, *Palaeogeogr. Palaeoclimatol. Palaeoecol.*, **5**, 77–86.

Mesolella, K. J., Matthews, R. K., Broecker, W. S. and Thurber, D. L. (1969) The astronomical theory of climatic change: Barbados data, *J. Geol.*, **77**, 250–74.

Metson, A. J. (1961) Methods of Chemical Analysis for Soil Survey Samples, *New Zealand Dept. Sci. and Industr. Res., Soil Bureau Bull.*, **12**.

Michael, H. N. and Ralph, E. K (1972) Discussion of radiocarbon dates obtained from precisely dated *Sequoia* and bristlecone-pine samples, *Proc. 8th Internat. Conf. on Radiocarbon Dating*, Roy. Soc. NZ, Wellington, pp. A11–A27.

Miller, C. D. (1973) Chronology of Neoglacial deposits in the northern Sawatch Range, Colorado, *Arct. Alp. Res.*, **5**, 373–84.

Miller, G. H., Bradley, R. S. and Andrews, J. T. (1975) The glaciation level and lowest equilibrium line altitude in the high Canadian Arctic: maps and climatic interpretation, *Arct. Alp. Res.*, **7**, 155–68.

Miller, G. H. and Hare, P. E. (1980) Amino acid geochronology: integrity of the carbonate matrix and potential of molluscan fossils, In P. E. Hare, T. C. Hoering and K. King Jr (eds.), *Biogeochemistry of Amino Acids*, Wiley, New York and Chichester, 415–43.

Miller, G. H., Hollin, J. T. and Andrews, J. T. (1979) Aminostratigraphy of UK Pleistocene deposits, *Nature*, **281**, 539–43.

Miller, H. (1884) On boulder glaciation, *Proc. R. Phys. Soc. Edinb.*, **8**, 156–89.

Milliman, J. D. and Emery, K O. (1968) Sea-levels during the past 35 000 years, *Science*, **162**, 1121–3.

Mitchell, G. F. (1942) A composite pollen diagram from Co. Meath, Ireland, *New Phytol.*, **41**, 257–61.

Mitchell, G. F. (1973) Fossil pingos in Camaross Townland, Co. Wexford, *Proc. R. Ir. Acad.*, **73B**, 269–82.

Mitchell, G. F. (1976) *The Irish Landscape*, Collins, London.

Mitchell, G. F. (1977) Raised beaches and sea-levels, In F. W. Shotton (ed.), *British Quaternary Studies – Recent Advances*, Oxford Univ. Press, 167–86.

Mitchell, G. F., Penny, L F., Shotton, F. W. and West, R G. (1973) A correlation of Quaternary deposits in the British Isles, *Geol. Soc. Lond., Spec. Rep.*, **4**, 1–99.

Mitchell, G. F. and West, R G. (eds.) (1977) The changing environmental conditions in Great Britain and Ireland during the Devensian (last) cold stage, *Phil. Trans. R. Soc. Lond.*, **B 280**, 103–374.

Monroe, W. H. (1970) A glossary of karst terminology, *Geol. Surv. Water Supply Pap.* **1899K**, US Govt. Printing Office, Washington.

Moore, P. D. (1977) Vegetational history, *Cambria*, **4**, 73–83.

Moore, P. D. (1980) The reconstruction of the Lateglacial environment: some problems associated with the interpretation of pollen data, In J. J. Lowe, J. M. Gray and J. E. Robinson (eds.), *Studies in the Lateglacial of North-west Europe*, Pergamon, Oxford and New York, 151–5.

Moore, P. D. and Bellamy, D. J. (1974) *Peatlands*, Elek Science, London.

Moore, P. D. and Webb, J. A. (1978) *An Illustrated Guide to Pollen Analysis*, Hodder and Stoughton, London.

Morgan, A. (1973) Late Pleistocene environmental changes indicated by fossil insect faunas of the English Midlands, *Boreas*, **2**, 173–212.

Morgan, A. V. (1973) The Pleistocene geology of the area north and west of Wolverhampton, Staffordshire, England, *Phil. Trans. R. Soc. Lond.*, **B 265**, 233–97.

Morgan, M. A. (1969) A Pleistocene fauna and flora from Great Billing, Northamptonshire, England, *Opuscula Entom.*, **34**, 109–29.

Morgan, V. I. and Budd, W. F. (1975) Radio-echo sounding of the Lambert Glacier basin, *J. Glaciol.*, **15**, 103–11.

Morley, J. J. and Hays, J. D. (1979) Comparison of glacial and interglacial oceanographic conditions in the South Atlantic from variations in calcium carbonate and radiolarian distribution, *Quat. Res.*, **12**, 396–408.

Mörner, N-A. (1976) Eustasy and geoid changes, *J. Geol.*, **84**, 123–51.

Mörner, N-A. (1979) The deglaciation of southern Sweden: a multi-parameter consideration, *Boreas*, **8**, 189–98.

Mörner, N-A. (1980a) The Fennoscandian uplift: geological data and their geodynamical implication, In N-A. Mörner (ed.), *Earth Rheology, Isostasy and Eustasy*, Wiley, New York and Chichester, 251–84.

Mörner, N-A. (1980b) A 10 700 years' palaeotemperature record from Gotland and Pleistocene/Holocene boundary events in Sweden, *Boreas*, **9**, 283–8.

Morrison, R. B. (1964) Lake Lahontan – geology of the southern Carson Desert, *US Geol. Surv. Prof. Pap.*, **424**, D 111–14.

Morrison, R. B. (1965) Quaternary geology of the Great Basin, In H. E. Wright and D. G. Frey (eds.), *The Quaternary of the United States*, Princeton Univ. Press, 265–86.

Morrison, R. B. (1967) Principles of Quaternary soil stratigraphy, In R. B. Morrison and H. E. Wright (eds.), *Quaternary Soils*, Proc. INQUA VIIth Congress, Desert Research Unit, Reno, Neveda, 1–69.

Morrison, R. B. (1968a) Pluvial lakes, In R. W. Fairbridge (ed.), *Encyclopaedia of Geomorphology*, Reinhold, New York, 873–83.

Morrison, R. B. (1968b) Means of time-stratigraphic subdivision and long-distance correlation of Quaternary successions, In R. B. Morrison and H. E. Wright Jr (eds.), *Means of Correlation of Quaternary Successions*, Univ. of Utah Press, Salt Lake City, 1–113.

Morrison, R. B. (1978) Quaternary soil stratigraphy – concepts, methods and problems, In W. C. Mahaney (ed.), *Quaternary Soils*, GeoAbstracts, Norwich, 77–108.

Morrison, R. B. and Frye, J. C. (1965) Correlation of Middle and Late Quaternary successions of the Lake Lahontan, Lake Bonneville, Rocky Mountain (Wasatch Range), southern Great Plains and eastern Midwest areas, *Nevada Bureau of Mines Rep.*, **9**, 1–45.

Morzadec-Kerfourn, M. T. (1975) Palynology of the Quaternary sediments in borehole V050, Appendix in Destombes, J. P., Shephard-Thorn, E. R and Redding, J. H., A buried valley system in the Strait of Dover, *Phil. Trans. R. Soc. Lond.*, **A 279**, 243–56.

Mosley-Thompson, E. and Thompson, L. G. (1982) Nine centuries of microparticle deposition at the South Pole, *Quat. Res.*, **17**, 1–13.

Mottershead, D. N. (1977) The Quaternary evolution of the south coast of England, In C. Kidson and M. J. Tooley (eds.), *The Quaternary History of the Irish Sea*, Seel House Press, Liverpool, 299–320.

Mottershead, D. N. (1980) Lichenometry – some recent applications, In R. A. Cullingford, D. A. Davidson and J. Lewin (eds.), *Timescales in Geomorphology*, Wiley, Chichester and New York, 95–108.

Mottershead, D. N. and Collin, R. L. (1976) A study of Flandrian glacier fluctuations in Tunsbergdalen, southern Norway, *Norsk Geol. Tidsskr.*, **56**, 413–36.

Mottershead, D. N. and White, I. D. (1972) The lichenometric dating of glacier recession, Tunbergdalbre, southern Norway, *Geogr. Annlr.*, **54A**, 47–52.

Muller, F. (1968) Pingos, modern, In R. W. Fairbridge (ed.), *Encyclopaedia of Geomorphology*, Reinhold, New York, 845–6.

Muller, R. A. (1977) Radioisotope dating with a cyclotron, *Science*, **196**, 489–94.

Muller, S. W. (1947) *Permafrost of Permanently Frozen Ground and Related Engineering Problems*, Ann. Arbor, Michigan.

Naeser, C. W. (1979) Fission-track dating and geologic annealing of fission tracks, In E. Jäger and J. C. Hunziker (eds.), *Lectures in Isotope Geology*, Springer-Verlag, Berlin and New York, 154–69.

Nagy, J. and Ofstad, K. (1980) Quaternary foraminifera and sediments in the Norwegian Channel, *Boreas*, **9**, 39–52.

NASA (1976) *Landsat Data User's Handbook*, NASA Goddard Space Flight Center, Greenbelt, Maryland.

Neale, J. W. (1964) Some factors influencing the distribution of recent British Ostracoda, *Pubbl. Staz. Zool. Napoli* (Suppl.), 247–307.

Neale, J. W. (ed.) (1969) *The Taxonomy, Morphology and Ecology of Recent Ostracoda*, Oliver and Boyd, Edinburgh.

Nelson, D. E., Korteling, R. G. and Stott, W. R. (1977) Carbon-14: direct detection at natural concentrations, *Science*, **198**, 507–8.

Nelson, J. G. (1966) Man and geomorphic process in the Chemung Valley, New York and Pennsylvania, *Ann. Assoc. Amer. Geogr.*, **56**, 24–32.

Neumann, A. C. and Moore, W. S. (1975) Sea level events and Pleistocene coral ages in the northern Bahamas, *Quat. Res.*, **5**, 215–24.

Newey, W. W. (1970) Pollen analysis of Late-Weichselian deposits at Corstorphine, Edinburgh, *New Phytol.*, **69**, 1167–77.

Nilsson, E. (1968) Södra Sveriges senkvartära historia (English summary), *Küngl. Svenska Vetensk. Akad. Hanlingar, Ser. 4*, 12, 117 pp.

Nilsson, T. (1964) Standard pollendiagramme und C^{14} datierungen aus dem Ageröds Mosse im Mittleren Schonen, *Lunds Univ. Arsskrift*, **59**, 7.

Norton, P. E. P. (1967) Marine molluscan assemblages in the early Pleistocene of Sidestrand, Bramerton and the Royal Society borehole at Ludham, *Phil. Trans. R. Soc. Lond.*, **B 253**, 161–200.

Norton, P. E. P. (1977) Marine mollusca in the East Anglian Pleistocene, In F. W. Shotton (ed.), *British Quaternary Studies – Recent Advances*, Oxford Univ. Press, 43–53.

Oakley, K. P. (1969) Analytical methods of dating bones, In D. Brothwell and E. Higgs (eds.), *Science in Archaeology*, Thames and Hudson, London, 35–45.

Oakley, K. P. (1980) Relative dating of the fossil hominids of Europe, *Bull. Brit. Museum (Nat. Hist.), Geol. Series*, **34**, 1–63.

Oerlemans, J. (1981) Modelling of Pleistocene European ice sheets: some experiments with simple mass balance parameterisation, *Quat. Res.*, **15**, 77–85.

Oeschger, H., Riesen, T. and Lerman, J. C. (1970) Bern Radiocarbon Dates, VII, *Radiocarbon*, **12**, 358

Olausson, E. (1965) Evidence of climatic changes in North Atlantic deep-sea cores with remarks on isotopic palaeotemperature analysis, *Progr. Oceanogr.*, **3**, 221–52.

Oldfield, F. (1978) Lakes and their drainage basins as units of sediment-based ecological study, *Progr. Phys. Geogr.*, **1**, 460–504.

Oldfield, F. (1981) Peats and lake sediments: formation, stratigraphy, description and nomenclature, In A. Goudie (ed.), *Geomorphological Techniques*, Allen and Unwin, London and Boston, 306–27.

Olsson, I. U. (1968) Modern aspects of radiocarbon datings, *Earth-Sci. Rev.*, **4**, 203–18.

Olsson, I. U. (1979) Radiocarbon dating, In B. E. Berglund (ed.), *Palaeohydrological Changes in the Temperate Zone in the last 15 000 years* (vol. II), IGCP Project 158, Dept. of Quaternary Geology, Lund, Sweden, 1–38.

Olsson, I. U. and Blake, W. (1961) Problems of radiocarbon dating of raised beaches based on experience in Spitsbergen, *Norsk Geogr. Tidsskr.*, **18**, 47–64.

Orford, J. D. (1981) Particle form, In A. Goudie (ed.), *Geomorphological Techniques*, Allen and Unwin, London and Boston, 86–90.

Osborne, P. J. (1971) An insect fauna from the Roman site at Alcester, Warwickshire, *Britannia*, **2**, 156–65.

Osborne, P. J. (1972) Insect faunas of Late Devensian and Flandrian age from Church Stretton, Shropshire, *Phil. Trans. R. Soc. Lond.*, **B 263**, 327–67.

O'Sullivan, P. E., Oldfield, F. and Battarbee, R. W. (1973) Preliminary studies of Lough Neagh sediments. I. Stratigraphy, chronology and pollen analysis, In H. J. B. Birks and R. G. West (eds.), *Quaternary Plant Ecology*, Blackwell, Oxford, 267–78.

Parmenter, C. and Folger, D. W. (1974) Eolian biogenic detritus in deep-sea sediments: a possible index of equatorial aridity, *Science*, **185**, 695–8.

Paterson, I. B. (1974) The supposed Perth Readvance in the Perth district, *Scott. J. Geol.*, **10**, 53–66.

Paterson, W. S. B. (1981) *The Physics of Glaciers*, (2nd ed.) Pergamon, Oxford and New York.

Paterson, W. S. B. *et al.* (1977) An oxygen-isotope climatic record from the Devon Island ice cap, arctic Canada, *Nature*, **266**, 508–11.

Pawluk, S. (1978) The pedogenic profile in the stratigraphic section, In W. C. Mahaney (ed.), *Quaternary Soils*, GeoAbstracts, Norwich, 61–76.

Peach, A. M. (1909) Boulder distribution from Lennoxtown, Scotland, *Geol. Mag.*, **46**, 26–31.

Peacock, J. D. (1970) Some aspects of the glacial geology of west Inverness-shire, *Bull. Geol. Surv. GB*, **33**, 43–56.

Peacock, J. D. (1971) Marine shell radiocarbon dates and the chronology of deglaciation in western Scotland, *Nature*, **230**, 43–5.

Peacock, J. D., Graham, D. K., Robinson, J. E. and Wilkinson, I. (1977) Evolution and chronology of Lateglacial marine environments at Lochgilphead, Scotland, In J. M. Gray and J. J. Lowe (eds.), *Studies in the Scottish Lateglacial Environment*, Pergamon, Oxford and New York, 89–100.

Peacock, J. D., Graham, D. K. and Wilkinson, I. P. (1978) Late-glacial and Post-glacial marine environments at Ardyne, Scotland, and their significance in the interpretation of the history of the Clyde Sea area, *Rep. Inst. Geol. Sci.*, **78/17**, 1–24.

Pearson, G. W., Pilcher, J. R., Baillie, M. G. L and Hillam, J. (1977) Absolute radiocarbon-dating using a low altitude European tree-ring calibration, *Nature*, **270**, 25–8.

Pearson, R. G. (1962) The Coleoptera from a detritus deposit of full-glacial age at Colney Heath near St Albans, *Proc. Linn. Soc. Lond.*, **173**, 37–55.

Peck, R. M. (1973) Pollen budget studies in a small Yorkshire catchment, In H. J. B. Birks and R. G. West (eds.), *Quaternary Plant Ecology*, Blackwell, Oxford, 43–60.

Peck, R. M. (1974) A comparison of four absolute pollen preparation techniques, *New Phytol.*, **73**, 567–87.

Peltier, W. R. (1981) Ice age geodynamics, *Ann. Rev. Earth Planet. Sci.*, **9**, 199–226.

Peltier, W. R., Farrell, W. E. and Clark, J. A. (1978) Glacial isostasy and relative sea level: a global finite element model, *Tectonophysics*, **50**, 81–110.

Penck, A. and Bruckner, E. (1909) *Die Alpen im Eiszeitalter*, Tachnitz, Leipzig.

Peng, T-H., Goddard, J. G. and Broecker, W. S. (1978) A direct comparison of ^{14}C and ^{230}Th ages at Searles Lake, California, *Quat. Res.*, **9**, 319–29.

Pennington, W. (1970) Vegetation history in the north-west of England: a regional synthesis, In. D. Walker and R. G. West (eds.), *Studies in the Vegetation History of the British Isles*, Cambridge Univ. Press, 41–79.

Pennington, W. (1975) A chronostratigraphic comparison of Late-Weichselian and Late-Devensian subdivisions, illustrated by two radiocarbon-dated profiles from western Britain, *Boreas*, **4**, 157–71.

Pennington, W. (1977a) Lake sediments and the Lateglacial environment in northern Scotland, In J. M. Gray and J. J. Lowe (eds.), *Studies in the Scottish Lateglacial Environment*, Pergamon, Oxford and New York, 119–41.

Pennington, W. (1977b) The Late Devensian flora and vegetation of Britain, *Phil. Trans. R. Soc. Lond.*, **B 280**, 247–71.

Pennington, W. (1978) Quaternary geology, In F. Moseley (ed.), *Geology of the Lake District*, Yorks. Geol. Soc., Occ. Publ. 3 Leeds, 207–25.

Pennington, W. (1980) Modern pollen samples from west Greenland and the interpretation of pollen data from the British Late-Glacial (Late Devensian), *New Phytol.*, **84**, 171–201.

Pennington, W., Cambray, R. S., Eakins, J. D. and Harkness, D. D. (1976) Radionuclide dating of the recent sediments of Blelham tarn, *Freshwater Biol.*, **6**, 317–31.

Pennington, W., Cambray, R. S. and Fisher, E. M. (1973) Observations on lake sediment using fallout ^{137}Cs as a tracer, *Nature*, **242**, 324–6.

Pennington, W., Haworth, E. Y., Bonny, A. P. and Lishman, J. P. (1972) Lake sediments in northern Scotland, *Phil. Trans. R. Soc. Lond.*, **B 264**, 191–294.

Pennington, W. and Lishman, J. P. (1971) Iodine in lake sediments in northern England and Scotland, *Biol. Rev.*, **46**, 279–313.

Penny, L. F., Coope, G. R. and Catt, J. A. (1969) Age and insect fauna of the Dimlington Silts, East Yorkshire, *Nature*, **224**, 65–7.

Péwé, T. L. (1959) Sand wedge polygons (tesselations) in the McMurdo Sound region, Antarctica, *Amer. J. Sci.*, **257**, 545–52.

Péwé, T. L. (1966) Palaeoclimatic significance of fossil ice wedges, *Biul. Peryglac.*, **15**, 65–73.

Péwé, T. L. (1969) *The Periglacial Environment*, McGill-Queen's Univ. Press, Montreal.

Péwé, T. L. (1973) Ice wedge casts and past permafrost distribution in North America, *Geoforum*, **15**, 15–26.

Phillips, L. (1974) Vegetational history of the Ipswichian/Eemian Interglacial in Britain and continental Europe, *New Phytol.*, **73**, 589–604.

Phillips, L. M (1976) Pleistocene vegetational history and geology in Norfolk, *Phil. Trans. R. Soc. Lond.*, **B 275**, 215–86.

Pias, J. (1970) Les formations sédimentaires Tertiaires et Quaternaires de la Cuvette Tehadienne et les sols qui en dérivent, *Office de la Recherche Scientifique et Technique Outre-Mer*, Mem. 43.

Pierce, K. L., Obradovich, J. D. and Friedman, I. (1976) Obsidian hydration dating and correlation of Bull Lake and Pinedale Glaciations near West Yellowstone, Montana, *Bull. Geol. Soc. Amer.*, **87**, 703–10.

Pigott, C. D. (1962) Soil formation and development on the Carboniferous limestone of Derbyshire. I. Parent materials, *J. Ecol.*, **50**, 145–56.

Pilcher, J. R. (1973) Tree-ring research in Ireland, *Tree-Ring Bull.*, **33**, 1–5.

Pilcher, J. R., Hillam, J., Baillie, M. G. L. and Pearson, G. W. (1977) A long sub-fossil oak tree-ring chronology from the north of Ireland, *New Phytol.*, **79**, 713–29.

Pirazzoli, P. (1978) High stands of Holocene sea levels in the north east Pacific, *Quat. Res.*, **10**, 1–29.

Pitcher, W. S., Shearman, D. J. and Pugh, D. C. (1954) The loess of Pegwell Bay, and its associated frost soils, *Geol. Mag.*, **91**, 308–14.

Pohl, F. (1937) Die pollenerzeugnung der Windbluter, *Botanisch. Centrablatt.*, **56A**, 365–470.

Pokorný, V. (1978) Ostracodes, In B. U. Haq and A. Boersma (eds.), *Introduction to Marine Micropalaeontology*, Elsevier, Amsterdam, 109–49.

Poole, E. G. and Whiteman, A. J. (1961) The glacial drifts of the southern part of the Cheshire-Shropshire plain, *Q. J. Geol. Soc. Lond.*, **117**, 91–130.

Porter, S. C. (1964) Late Pleistocene glacial chronology of north-central Brooks Range, Alaska, *Amer. J. Sci.*, **262**, 446–60.

Porter, S. C. (1970) Quaternary glacial record in Swat Kohistan, West Pakistan, *Bull. Geol. Soc. Amer.*, **81**, 1421–46.

Porter, S. C. (1975) Equilibrium-line altitudes of late Quaternary glaciers in the Southern Alps, New Zealand, *Quat. Res.*, **5**, 27–47.

Porter, S. C. and Denton, G H. (1967) Chronology of Neoglaciation in the North American cordillera, *Amer. J. Sci.*, **265**, 177–210.

Powers, M. C. (1953) A new roundness scale for sedimentary particles, *J. Sed. Petrol.*, **23**, 117–19.

Preece, R. C. (1980) The biostratigraphy and dating of a postglacial slope deposit at Gore Cliff, near Blackgang, Isle of Wight, *J. Archaeol. Sci.*, **7**, 255–65.

Preece, R. C. (1981) The value of shell microsculpture as a guide to the identification of land mollusca from Quaternary deposits, *J. Conch.*, **30**, 331–7,

Prell, W. L. and Hays, J. D. (1976) Late Pleistocene faunal and temperature patterns of the Columbia Basin, Caribbean Sea, *Geol. Soc. Amer. Mem.*, **145**, 201–20.

Prest, V. K. (1970) Quaternary geology of Canada, In R. J. W. Douglas (ed.), *Geology and Economic Minerals of Canada*, *Geol. Surv. Can.*, *Rep.* **1**, 675–764.

Pullar, W. A. (1965) Chronology of flood-plains, fans and terraces in the Gisborne and Bay of Plenty districts, *Proc. 4th NZ Geogr. Conf.*, 77–81.

Pullar, W. A., Pain, C. F. and Johns, R. J. (1967) Chronology of terraces floodplains, fans and dunes in the Whakatone Valley, *Proc. 5th NZ Geogr. Conf.*, 175–80.

Rackham, D. J. (1978) Evidence for changing vertebrate communities in the Middle Devensian, *Quat. Res. Assoc. (GB) Newsl.*, **25**, 1–3.

Ralph, E. K., Michael, H. N. and Han, M. C. (1973) Radiocarbon dates and reality, *MASCA Newsletter*, **9**, 1–18.

Rampton, V. (1970) Neoglacial fluctuations of the Natazhat and Klutlan Glaciers, Yukon Territory, Canada, *Can. J. Earth Sci.*, **7**, 1236–63.

Ramsay, A. T. S. (1977) *Ocean Micropalaeontology*, Academic Press, London and New York (2 vols.).

Ramsden, J. and Westgate, J. A. (1971) Evidence for reorientation of a till fabric in the Edmonton area, Alberta, In R. P. Goldthwait (ed.), *Till: a Symposium*, Ohio State Univ. Press, 335–44.

Reed, B., Galvin, C. J. and Miller, J. P. (1962) Some aspects of drumlin geometry, *Amer. J. Sci.*, **260**, 200–10.

Reiche, P. (1938) An analysis of cross-lamination: the Coconino Sandstone, *J. Geol.*, **46**, 905–32.

Reid, C. (1899) *The Origin of the British Flora*, Dulau, London.

Reider, R. C. (1975) Morphology and genesis of soils on the Prairie Divide deposit (pre-Wisconsin), Front Range, Colorado, *Arct. Alp. Res.*, **7**, 353–72.

Reineck, H-E. and Singh, I. B. (1973) *Depositional Sedimentary Environments*, Springer-Verlag, Berlin and New York.

Reynoulds, S. G. and Aldous, K. (1970) *Atomic Absorption Spetroscopy*, Griffin, London.

Rice, R. J. (1977) *Fundamentals of Geomorphology*, Longman, London and New York.

Richardson, J. L. and Richardson, A. E. (1972) History of an African rift lake and its climatic implications, *Ecol. Monogr.*, **42**, 499–534.

Richmond, G. M. (1965) Glaciation of the Rocky Mountains, In H. E. Wright and D. G. Frey (eds.), *The Quaternary of the United States*, Princeton Univ. Press, Princeton, NJ, 217–30.

Richmond, G. M. (1970) Comparison of the Quaternary stratigraphy of the Alps and Rocky Mountains, *Quat. Res.*, **1**, 3–28.

Richmond, G. M. (1977) Extensive glaciers in the Yellowstone National Park *ca.* 114 000 and *ca.* 88 000 years ago, *Xth INQUA Congr., Birmingham, 1977, Abstracts*, p. 382.

Richter, K. (1936) Gefugestudien im Engaebrae, Fondalsbrae, und ihren Vorlandsedimenten, *Zeits. Gletscherkunde*, **24**, 22–30.

Robin, G. de Q. (1977) Ice cores and climatic change, *Phil. Trans. R. Soc. Lond.*, **B 280**, 143–68.

Robin, G. de Q., Drewry, D. J. and Meldrum, D. T. (1977) International studies of ice sheet and bedrock, *Phil. Trans. R. Soc. Lond.*, **B 279**, 185–96.

Robinson, J. E. (1980) The marine ostracod record from the Lateglacial period in Britain and NW Europe: a review, In J. J. Lowe, J. M. Gray and J. E. Robinson (eds.), *Studies in the Lateglacial of North-west Europe*, Pergamon, Oxford and New York, 115–22.

Robinson, M. and Ballantyne, C. K. (1979) Evidence for a glacial advance predating the Loch Lomond Advance in Wester Ross, *Scott. J. Geol.*, **15**, 271–7.

Rognon, P. (1980) Pluvial and arid phases in the Sahara: the role of non-climatic factors, In E. M. van Zinderen Bakker and J. A. Coetzee (eds.), *Palaeoecology of Africa*, Balkema, Rotterdam, 45–62.

Rolfe, W. D. I. (1966) Woolly rhinoceros from the Scottish Pleistocene, *Scott. J. Geol.*, **2**, 253–8.

Ronai, A. (1965) Neotectonic subsidences in the Hungarian Basin, *Geol. Soc. Amer. Spec. Pap.*, **84**, 219–32.

Rose, J. (1974) Small scale spatial variability of some sedimentary properties of lodgement and slumped till, *Proc. Geol. Assoc.*, **85**, 223–37.

Rose, J. (1981) Raised shorelines, In A. Goudie (ed.), *Geomorphological Techniques*, Allen and Unwin, London and Boston, 327–41.

Rose, J., Allen, P. and Hey, R. W. (1976) Middle Pleistocene stratigraphy in southern East Anglia, *Nature*, **263**, 492–4.

Rose, J. and Letzer, J. M. (1975) Drumlin measurements: a test of the reliability of data derived from 1 : 25 000 scale topographic maps, *Geol. Mag.*, **112**, 361–71.

Rose, J., Turner, C., Coope, G. R. and Bryan, M. D. (1980) Channel changes in a lowland river catchment over the last 13 000 years. In R. A. Cullingford, D. A. Davidson and J. Lewin (eds.), *Timescales in Gemorphology*, Wiley,

Chichester and New York, 159–75.

Rottländer, R. C. A. (1976) Variations in the chemical composition of bone as an indicator of climatic change, *J. Archaeol. Sci.*, **3**, 83–6.

Round, F. E. (1964) The diatom sequence in lake deposits: some problems of interpretation, *Verh. Int. Verein. Limnol.*, **15**, 1012–20.

Rowlands, B. M. (1971) Radiocarbon evidence of the age of an Irish Sea glaciation in the Vale of Clwyd, *Nature*, **230**, 9–11.

Ruddiman, W. F. and McIntyre, A. (1973) Time-transgressive deglacial retreat of polar waters from the North Atlantic, *Quat. Res.*, **3**, 117–30.

Ruddiman, W. F. and McIntyre, A. (1976) Northeast Atlantic palaeoclimatic changes over the past 600 000 years, *Geol. Soc. Amer. Mem.*, **145**, 111–46.

Ruddiman, W. F. and McIntyre, A. (1981) The North Atlantic during the last deglaciation, *Palaeogeogr. Palaeoclimatol. Palaeoecol.*, **35**, 145–214.

Ruddiman, W. F., McIntyre, A., Niebler-Hunt, V. and Durazz, J. T. (1980) Oceanic evidence for the mechanism of rapid Northern Hemisphere glaciation, *Quat. Res.*, **13**, 33–64.

Ruddiman, W. F., Sancetta, C. D. and McIntyre, A. (1977) Glacial/interglacial response rate of subpolar North Atlantic water to climatic change: the record in ocean sediments, *Phil. Trans. R. Soc. Lond.*, B **280**, 119–42.

Ruellan, A. (1971) The history of soils: some problems of definition and interpretation, In. D. H. Yaalon (ed.), *Palaeopedology*, Internat. Soc. Soil Sci. and Israel Univ. Press, Jerusalem, 3–12.

Ruggieri, C. (1971) Ostracodes as cold climate indicators in the Italian Quaternary, In H. J. Oertli (ed.), *Palaéoécologie Ostracodes, Pau 1970, Bull. Centre Rech., Pau-SNPA* **5**,285–93.

Ruhe, R. V. (1965) Quaternary paleopedology, In H. E. Wright and D. G. Frey (eds.), *The Quaternary of the United States*, Princeton Univ. Press, Princeton, NJ, 755–64.

Ruhe, R. V. (1970) Soils, paleosols and environment, In W. Dort and J. Knox Jones (eds.), *Pleistocene and Recent Environments of the Central Great Plains*, Univ. Kansas Press, 37–52.

Rutter, N. W. (1969) Comparison of moraines formed by normal and surging glaciers, *Can. J. Earth Sci.*, **6**, 991–9.

Rymer, L. (1978) The use of uniformitarianism and analogy in palaeoecology, In. D. Walker and J. C. Guppy (eds.), *Biology and Quaternary Environments*, Australian Acad. Sci., Canberra, 245–58.

Ryvarden, L. (1971) Studies in seed dispersal. I. Trapping of diaspores in the alpine zone of Finse, Norway, *Norw. J. Bot.*, **18**, 215–26.

Saarnisto, M. (1974) The deglaciation history of the Lake Superior region and its climatic implications, *Quat. Res.*, **4**, 316–39.

Saarnisto, M. (1979) Studies of annually laminated lake sediments, In B. E. Berglund (ed.), *Palaeohydrological Changes in the Temperate Zone in the last 15 000 years* (vol. II), IGCP Project 158, Dept. of Quaternary Geology, Lund, Sweden, 61–80.

Saarnisto, M., Huttunen, P. and Tolonen, K. (1977) Annual lamination of sediments in Lake Lovojärvi, southern Finland, during the past 600 years, *Ann. Bot. Fenn.*, **14**, 35–45.

Saarnthein, M. (1978) Sand deserts during glacial maximum and climatic optimum, *Nature*, **272**, 43–6.

Sancetta, C., Imbrie, J. and Kipp, N. G. (1973) Climatic record of the past 130 000 years in North Atlantic deep-sea core V23–82: correlation with the terrestrial record, *Quat. Res.*, **3**, 110–16.

Sangster, A. G. and Dale, H. M. (1964) Pollen preservation of under-represented species in fossil spectra, *Can. J. Bot.* **42**, 437–49.

Saunders, G. E. (1968a) Glaciation of possible Scottish Readvance age in north-west Wales, *Nature*, **218**, 76–8.

Saunders, G. E. (1968b) A fabric analysis of the ground moraine deposits of the Lleyn Peninsula of south-west Caernarvonshire, *Geol. J.*, **6**, 105–18.

Sauramo, M. (1918) Geochronologische Studien über die spätglaciale Zeit in Südfinnland, *Commn. Geol. Finlande Bull.*, **50**, 44 pp.

Sauramo, M. (1923) Studies on the Quaternary varve sediments in southern Finland, *Commn. Geol. Finlande Bull.*, **60**, 164 pp.

Savigear, R. A G. (1965) A technique of morphological mapping. *Ann. Assoc. Amer. Geogr.*, **55**, 514–39.

Scharpenseel, H. W. (1971) Radiocarbon dating of soils – problems, troubles, hopes, In D. A. Yaalon (ed.), *Palaeopedology*, Internat. Soc. Soil Sci., and Israel Univ. Press, Jerusalem, 77–87.

Scharpenseel, H. W. and Schiffman, H. (1977) Radiocarbon dating of soils — a review, *Zeit. für Pflanzenernährung Düngen und Bodenkunde*, **140**, 159–74.

Schlüchter, Ch. (ed.) (1979) *Moraines and Varves*, Proc. INQUA Symp. Zurich, 1978, Balkema, Rotterdam.

Schmid, E. (1969) Cave sediments and prehistory. In D. Brothwell and E. Higgs (eds.), *Science in Archaeology*, Thames and Hudson, London, 151–66.

Schneekloth, H. (1968) The significance of the limiting horizon for the chronostratigraphy of raised bogs: results of a critical investigation, *Proc. 3rd Internat. Peat Congr., Quebec*, 116.

Schott, W. (1935) Die foraminiferen in dem äquatoriales teil des Atlantischen Ozeans, *Deutsch. Atlant. Exped. Meteor 1925–1927, Wiss. Ergebn.*, **3**, 43–134.

Schulman, E. (1954) Longevity under adversity in conifers, *Science*, **119**, 396–9.

Schulman, E. (1958) Bristlecone pine, oldest known living thing, *Nat. Geogr. Mag.*, **113**, 354–72.

Schwarcz, H. P., Harmon, R. S., Thompson, P. and Ford, D. C. (1976) Stable isotope studies of fluid inclusions in speleothems and their palaeoclimatic significance, *Geochim. et Cosmochim. Acta*, **40**, 657–65.

Schweingrüber, F. H. *et al.* (1978) The X-ray technique as applied to dendroclimatology, *Tree-Ring Bull.*, **38**, 61–91.

Shackleton, N. J. (1967) Oxygen isotope analyses and Pleistocene temperatures reassessed, *Nature*, **215**, 15–17.

Shackleton, N. J. (1969a) Marine mollusca in archaeology, In D. Brothwell and E. Higgs (eds.), *Science and Archaeology*, Thames and Hudson, London, 407–14.

Shackleton, N. J. (1969b) The last interglacial in the marine and terrestrial record, *Proc. R. Soc. Lond.*, B **174**, 135–54.

Shackleton, N. J. (1975) The stratigraphic record of deep-sea cores and its implications for the assessment of glacials, interglacials and interstadials in the Mid-Pleistocene, In K. W. Butzer and G. L. Isaac (eds.), *After the Australopithecines*, Mouton Press, The Hague, 1–24.

Shackleton, N. J. (1977) The oxygen isotope record of the Late Pleistocene, *Phil. Trans. R. Soc. Lond.*, B **280**, 169–82.

Shackleton, N. J. and Opdyke, N.D. (1973) Oxygen isotope and palaeomagnetic stratigraphy of equatorial Pacific core V28–238: oxygen isotope temperatures and ice volumes on a 10^5 and 10^6 year scale, *Quat. Res.*, **3**, 39–55.

Shackleton, N. J. and Opdyke, N. D. (1976) Oxygen isotope and palaeomagnetic stratigraphy of Pacific core V28–239, Late Pliocene to Late Holocene, *Geol. Soc. Amer. Mem.*, **145**, 449–64.

Shackleton, N. J. and Opdyke, N. D. (1977) Oxygen isotope and palaeomagnetic evidence for early Northern Hemisphere glaciation, *Nature*, **261**, 547–50.

Shakesby, R. A. (1978) Dispersal of glacial erratics from Lennoxtown, Stirlingshire, *Scott. J. Geol.*, **14**, 81–6.

Shaw, J. and Archer, J. (1978) Winter turbidity current deposits in late Pleistocene glaciolacustrine varves, Okanagan Valley, British Columbia, Canada, *Boreas*, **7**, 123–30.

Shepard, F. P. (1963) Thirty-five thousand years of sea level, *Essays in Marine Geology*, Univ. California Press, 1–10.

Shepard, F. P. and Curray, J. R. (1967) Carbon-14 determination of sea-level changes in stable areas, *Progr. Oceanogr.*, **4**, 283–91.

Shilts, W. W. (1980) Flow patterns in the central North American ice sheet, *Nature*, **286**, 213–18.

Shotton, F. W. (1967) The problems and contributions of methods of absolute dating within the Pleistocene period, *Q. J. Geol. Soc. Lond.*, **122**, 357–83.

Shotton, F. W. (1972) An example of hard-water error in radiocarbon dating of vegetable matter, *Nature*, **240**, 460–1.

Shotton, F. W. (ed.) (1977a) *British Quaternary Studies – Recent Advances*, Oxford Univ. Press, Oxford.

Shotton, F. W. (1977b) The Devensian Stage: its development, limits and substages, *Phil. Trans. R. Soc. Lond.*, B **280**, 107–18.

Shotton, F. W., Blundell, P. J. and Williams, R. E. G.(1970) Birmingham University radiocarbon dates IV, *Radiocarbon*, **12**, 385–99.

Shotton, F. W. and Williams, R. E. G. (1971) Birmingham University radiocarbon dates VI, *Radiocarbon*, **13**, 141–56.

Shotton, F. W. and Williams, R. E. G. (1973) Birmingham University radiocarbon dates VII, *Radiocarbon*, **15**, 451–68.

Simpson, I. M. and West, R. G. (1958) On the stratigraphy and palaeobotany of a late-Pleistocene organic deposit at Chelford, Cheshire, *New Phytol.*, **57**, 239–50.

Sissons, J. B. (1966) Relative sea-level changes between 10 000 and 8 000 BP in part of the Carse of Stirling, *Trans. Inst. Brit. Geogr.*, **39**, 19–29.

Sissons, J. B. (1967a) *The Evolution of Scotland's Scenery*, Oliver and Boyd, Edinburgh.

Sissons, J. B. (1967b) Glacial stages and radiocarbon dates in Scotland, *Scott. J. Geol.*, **3**, 375–81.

Sissons, J. B. (1972) The last glaciers in part of the south-east Grampians, *Scott. Geogr. Mag.*, **88**, 168–81.

Sissons, J. B. (1973) Delimiting the Loch Lomond Readvance in the eastern Grampian Highlands, *Scott. Geogr. Mag.*, **89**, 138–9.

Sissons, J. B. (1974a) A Late-glacial ice cap in the central Grampians, Scotland, *Trans. Inst. Brit. Geogr.*, **62**, 95–114.

Sissons, J. B. (1974b) Lateglacial marine erosion in Scotland, *Boreas*, **3**, 41–8.

Sissons, J. B. (1974c) The Quaternary in Scotland: a review, *Scott. J. Geol.*, **10**, 311–37.

Sissons, J. B. (1975) A fossil rock glacier in Wester Ross, *Scott. J. Geol.*, **11**, 83–6.

Sissons, J. B. (1976) *The Geomorphology of the British Isles: Scotland*, Methuen, London and New York.

Sissons, J. B. (1977a) The Loch Lomond Readvance in the northern mainland of Scotland, In J. M. Gray and J. J. Lowe (eds.), *Studies in the Scottish Lateglacial Environment*, Pergamon, Oxford and New York, 45–59.

Sissons, J. B. (1977b) The Loch Lomond Readvance in southern Skye and some palaeoclimatic implications, *Scott. J. Geol.*, **13**, 23–36.

Sissons, J. B. (1979a) The Loch Lomond Stadial in the British Isles, *Nature*, **280**, 199–202.

Sissons, J. B. (1979b) Palaeoclimatic inferences from former glaciers in Scotland and the Lake District, *Nature*, **278**, 518–21.

Sissons, J. B. (1980) The Loch Lomond Advance in the Lake District, northern England, *Trans. R. Soc. Edinb.: Earth Sciences*, **71**, 13–27.

Sissons, J. B. (1981) The last Scottish ice-sheet: facts and speculative discussion, *Boreas*, **10**, 1–17.

Sissons, J. B. and Grant, A. J. H. (1972) The last glaciers in the Lochnagar area, Aberdeenshire, *Scott. J. Geol.*, **8**, 85–93.

Sissons, J. B. and Smith, D. E. (1965) Raised shorelines associated with the Perth Readvance in the Forth Valley and their relation to glacial isostasy, *Trans. R. Soc. Edinb.*, **66**, 143–68.

Sissons, J. B., Smith, D. E. and Cullingford, R. A. (1966) Lateglacial and postglacial shorelines in south-east Scotland, *Trans. Inst. Brit. Geogr.*, **39**, 9–18.

Sissons, J. B. and Sutherland, D. G. (1976) Climatic inferences from former glaciers in the south-east Grampian Highlands, Scotland, *J. Glaciol.*, **17**, 325–46.

Sissons, J. B. and Walker, M. J. C. (1974) Lateglacial site in the central Grampian Highlands, *Nature*, **249**, 822–4.

Smalley, I. J. and Unwin, D. J. (1968) The formation and shape of drumlins and their distribution and orientation in drumlin fields, *J. Glaciol.*, **7**, 377–90.

Smith, D. E., Sissons, J. B. and Cullingford, R. A. (1969) Isobases for the Main Perth raised shoreline in south-east Scotland as determined by trend-surface analysis, *Trans. Inst. Brit. Geogr.*, **46**, 45–52.

Smith, G. I. (1968) Late Quaternary geologic and climatic history of Searles Lake, south-eastern California, In R. B. Morrison and H. E. Wright (eds.), *Means of Correlation of Quaternary Successions*, Univ. Utah Press, Salt Lake City, 293–310.

Smith, G. I. (1974) Quaternary deposits in south-western Afghanistan, *Quat. Res.*, **4**, 39–52.

Smith, G. I. (1976) Palaeoclimatic record in the upper Quaternary sediments of Searles Lake, California, USA, In S. Horie (ed.), *Palaeolimnology of Lake Biwa and the Japanese Pleistocene*, vol. 4, Contribution to the 'Palaeolimnology of Lake Biwa and the Japanese Pleistocene', no. 155, Kyoto Univ., Otsu, 577–604.

Smith, H. T. U. (1949) Physical effects of Pleistocene climatic changes in non-glaciated areas: eolian phenomena, frost action and stream terracing, *Bull. Geol. Soc. Amer.*, **60**, 1485–516.

Smith, H. T. U. (1964) Periglacial eolian phenomena in the United States, *Rep. 6th INQUA Congr. (Warsaw, 1961)*, Lodz, Poland, **4**, 177–86.

Smith, H. T. U. (1965) Dune morphology and chronology in central and western Nebraska, *J. Geol.*, **73**, 557–78.

Sneed, E. D. and Folk, R L. (1958) Pebbles in lower

Colorado River, Texas: a study in particle morphogenesis, *J. Geol.*, **66**, 114–50.

Sohn, I. G., Berdan, J. M. and Peck, R. E. (1965) Ostracods, In B. Kummel and D. Raup (eds.), *Handbook of Palaeontological Techniques*, Freeman, San Francisco, 75–89.

Sorensen, C. J. (1977) Reconstructed Holocene bioclimates, *Ann. Assoc. Amer. Geogr.*, **67**, 214–22.

Sorensen, C. J., Knox, J. C., Larsen, J. A. and Bryson, R. A. (1971) Paleosols and the forest border in Keewatin, *Quat. Res.*, **1**, 468–73.

Sparks, B. W. (1957) The non-marine mollusca of the interglacial deposits at Bobbitshole, Ipswich, *Phil. Trans. R. Soc. Lond.*, **B 241**, 33–44.

Sparks, B. W. (1961) The ecological interpretation of Quaternary non-marine mollusca, *Proc. Linn. Soc. Lond.*, **172**, 71–80.

Sparks, B. W. (1964) Non-marine mollusca and Quaternary ecology, *J. Anim. Ecol.*, **33**, (suppl.), 87–98.

Sparks, B. W. (1969) Non-marine mollusca and archaeology, In D. Brothwell and E. Higgs (eds.), *Science in Archaeology*, Thames and Hudson, London.

Sparks, B. W. and West, R. G. (1959) The palaeoecology of the interglacial deposits at Histon Road, Cambridge, *Eiszeit, Gegenw.*, **10**, 123–43.

Sparks, B. W. and West, R. G. (1970) Late Pleistocene deposits at Wretton, Norfolk. I. Ipswichian interglacial deposits, *Phil. Trans. R. Soc. Lond.*, **B 258**, 1–30.

Sparks, B. W. and West, R G. (1972) *The Ice Age in Britain*, Methuen, London and New York.

Sparks, B. W., Williams, R. B. and Bell, F. G. (1972) Presumed ground-ice depressions in East Anglia, *Proc. R. Soc. Lond.*, **A 327**, 329–43.

Spencer, P. J. (1975) Habitat change in coastal sand-dune areas: the molluscan evidence, In J. G. Evans, S. Limbrey and H. Cleare (eds.), *The Effect of Man on the Landscape: the Highland Zone*, CBA Res. Rep., **11**, 96–103.

Spicer, R. A. (1981) The sorting and deposition of allochthonous plant material in a modern environment at Silwood Lake, Silwood Park, Berkshire, England, *US Geol. Surv. Prof. Pap.*, **1143**, 1–77.

Stalker, A. Mac. (1977) The probable extent of classical Wisconsin ice in southern and central Alberta, *Can. J. Earth Sci.*, **14**, 2614–9.

Stearns, C. E. and Thurber, D. L. (1965) ^{230}Th-^{234}U dates of late Pleistocene marine fossils from the Mediterranean and Mallorcan littorals, *Quaternaria*, **7**, 29–42.

Stephens, N. and McCabe, A. M. (1977) Late-Pleistocene ice movements and patterns of Late- and Post-glacial shorelines on the coast of Ulster (Ireland), In C. Kidson and M. J. Tooley (eds.), *The Quaternary History of the Irish Sea*, Seel House Press, Liverpool, 179–98.

Stieglitz, R. D. *et al.* (1978) Pre-Twocreekan age of the type Valders till, Wisconsin: comment by R. D. Stieglitz, J. M. Moran and D. P. Quigley. *Geology*, **6**, 136.

Stockton, C. W. and Meko, D. M. (1975) A long-term history of drought occurrence in western United States as inferred from tree rings, *Weatherwise*, **28**, 245–9.

Straw, A. (1963) Some observations on the 'Cover sands' of north Lincolnshire, *Trans. Lincs. Nat. Un.*, **15**, 260–9.

Straw, A. (1979) An early Devensian glaciation in eastern England, *Quat. Res. Assoc. (GB) Newsl.*, **28**, 18–24.

Straw, A. (1980) An early Devensian glaciation in eastern England reiterated, *Quat. Res. Assoc. (GB) Newsl.*, **31**, 18–23.

Street, F. A. and Grove, A. T. (1976) Environmental and climatic implications of Late Quaternary lake-level fluctuations in Africa, *Nature*, **261**, 385–90.

Street, F. A. and Gove, A. T. (1979) Global maps of lake-level fluctuations since 30 000 yr BP, *Quat. Res.*, **12**, 83–118.

Stride, A. H. (1962) Low Quaternary sea levels, *Proc. Ussher Soc.*, **1**, 6–7.

Stromberg, B. (1972) Glacial striae in southern Hinlopenstretet and Kong Karls Land, Svalbard, *Geogr. Annlr.*, **54A**, 53–65.

Stuart, A. J. (1974) Pleistocene history of the British vertebrate fauna, *Biol. Rev.*, **49**, 225–66.

Stuart, A. J. (1975) The vertebrate fauna of the type Cromerian, *Boreas*, **4**, 63–76.

Stuart, A. J. (1976) The history of the mammalian fauna during the Ipswichian/last Interglacial in England, *Phil. Trans. R. Soc. Lond.*, **B 276**, 221–50.

Stuart, A. J. (1977) The vertebrates of the Last Cold Stage in Britain and Ireland, *Phil. Trans. R. Soc. Lond.*, **B 280**, 295–312.

Stuart, A. J. (1979) Pleistocene occurrences of the European pond tortoise (*Emys orbicularis* L.) in Britain, *Boreas*, **8**, 359–71.

Stuart, A. J. (1980) The vertebrate fauna from the Interglacial deposits at Sugworth, near Oxford, *Phil. Trans. R. Soc. Lond.*, **B 289**, 87–97.

Stuart, A. J. (1982) *Pleistocene Vertebrates in the British Isles*, Longman, London and New York.

Stuiver, M. (1970) Long-term C14 variations, In I. U. Olsson (ed.), *Radiocarbon Variations and Absolute Chronology*, Wiley, New York and London, 197–213.

Suess, H. E. (1965) Secular variations in the cosmic-ray produced carbon-14 in the atmosphere and their interpretations, *J. Geophys. Res.*, **70**, 5937–50.

Suess, H. E. (1970a) The three causes of the secular C14 fluctuations, their amplitudes and time constants, In I. U. Olsson (ed.), *Radiocarbon Variations and Absolute Chronology*, Wiley, New York and London, 595–605.

Suess, H. E. (1970b) Bristlecone-pine calibration of the radiocarbon time-scale 5 000 BC to the present, In I. U. Olsson (ed.), *Radiocarbon Variations and Absolute Chronology*, Wiley, New York and London, 303–11.

Sugden, D. E. (1970) Landforms of deglaciation in the Cairngorm Mountains, *Trans. Inst. Brit. Geogr.*, **45**, 79–92.

Sugden, D. E. (1977) Reconstruction of the morphology, dynamics, and thermal characteristics of the Laurentide Ice Sheet at its maximum, *Arct. Alp. Res.*, **9**, 21–47.

Sugden, D. E. (1978) Glacial erosion by the Laurentide ice sheet, *J. Glaciol.*, **20**, 367–92.

Sugden, D. E. and John, B. S. (1973) The ages of glacier fluctuations in the South Shetland Islands, Antarctica, In E. M. van Zinderen Bakker (ed.), *Palaeoecology of Africa, the Surrounding Islands, and Antarctica*, Balkema, Cape Town, 139–59.

Sugden, D. E. and John, B. S. (1976) *Glaciers and Landscape*. Edward Arnold, London.

Sutcliffe, A. J. (1960) Joint Mitnor Cave, Buckfastleigh, *Trans. Proc. Torquay Nat. Hist. Soc.*, **13**, 1–26.

Sutcliffe, A. J. (1975) A hazard in the interpretation of Glacial-Interglacial sequences, *Quat. Res. Assoc. (GB) Newsl.*, **17**, 1–3.

Sutcliffe, A. J. (1976) The British Glacial-Interglacial sequence, *Quat. Res. Assoc. (GB) Newsl.*, **18**, 1–7.

Sutcliffe, A. J. (1981) Progress report on excavations in Minchin Hole, Gower, *Quat. Res. Assoc. (GB) Newsl.*, **33**, 1–17.

Sutcliffe, A. J. and Bowen, D. Q. (1973) Preliminary report on excavations at Minchin Hole, April-May, 1973, *Newsl. William Pengelly Cave Studies Trust*, **21**, 12–25.

Sutherland, D. G. (1980) Problems of radiocarbon dating of deposits from newly deglaciated terrain: examples from the Scottish Lateglacial, In J. J. Lowe, J. M, Gray and J. E. Robinson (eds.), *Studies in the Lateglacial of North-west Europe*, Pergamon, Oxford and New York, 139–49.

Sutherland, D. G. (1981a) The high-level marine shell beds of Scotland and the build-up of the last Scottish ice sheet, *Boreas*, **10**, 247–54.

Sutherland, D. G. (1981b) The raised shorelines and deglaciation of the Loch Long/Loch Fyne area, western Scotland, Unpubl. PhD Thesis, Univ. of Edinburgh.

Swain, A. N. (1973) A history of fire and vegetation in north-eastern Minnesota as recorded in lake sediments, *Quat. Res.*, **3**, 383–96.

Swain, F. M. (ed.) (1977) *Stratigraphic Micropalaeontology of Atlantic Basin and Borderlands*, Elsevier, Amsterdam.

Sweeting, M. M. (1972) *Karst Landforms*, MacMillan, New York and London.

Switsur, V. R. (1973) The radiocarbon calendar recalibrated, *Antiquity*, **XLVII**, 131–7.

Synge, F. M. (1977) Records of sea levels during the Late Devensian, *Phil. Trans. R. Soc. Lond.*, **B 280**, 211–28.

Synge, F. M. (1979) Quaternary glaciation in Ireland, *Quat. Res. Assoc. (GB) Newsl.*, **28**, 1–18.

Szabo, B. J. (1980) Results and assessment of uranium-series dating of vertebrate fossils from Quaternary alluviums in Colorado, *Arct. Alp. Res.*, **12**, 95–100.

Szabo, B. J. and Butzer, K. W. (1979) Uranium-series dating of lacustrine limestones from pan deposits with Final Acheulian Assemblage at Rooidam, Kimberley District, South Africa, *Quat. Res.*, **11**, 257–60.

Szabo, B. J. and Rosholt, J. N. (1969) Uranium-series dating of Pleistocene molluscan shells from southern California – an open system model, *J. Geophys. Res.*, **74**, 3253–60.

Talbot, M. R. (1980) Environmental responses to climatic change in the West African Sahel over the past 20 000 years, In M. A. J. Williams and H. Faure (eds.), *The Sahara and the Nile*, Balkema, Rotterdam, 37–62.

Tarling, D. H. (1971) *Principles and Applications of Palaeomagnetism*, Chapman and Hall, London (distr. USA by Barnes and Noble Inc.).

Tauber, H. (1965) Differential pollen dispersal and the interpretation of pollen diagrams, *Danm. Geol. Unders. Ser. II*, **89**, 1–69.

Tauber, H. (1970) The Scandinavian varve chronology and C14 dating, In I. U. Olsson (ed.), *Radiocarbon Variations and Absolute Chronology*, Wiley, New York and London, 173–95.

Thomas, G. S. P. (1976) The Quaternary stratigraphy of the Isle of Man, *Proc. Geol. Assoc.*, **87**, 307–23.

Thomas, G. S. P. (1977) The Quaternary of the Isle of Man, In C. Kidson and M. J. Tooley (eds.), *The Quaternary History of the Irish Sea*, Seel House Press, Liverpool, 155–78.

Thomas, M. F. (1974) *Tropical Geomorphology*, MacMillan, London and New York.

Thompson, M. M. (ed.) (1966) *Manual of Photogrammetry* (2 vols.), Amer. Soc. Photogramm., Virginia.

Thompson, R. (1978a) Resistivity investigation of an infilled kettle hole, *Quat. Res.*, **9**, 231–7.

Thompson, R. (1978b) European palaeomagnetic secular variation 13 000 – 0 BP, *Pol. Arch. Hydrobiol.*, **25**, 413–18.

Thompson, R. (1979) Palaeomagnetic correlation and dating, In B. E. Berglund (ed.), *Palaeohydrological Changes in the Temperate Zone in the last 15 000 years*, IGCP Project 158, Dept. Quat. Geology, Lund, Sweden, 39–59.

Thompson, R., Battarbee, R. W., O'Sullivan, P. E. and Oldfield, F. (1975) Magnetic susceptibility of lake sediments, *Limnol. Oceanogr.*, **20**, 687–98.

Thompson, R. and Turner, G. M. (1979) British geomagnetic master curve 10 000 – 0 yr BP from dating European sediments, *Geophys. Res. Letters*, **6**, 249–52.

Thomson, M. E. and Eden, R. A. (1977) Quaternary deposits of the central North Sea. 3. The Quaternary sequence in the west-central North Sea, *Rep. Inst. Geol. Sci. Lond.*, **77/12**, 1–18.

Thorp, P. W. (1981) A trimline method for defining the upper limit of Loch Lomond Advance glaciers: examples from the Loch Leven and Glen Coe areas, *Scott. J. Geol.*, **17**, 49–64.

Thorpe, P. M. (1980) Radiocarbon dating of tufa nodules from the base of the Blashenwell tufa, Dorset, *J. Archaeol. Sci.*, **7**, 361–2.

Tinsley, J. (1950) The determination of organic carbon in soils by chromatic mixtures, *Trans. 4th Internat. Congr. Soil Sci.*, **1**, 161–4.

Todd, R., Low, D. and Mello, J. M. (1965) Smaller foraminifera, In B. Kummel and D. Raup (eds.), *Handbook of Palaeontological Techniques*, Freeman, San Francisco, 14–20.

Tooley, M. J. (1978) *Sea-Level Fluctuations in North-west England During the Flandrian Stage*, Clarendon Press, Oxford.

Tooley, M. J. (1981) Methods of reconstruction, In I. G. Simmons and M. J. Tooley (eds.), *The Environment in British Prehistory*, Duckworth, London, 1–48.

Townshend, J. R. G. (1981) *Terrain Analysis and Remote Sensing*, Allen and Unwin, London and Boston.

Trenhaile, A. S. (1975) The morphology of a drumlin field, *Ann. Assoc. Amer. Geogr.*, **65**, 297–312.

Tricart, J. (1975) Influence des oscillations climatiques recentes sur le modelé en Amazonie orientale (Région de Santarem) d'après les images de radar latéral, *Zeit. für Geomorph.*, **19**, 140–63.

Troughton, J. H. (1972) Carbon isotope fractionation by plants, *Proc. 8th Internat. Conf. on Radiocarbon Dating*, Roy. Soc. New Zealand, Wellington, pp. E40–E57.

Turner, C. (1970) The middle Pleistocene deposits at Marks Tey, Essex, *Phil. Trans. R. Soc. Lond.*, **B 257**, 373–440.

Turner, J. and Kershaw, A. P. (1973) A Late- and Post-glacial pollen diagram from Cranberry Bog, near Beamish, County Durham, *New Phytol.*, **72**, 915–28.

Tyldesley, J. B. (1973) Long-range transmission of tree pollen to Shetland. I. Sampling and trajectories, *New Phytol.*, **72**, 175–81.

Unwin, D. J. (1975) The nature and origin of the corrie moraines of Snowdonia, *Cambria*, **2**, 20–33.

Urey, H. C. (1947) The thermodynamic properties of isotopic substances, *J. Chem. Soc.* 1974, 562–81.

Valentine, K. W. G. and Dalrymple, J. B. (1975) The identification, lateral variation and chronology of two buried palaeocatenas at Woodhall Spa and West Runton, England, *Quat. Res.*, **4**, 551–90.

Valentine, K. W. G. and Dalrymple, J. B. (1976) Quaternary buried palaeosols: a critical review, *Quat. Res.*, **6**, 209–22.

Van der Werff, A. and Huls, H. (1958–74) *Diatomeenflora van Nederland*, (8 parts), publ. privately by A. Van der Werff, Westziide, 13A, De Hoef (V). The Netherlands.

Van Donk, J. (1976) An ^{18}O record of the Atlantic Ocean for the entire Pleistocene, *Geol. Soc. Amer. Mem.*, **145**, 147–64.

Vasari, Y. (1977) Radiocarbon dating of the Lateglacial and early Flandrian vegetational succession in the Scottish Highlands and the Isle of Skye, In J. M. Gray and J. J. Lowe (eds.), *Studies in the Scottish Lateglacial Environment*, Pergamon, Oxford and New York, 143–62.

Vasari, Y. and Vasari, A. (1968) Late and post-glacial macrophytic vegetation in the lochs of Northern Scotland, *Acta Bot. Fenn.*, **80**, 1–120.

Velichko, A. A. (1975) Paragenesis of a cryogenic (periglacial) zone, *Biul. Peryglac.*, **24**, 89–110.

Verstappen, H. Th. (1977) *Remote Sensing in Geomorphology*, Elsevier, Amsterdam and New York.

Virkkala, K. (1951) Glacial geology of the Suomussalmi area, east Finland, *Bull. Commn. Geol: Finl.*, **155**, 1–66.

Virkkala, K. (1960) On the striations and glacier movements in the Tampere region, southern Finland, *Bull. Commn. Geol. Finl.*, **188**, 159–76.

Virkkala, K. (1963) On ice-marginal features in southwestern Finland, *Bull. Commn. Geol. Finl.*, **210**, 1–76.

Vita-Finzi, C. (1969) *The Mediterranean Valleys: Geological Changes in Historical Times*, Cambridge Univ. Press, Cambridge.

Vita-Finzi, C. (1973) *Recent Earth History*, MacMillan, London and New York.

Vogel, J. C. (1970) C14 trends before 6000 BP, In I. U. Olsson (ed.), *Radiocarbon Variations and Absolute Chronology*, Wiley, New York and London, 313–25.

Vogel, J. C. and Kronfeld, J. (1980) A new method for dating peat, *S. Afr. J. Sci.*, **76**, 557–8.

Vogel, J. C. and Zagwijn, W. H. (1967) Groningen radiocarbon dates VI, *Radiocarbon*, **9**, 63–106.

Von Post, L. (1916) Forest tree pollen in south Swedish peat bog deposits (transl. into Engl. by M. B. Davis and K. Faegri, 1967), *Pollen Spores*, **9**, 375–401.

Von Weymarn, J. and Edwards, K. J. (1973) Interstadial site on the Island of Lewis, Scotland, *Nature*, **246**, 473–4.

Vorren, T. O. (1977) Weichselian ice movement in South Norway and adjacent areas, *Boreas*, **6**, 247–57.

Waechter, J. d' A., Newcomer, M. H. and Conway, B. W. (1970) Swanscombe (Kent-Barnfield Pit) 1971, *Proc. R. Anthropol. Inst. for 1970*, 43–64.

Wagner, G. A. (1979) Archaeomagnetic dating, In E. Jäger and J. C. Hunziker (eds.), *Lectures in Isotope Geology*, Springer-Verlag, Berlin and New York, 178–88.

Wagstaff, J. M. (1981) Buried assumptions: some problems in the interpretation of the 'Younger Fill' raised by recent data from Greece, *J. Archaeol. Sci.*, **8**, 247–64.

Walcott, R. J. (1970) Isostatic response to loading of the crust in Canada, *Can. J. Earth Sci.*, **7**, 716–26.

Walcott, R. J. (1972) Past sea levels, eustasy and deformation of the earth, *Quat. Res.*, **2**, 1–14.

Walker, D. (1966) The late Quaternary history of the Cumberland lowland, *Phil. Trans. R. Soc. Lond.*, **B 251**, 1–210.

Walker, D. (1970) Direction and rate in some British post-glacial hydroseres, In D. Walker and R. G. West (eds.), *Studies in the Vegetational History of the British Isles*, Cambridge Univ. Press, 117–39.

Walker, D. and Godwin, H. (1954) Lake stratigraphy, pollen analysis and vegetational history, In J. G. D. Clark (ed.), *Excavations at Star Carr*, Cambridge Univ. Press, London, 25–69.

Walker, M. J. C. (1975a) Two Lateglacial pollen diagrams from the eastern Grampian Highlands, Scotland, *Pollen Spores*, **17**, 67–92.

Walker, M. J. C. (1975b) Lateglacial and Early Postglacial environmental history of the central Grampian Highlands, Scotland *J. Biogeogr.*, **2**, 265–84.

Walker, M. J. C. (1977) Corrydon: a Lateglacial profile from Glenshee, south-east Grampian Highlands, *Pollen Spores*, **19**, 391–406.

Walker, M. J. C. (1980) Late-Glacial history of the Brecon Beacons, South Wales, *Nature*, **287**, 133–5.

Walker, M. J. C. (1982a) The Late-glacial and early Flandrian deposits at Traeth Mawr, Brecon Beacons, South Wales, *New Phytol.*, **90**, 177–94.

Walker, M. J. C. (1982b) Early and mid-Flandrian environmental history of the Brecon Beacons, South Wales, *New Phytol.*, **91**, 147–65.

Walker, M. J. C. and Lowe, J. J. (1977) Postglacial environmental history of Rannoch Moor, Scotland. I. Three pollen diagrams from the Kingshouse area, *J. Biogeogr.*, **4**, 333–51.

Walker, M. J. C. and Lowe, J. J. (1979) Postglacial environmental history of Rannoch Moor, Scotland. II. Pollen diagrams and radiocarbon dates from the Rannoch Station and Corrour areas, *J. Biogeogr.*, **6**, 349–62.

Walker, M. J. C. and Lowe, J. J. (1982) Lateglacial and early Flandrian chronology of the Isle of Mull, Scotland, *Nature*, **296**, 558–61.

Walker, R. (1978) Diatom and pollen studies of a sediment profile from Melynllyn, a mountain tarn in Snowdonia, North Wales, *New Phytol.*, **81**, 791–804.

Walkley, A. and Black, L. A. (1934) An examination of the Detjareff Method for determining soil organic matter and a proposed modification of the chromic acid titration method, *Soil Sci.*, **37**, 29–38.

Warren, A. (1970) Dune trends and their implications in the central Sudan, *Zeit. für Geomorph.*, Suppl., **10**, 154–80.

Washbourn-Kamau, C. K. (1971) Late Quaternary lakes in the Nakuru-Elementeita Basin, Kenya, *Geogr. J.*, **137**, 522–35.

Washburn, A. L. (1956) Classification of patterned ground and a review of suggested origins, *Bull. Geol. Soc. Amer.*, **67**, 823–65.

Washburn, A. L. (1973) *Periglacial Processes and Environments*, Edward Arnold, London.

Washburn, A. L. (1979) *Geocryology: A Survey of Periglacial Processes and Environments*, Edward Arnold, London.

Waters, R. S. (1958) Morphological mapping, *Geography*, **43**, 10–17.

Watkins, N. D. (1972) Review of the development of the geomagnetic time-scale and discussion of prospects for its finer definition, *Bull. Geol. Soc. Amer.*, **83**, 551–74.

Watkins, T. (ed.) (1975) *Radiocarbon: Calibration and Prehistory*, Edinb. Univ. Press, Edinburgh.

Watson, A. (1979) Gypsum crusts in deserts, *J. Arid Envs.*, **2**, 3–20.

Watson, E. (1969) The slope deposits of the Nant Iago

Valley near Cader Idris, Wales, *Biul. Peryglac.*, **18**, 95–113.

Watson, E. (1972) Pingos of Cardiganshire and the latest ice limit, *Nature*, **238**, 343–4.

Watson, E. (1977) The periglacial environment of Britain during the Devensian, *Phil. Trans. R. Soc. Lond.*, **B 280**, 183–98.

Watson, E. and Watson, S. (1971) Vertical stones and analogous structures, *Geogr. Annlr.*, **53A**, 107–14.

Watson, E. and Watson, S. (1974) Remains of pingos in the Cletwr basin, southwest Wales, *Geogr. Annlr.*, **56A**, 213–25.

Watts, W. A. (1963) Late-glacial pollen zones in western Ireland, *Irish Geogr.*, **4**, 367–76.

Watts, W. A. (1967) Late-glacial plant macrofossils from Minnesota, In E. J. Cushing and H. E. Wright (eds.), *Quaternary Palaeoecology*, Yale Univ. Press, New Haven, 89–97.

Watts, W. A. (1977) The Late Devensian vegetation of Ireland, *Phil. Trans. R. Soc. Lond.*, **B 280**, 273–93.

Watts, W. A. (1978) Plant macrofossils and Quaternary palaeoecology, In D. Walker and J. C. Guppy (eds.), *Biology and Quaternary Environments*, Australian Acad. Sci., Canberra, 53–68.

Watts, W. A. and Winter, T. C. (1966) Plant macrofossils from Kirchner Marsh, Minnesota – a palaeoecological study, *Bull. Geol. Soc. Amer.*, **77**, 1339–60.

Webb, T. III, and Bryson, R. A. (1972) Late- and postglacial climatic change in the northern midwest USA: quantitative estimates derived from fossil pollen spectra by multivariate statistical techniques, *Quat. Res.*, **2**, 70–115.

Welten, M. (1944) Pollenanalytische, stratigraphische und geochronologische Untersuchungen aus dem Faulenseemoos bei Spiez, *Veröff. Geobot. Inst. Rübel*, **21**, 1–201.

Werner, D. (ed.) (1977) *The Biology of Diatoms*, Blackwell, Oxford.

West, R. G. (1957) Interglacial deposits at Bobbitshole, Ipswich, *Phil. Trans. R. Soc. Lond.*, **B 241**, 1–31.

West, R. G. (1961) Vegetational history of the early Pleistocene of the Royal Society borehole at Ludham, Norfolk, *Proc. R. Soc. Lond.*, **B 155**, 437–53.

West, R. G. (1977a) *Pleistocene Geology and Biology*, Longman, London and New York.

West, R. G. (1977b) Early and Middle Devensian flora and vegetation, *Phil. Trans. R. Soc. Lond.*, **B 280**, 229–46.

West, R. G. (1980) *The Pre-glacial Pleistocene of the Norfolk and Suffolk Coasts*, Cambridge Univ. Press, London.

West, R. G., Dickson, C. A. *et al.* (1974) Late Pleistocene deposits at Wretton, Norfolk. II. Devensian deposits, *Phil. Trans. R. Soc. Lond.*, **B 267**, 337–420.

West, R. G. and Donner, J. J. (1956) The glaciation of East Anglia and the East Midlands: a differentiation based on stone orientation measurement of tills, *Q. J. Geol. Soc. Lond.*, **112**, 146–84.

West, R. G., Funnell, B. M. and Norton, P. E. P. (1980) An early Pleistocene cold marine episode in the North Sea: pollen and faunal assemblages at Covehithe, Suffolk, England, *Boreas*, **9**, 1–10.

West, R. G. and Norton, P. E. P. (1974) The Icenian Crag of southeast Suffolk, *Phil. Trans. R. Soc. Lond.*, **B 269**, 1–28.

West, R. G. and Sparks, B. W. (1960) Coastal interglacial deposits of the English Channel, *Phil. Trans. R. Soc. Lond.*, **B 243**, 95–133.

Western, A. C. (1969) Wood and Charcoal in Archaeology, In D. Brothwell and E. Higgs (eds.), *Science in Archaeology*, Thames and Hudson, London.

Westgate, J. A. and Gold, C. M. (1974) *World Bibliography and Index of Tephrochronology*, Univ. Alberta Press.

Westgate, J. A., Smith, D. G. W. and Tomlinson, M. (1970) Late Quaternary tephra layers in south-western Canada, In R. A. Smith and J. W. Smith (eds.), *Early Man and Environment in Northwest North America*, Univ. Calgary Press, 13–33.

Whillans, I. N. (1976) Radio-echo layers and the recent stability of the West Antarctic ice sheet, *Nature*, **264**, 152–5.

White, S. E. (1971) Rock glacier studies in the Colorado Front Range, 1961–1968, *Arct. Alp. Res.*, **3**, 43–64.

White, S. E. (1976) Rock glaciers and block fields, review and new data, *Quat. Res.*, **6**, 77–98.

Whitehead, P. F. (1977) A note on *Picea* in the Chelfordian Interstadial organic deposit at Chelford, Cheshire, *Quat. Res. Assoc. (GB) Newsl.*, **23**, 8–10.

Whittington, R. J. (1977) A late-glacial drainage pattern in the Kish Bank area and post-glacial sediments in the Central Irish Sea, In C. Kidson and M. J. Tooley (eds.), *The Quaternary History of the Irish Sea*, Seel House Press, Liverpool, 55–68.

Whittow, J. B. and Ball, D. F. (1970) North-west Wales, In C. A. Lewis (ed.), *The Glaciations of Wales and Adjoining Regions*, Longman, London.

Wigley, T. M. L. (1981) Climate and palaeoclimate: what we can learn about solar luminosity variations, *Solar Physics*, **74**, 435–71.

Wilcox, R. E. (1965) Volcanic ash chronology, In H. E. Wright and D. G. Frey (eds.), *The Quaternary of the United States*, Princeton Univ. Press, Princeton, NJ, 807–16.

Wilding, L. P. and Drees, L. R. (1969) Biogenic opal in soils as an index of vegetative history in the Prairie Peninsula, In R. E. Bergstrom (ed.), *The Quaternary of Illinois*, Univ. Illinois, Coll. of Agric., Spec. Publ., **4**.

Williams, G. E. (1973) Late Quaternary piedmont sedimentation, soil formation and palaeoclimates in Australia, *Zeit. für Geomorph.*, **17**, 102–25.

Williams, G. E. and Polach, H. A. (1969) The evaluation of ^{14}C ages for soil carbonates from the arid zone, *Earth Planet. Sci. Lett.*, **4**, 240–2.

Williams, J. and Barry, R. G. (1974) Ice age experiments with the NCAR General Circulation Model: conditions in the vicinity of the northern continental ice sheets, In G. Weller and S. A. Bowling (eds.), *Climate of the Arctic*, Univ. Alaska Press, Fairbanks, 143–9.

Williams, J., Barry, R. G. and Washington, W. M. (1974) Simulation of the atmospheric circulation using the NCAR global circulation model with ice age boundary conditions, *J. Appl. Meteor.*, **13**, 305–17.

Williams, R. B. G. (1975) The British climate during the Last Glaciation: an interpretation based on periglacial phenomena, In A. E. Wright and F. Moseley (eds.), *Ice Ages: Ancient and Modern*, Seel House Press, Liverpool, 95–120.

Willman, H. B. and Frye, J. C. (1970) Pleistocene stratigraphy of Illinois, *Illinois Geol. Surv. Bull.*, **94**.

Willman, H. B., Glass, H. D. and Frye, J. C. (1963) Mineralogy of glacial tills and their weathering profile in Illinois. Part I. Glacial tills, *Illinois Geol. Surv. Circ.*, **347**.

Willman, H. B., Glass, H. D. and Frye, J. C. (1966) Mineralogy of glacial tills and their weathering profiles in

Illinois. Part II. Weathering profiles, *Illinois Geol. Surv. Circ.*, **400.**

Winterhalter, B. (1972) On the geology of the Bothnian Sea, an epeiric sea that has undergone Pleistocene glaciation, *Bull. Geol. Surv. Finl.*, **258**, 1–66.

Wintle, A. G. (1981) Thermoluminescence dating of Late Devensian loesses in southern England, *Nature*, **289**, 479–81.

Wintle, A. G. and Huntley, D. J. (1979) Thermoluminescence dating of a deep-sea sediment core, *Nature*, **279**, 710–12.

Wintle, A. G. and Huntley, D. J. (1982) Thermoluminescence dating of sediments, *Quat. Sci. Rev.*, **1**, 31–53.

Wise, S. M. (1980) Caesium-137 and lead-210: a review of the techniques and some applications in geomorphology, In R. A. Cullingford, D. A. Davidson and J. Lewin (eds.), *Timescales in Geomorphology*, Wiley, Chichester and New York, 109–27.

Woillard, G. M. (1978) Grande Pile peat bog: a continuous pollen record for the past 140 000 years, *Quat. Res.*, **9**, 1–21.

Wolf, P. R. (1974) *Elements of Photogrammetry*, McGraw-Hill, New York.

Wollin, G., Ericson, D. B. and Ewing, M. (1971) Late Pleistocene climates recorded in Atlantic and Pacific deep-sea sediments, In K. K. Turekian (ed.), *The Late Cenozoic Glacial Ages*, Yale Univ. Press, New Haven, 199–214.

Wolstedt, P. (1958) *Das Eiszeitalter* (Bd. II), Enke, Stuttgart.

Woods, A. J. (1980) Geomorphology, deformation and chronology of marine terraces along the Pacific Coast of central Baja California, *Quat. Res.*, **13**, 346–64.

Wooldridge, S. W. (1938) The glaciation of the London Basin and the evolution of the Lower Thames drainage system, *Q. J. Geol. Soc. Lond.*, **94**, 627–67.

Wooldridge, S. W. and Linton, D. L. (1955) *Structure, Surface and Drainage in South-east England*, George Philip, London.

Worsley, P. (1967) Fossil frost wedge polygons at Congleton, Cheshire, England, *Geogr. Annlr.*, **48A**, 211–19.

Worsley, P. (1977) The Cheshire-Shropshire Plain, In D. Q. Bowen (ed.), *Wales and the Cheshire-Shropshire Lowland, Proc. Xth INQUA Conf. Birmingham, 1977*, Norwich, 53–64.

Worsley, P. (1980) Problems in radiocarbon dating the Chelford Interstadial of England, In R. A. Cullingford, D. A. Davidson and J. Lewin (eds.), *Timescales in Geomorphology*, Wiley, Chichester and New York, 289–304.

Wright, H. E. (1971) Retreat of the Laurentide ice sheet from 14 000 to 9 000 years ago, *Quat. Res.*, **1**, 316–30.

Wright, H. E. (1972) Quaternary history of Minnesota, In P. K. Sims and G. B. Morey (eds.), *Geology of Minnesota*, Minnesota Geol. Surv., 515–48.

Wright, H. E. (1973) Tunnel valleys, glacial surges and subglacial hydrology of the Superior lobe, Minnesota, *Geol. Soc. Amer. Mem.*, **136**, 251–76.

Wright, H. E. (1976) Ice retreat and revegetation in the western Great Lakes area, In W. C. Mahaney (ed.), *Quaternary Stratigraphy of North America*, Dowden, Hutchinson and Ross, Pennsylvania, 119–32.

Wright, H. E. (1980) Surge moraines of the Klutlan Glacier, Yukon Territory, Canada: origin, wastage, vegetation succession, lake development and application to the Late-glacial of Minnesota, *Quat. Res.*, **14**, 2–18.

Wright, H. E., Cushing, E. J. and Livingstone, D. A. (1965) Coring devices for lake sediment, In B. Kummel and D. Raup (eds.), *Handbook of Palaeontological Techniques*, Freeman, San Francisco, 494–520.

Wright, H. E. and Frey, D. G. (eds.) (1965) *The Quaternary of the United States*, Princeton Univ. Press, Princeton, NJ.

Wyckoff, R. W. G. (1980) Collagen in fossil bones, In P. E. Hare, T. C. Hoering and K. King Jr (eds.), *Biogeochemistry of Amino Acids*, Wiley, New York and Chichester, 17–22.

Yatsu, E. and Shimoda, S. (1981) X-ray diffraction of clay minerals, In A. S. Goudie (ed.), *Geomorphological Techniques*, Allen and Unwin, London and Boston, 110–14.

Zagwijn, W. H. (1975) Variations in climate as shown by pollen analysis, especially in the Lower Pleistocene of Europe, In A. E. Wright and F. Moseley (eds.), *Ice Ages: Ancient and Modern*, Seel House Press, Liverpool, 137–52.

Zeuner, F. E. (1945) *The Pleistocene Period: Its Climate, Chronology and Faunal Successions* (1st ed.), Hutchinson, London.

Zeuner, F. E (1959) *The Pleistocene Period* (2nd ed.), Hutchinson, London.

Zingg, T. (1935) Beitrag zür Schotteranalyse, *Schweg. Mineralog. und Petrog. Mitt. Bd.*, **15**, 39–114.

Zumberge, J. E., Engel, M. H. and Nagy, B. (1980) Amino acids in bristlecone pine: an evaluation of factors affecting racemization rates and palaeothermometry, In P. E. Hare, T. C. Hoering and K. King Jr. (eds.), *Biogeochemistry of Amino Acids*, Wiley, New York and Chichester, 503–25.

Index